ENGINEERING INTERVENTIONS IN AGRICULTURAL PROCESSING

Innovations in Agricultural and Biological Engineering

ENGINEERING INTERVENTIONS IN AGRICULTURAL PROCESSING

Edited by
Megh R. Goyal, PhD, PE
Deepak Kumar Verma

Apple Academic Press Inc.
3333 Mistwell Crescent
Oakville, ON L6L 0A2 Canada

Apple Academic Press Inc.
9 Spinnaker Way
Waretown, NJ 08758 USA

© 2018 by Apple Academic Press, Inc.
Exclusive worldwide distribution by CRC Press, a member of Taylor & Francis Group
No claim to original U.S. Government works
Printed in the United States of America on acid-free paper
International Standard Book Number-13: 978-1-77188-556-0 (Hardcover)
International Standard Book Number-13: 978-1-315-20737-7 (eBook)

All rights reserved. No part of this work may be reprinted or reproduced or utilized in any form or by any electric, mechanical or other means, now known or hereafter invented, including photocopying and recording, or in any information storage or retrieval system, without permission in writing from the publisher or its distributor, except in the case of brief excerpts or quotations for use in reviews or critical articles.

This book contains information obtained from authentic and highly regarded sources. Reprinted material is quoted with permission and sources are indicated. Copyright for individual articles remains with the authors as indicated. A wide variety of references are listed. Reasonable efforts have been made to publish reliable data and information, but the authors, editors, and the publisher cannot assume responsibility for the validity of all materials or the consequences of their use. The authors, editors, and the publisher have attempted to trace the copyright holders of all material reproduced in this publication and apologize to copyright holders if permission to publish in this form has not been obtained. If any copyright material has not been acknowledged, please write and let us know so we may rectify in any future reprint.

Trademark Notice: Registered trademark of products or corporate names are used only for explanation and identification without intent to infringe.

Library and Archives Canada Cataloguing in Publication

Engineering interventions in agricultural processing / edited by Megh R. Goyal, PhD, PE, Deepak Kumar Verma.
(Innovations in agricultural and biological engineering)
Includes bibliographical references and index.
Issued in print and electronic formats.
ISBN 978-1-77188-556-0 (hardcover).--ISBN 978-1-315-20737-7 (PDF)
1. Agricultural processing. I. Goyal, Megh Raj, editor II. Verma, Deepak Kumar, 1986-, editor
III. Series: Innovations in agricultural and biological engineering
S698.E54 2017   630.2   C2017-903235-6   C2017-903236-4

Library of Congress Cataloging-in-Publication Data

Names: Goyal, Megh Raj, editor. | Verma, Deepak Kumar, 1986- editor.
Title: Engineering interventions in agricultural processing / editors: Megh R. Goyal, PhD, PE; Deepak Kumar Verma.
Description: Waretown, NJ : Apple Academic Press, 2017. | Series: Innovations in agricultural & biological engineering | Includes bibliographical references and index.
Identifiers: LCCN 2017021024 (print) | LCCN 2017022126 (ebook) | ISBN 9781315207377 (ebook) | ISBN 9781771885560 (hardcover : alk. paper)
Subjects: LCSH: Agricultural processing.
Classification: LCC S698 (ebook) | LCC S698 .E54 2017 (print) | DDC 630.2--dc23
LC record available at https://lccn.loc.gov/2017021024

Apple Academic Press also publishes its books in a variety of electronic formats. Some content that appears in print may not be available in electronic format. For information about Apple Academic Press products, visit our website at **www.appleacademicpress.com** and the CRC Press website at **www.crcpress.com**

CONTENTS

List of Contributors .... *vii*

List of Abbreviations .... *xi*

List of Symbols .... *xvii*

Foreword 1 by R. K. Sivanappan .... *xxi*

Foreword 2 by Murlidhar Meghwal .... *xxiii*

Foreword 3 by Khursheed Alam Khan .... *xxv*

Preface 1 by Megh R. Goyal .... *xxvii*

Preface 2 by Deepak Kumar Verma .... *xxxi*

Warning/Disclaimer .... *xxxv*

About Senior Editor-in-Chief .... *xxxvii*

About Co-Editor .... *xxxix*

Book Endorsements .... *xli*

Books on Agricultural and Biological Engineering
from Apple Academic Press, Inc. .... *xliii*

Editorial .... *xlv*

PART I: AGRICULTURAL PROCESSING: INTERVENTIONS IN ENGINEERING TECHNOLOGIES 1

1. **Machine Vision Technology in Food Processing Industry: Principles and Applications—A Review** 3

   P. S. Minz, Charanjiv Singh, and I. K. Sawhney

2. **Cold Plasma Technology: An Emerging Non-Thermal Processing of Foods—A Review** 33

   R. Mahendran, C. V. Kavitha Abirami, and K. Alagusundaram

3. **Technology of Ohmic Heating in Meat Processing** 57

   Asaad Rehman Saeed Al-Hilphy

4. **Technology of Solid State Fermentation in Dairy Products: Production of β-Galactosidase from Mold *Aspergillus oryzae*** 71

Alaa Jabbar Abd Al-Manhal, Ali Khudhair Jaber Alrikabi, Gheyath H. Majeed, and Abdullah M.Alsalim

5. **Nano-Particle Based Delivery Systems: Applications in Agriculture** 107

Deepak Kumar Verma, Shikha Srivastava, Vipul Kumar, Bavita Asthir, Mukesh Mohan, and Prem Prakash Srivastav

6. **Green Synthesis of Silver Nanoparticles from Endophytic Fungus *Aspergillus niger*** 131

Poonam Rani and Vedpriya Arya

PART II: NOVEL PRACTICES IN AGRICULTURAL PROCESSING 145

7. **Practices in Bioleaching: A Review on Clean and Economic Alternative for Safe and Green Environment** 147

Sunita Devi, Bindu Devi, and Seema Verma

8. **Practices in Seed Priming: Quality Improvement of Oil Seed Crops** 179

Hasnain Nangyal and Nighat ZiaUdin

9. **Modified Pearl Millet Starch: A Review on Chemical Modification, Characterization and Functional Properties** 191

Mandira Kapri, Deepak Kumar Verma, Ajesh Kumar, Sudhanshi Billoria, Dipendra Kumar Mahato, Baljeet Singh Yadav, and Prem Prakash Srivastav

PART III: AGRICULTURAL PROCESSING: HEALTH BENEFITS OF MEDICINAL PLANTS 227

10. **Potential Health Benefits of Tea Polyphenols—A Review** 229

Brij Bhushan, Dipendra Kumar Mahato, Deepak Kumar Verma, Mandira Kapri, and Prem Prakash Srivastav

11. **Potential Health Benefits of Lemon Grass—A Review** 283

Nighat Zia Udin and Hasnain Nangyal

Glossary 295

Index 305

LIST OF CONTRIBUTORS

K. Alagusundaram, PhD
Deputy Director General (Agricultural Engineering), Division of Agricultural Engineering, ICAR, Krishi Anusandhan Bhawan II, New Delhi – 110012, India. Tel.: +91-11-25843415; E-mail: ddgengg@icar.org.in

Asaad Rehman Saeed Al-Hilphy, PhD
Assistant Professor, Department of Food Science, College of Agriculture, University of Basrah, Basra City, Iraq. Mobile: +00-96-47702696458; E-mail: aalhilphy@yahoo.co.uk

Alaa Jabbar Abd Al-Manhal, PhD
Assistant Professor, Department of Food Science, College of Agriculture, University of Basrah, Basra City, Iraq. Tel.: +964-7808785772; E-mail: alaafood_13@yahoo.com

Ali Khudhair Jaber Alrikabi, PhD
Assistant Professor, Department of Food Science, College of Agriculture, University of Basrah, Basra City, Iraq. Tel.: +964-7716733405; E-mail: alikhudhair2012@yahoo.com

Abdullah M. Alsalim, PhD
Lecturer, Department of Food Science, College of Agriculture, University of Basrah, Basra City, Iraq. Tel.: +964-7705504849; E-mail: abdullahphd@yahoo.com

Vedpriya Arya, PhD
Scientist E, Patanjali Herbal Research Department, Patanjali Yogpeeth, Phase I, Haridwar, Uttrakhand; Formerly Assistant Professor, Department of Biotechnology, Guru Nanak Girls College, Model Town, Ludhiana-141001, Punjab, India, Mobile: +91-7060472471; E-mail: ved.nano2008@gmail.com

Brij Bhushan, MSc
Senior Research Fellow, Indian Agricultural Research Institute, Pusa campus, New Delhi 110012, India. Mobile: +91-8586935562, +91-8802118902; E-mail: bhushanmishra909@gmail.com

Sudhanshi Billoria, MSc
Research Scholar, Department of Agricultural and Food Engineering, Indian Institute of Technology, Kharagpur – 721302, West Bengal, India. Mobile: +91-8768126479; E-mail: sudharihant@gmail.com

Bindu Devi, MSc
Assistant Professor (Microbiology), Department of Microbiology, College of Basic Sciences, CSK, Himachal Pradesh Agricultural University, Palampur – 176062, Himachal Pradesh, India. Mobile: +91-9882191747; E-mail: bindu.thakur111@gmail.com

Sunita Devi, PhD
Assistant Professor (Microbiology), Department of Basic Sciences, College of Forestry, University of Horticulture and Forestry- Nauni, Solan – 173230, Himachal Pradesh, India. Mobile: +91-9418014742; E-mail: sunitachamba@gmail.com

Megh R. Goyal, PhD, PE

Retired Professor in Agricultural and Biomedical Engineering, University of Puerto Rico – Mayaguez Campus; and Senior Technical Editor-in-Chief in Agriculture Sciences and Biomedical Engineering, Apple Academic Press Inc., PO Box 86, Rincon – PR – 00677, USA; E-mail: goyalmegh@gmail.com

Mandira Kapri, MSc

Junior Project Assistant, Department of Agricultural and Food Engineering, Indian Institute of Technology, Kharagpur – 721302, West Bengal, India. Mobile: +91-9051245266; E-mail: kaprimandira@gmail.com

C. V. Kavitha Abirami, PhD

Associate Professor, Department of Food Packaging and System Development, Indian Institute of Crop Processing Technology, Ministry of Food Processing Industries, Government of India, Pudukottai Road, Thanjavur – 613005, Tamil Nadu, India. Mobile: +91-9750968412; E-mail: kaviabi@iicpt.edu.in

Khursheed Alam Khan, PhD

Assistant Professor (Agricultural Engineering), College of Horticulture, Rajmata Vijayaraje Scindia Agriculture University (RVSKVV), Mandsaur – 458001, Gwalior, India, Mobile: +91-9425942903; E-mail: khan_undp@yahoo.ca

Ajesh Kumar, MTech

Research Scholar, Department of Agricultural and Food Engineering, Indian Institute of Technology, Kharagpur – 721302, West Bengal, Mobile: +91-8900479007; Telephone: +91-3222-281673; Fax: +91-3222-282224; E-mail: ajeshmtr@gmail.com

Vipul Kumar, PhD

Assistant Professor (Plant Protection), Department of Plant Protection, School of Agriculture, Lovely Professional University, Phagwara, Punjab – 144411, India. Mobile: +91 7525969981; E-mail: vipulpathology@gmail.com

Dipendra Kumar Mahato, MSc

Senior Research Fellow, Indian Agricultural Research Institute, Pusa Campus, New Delhi – 110012, India. Mobile: +91 9911891494, +91-9958921936; E-mail: kumar.dipendra2@gmail.com

R. Mahendran, PhD

Assistant Professor, Department of Food Packaging and System Development, Indian Institute of Crop Processing Technology, Ministry of Food Processing Industries, Government of India, Pudukottai Road, Thanjavur – 613005, Tamil Nadu, India. Mobile: +91-9750968418; E-mail: mahendran@iicpt.edu.in

Gheyath H. Majeed, PhD

Professor, Department of Food Science, College of Agriculture, University of Basrah, Basra City, Iraq. Tel.: +964-7801009828; E-mail: ghmajeed@yahoo.com

Murlidhar Meghwal, PhD

Assistant Professor, Department of Food Science and Technology, National Institute of Food Technology Entrepreneurship & Management, Kundli -131028, Sonepat, Haryana, India; Mobile: +91 9739204027; Email: murli.murthi@gmail.com

P. S. Minz, MTech

Scientist, Machine Vision Lab, Dairy Engineering Division, ICAR – National Dairy Research Institute, Karnal – 132001, Haryana, India. Mobile: +91-9992454593, Tel.: 0184-2259281; E-mail: psminz@gmail.com

List of Contributors

Mukesh Mohan, PhD
Associate Professor, Department of Agricultural Biochemistry, C. S. Azad University of Agriculture and Technology, Kanpur – 208002, India; E-mail: drmukeshmohan@rediffmail.com

Hasnain Nangyal, MSc
Research Scholar, Department of Botany, Faculty of Sciences, Hazara University, Mansehra, 32040 (Khyber Pakhtoonkhwa), Pakistan. Mobile: +92-3468982308, Tel.: +92-45618658, Fax: +92-419200764; E-mail: hasnain308@gmail.com

Nighat Nangyal, MSc
Research Scholar, Department of Biochemistry, University of Agriculture, Faisalabad 38040 (Punjab) Pakistan. Mobile: +92-3027060966, Fax: +92-419200764; E-mail: nighatzia72@gmail.com

Poonam Rani, MSc
Assistant Professor, Department of Biotechnology, Guru Nanak Girls College, Model Town, Ludhiana-141001, Punjab. Mobile: +91-9876326036; E-mail: vermarishi69@gmail.com

A.K. Sawhney, PhD
Emeritus Scientist, Dairy Engineering Division, ICAR – National Dairy Research Institute, Karnal-132001, Haryana, India. Mobile: +91-9896440784; E-mail: charanjiv_cjs@yahoo.co.in

Charanjiv Singh, PhD
Associate Professor, Department of Food Engineering and Technology, Sant Longowal Institute of Engineering and Technology (SLIET), Longowal – 148106, Sangrur, Punjab, India. Mobile: +91-9872980044; E-mail: charanjiv_cjs@yahoo.co.in

R. K. Sivanappan, PhD
Former Professor and Dean, College of Agricultural Engineering and Technology, Tamil Nadu Agricultural University (TNAU), Coimbatore. Current mailing address: Consultant, 14, Bharathi Park, 4th Cross Road, Coimbatore, TN – 641 043, India. E-mail: sivanappanrk@hotmail.com

Prem Prakash Srivastav, PhD
Associate Professor, Department of Agricultural and Food Engineering, Indian Institute of Technology, Kharagpur – 721302, West Bengal, India. Mobile: +91-9434043426. Tel.: +91-3222-283134, Fax: +91-3222-282224; E-mail: pps@agfe.iitkgp.ernet.in

Shikha Srivastava, MSc
Research Scholar, Department of Botany, Deen Dyal Upadhyay Gorakhpur University, Gorakhpur 273 009, Uttar Pradesh, India; E-mail: shikha.sriv13@gmail.com

Deepak Kumar Verma, MSc
Research Scholar, Agricultural and Food Engineering Department, Indian Institute of Technology, Kharagpur – 721302, West Bengal, India. Tel.: +91-3222-281673. Mobile: +91-7407170260, +91-9335993005. Fax: +91-3222-282224; E-mail: deepak.verma@agfe.iitkgp.ernet.in; rajadkv@rediffmail.com

Seema Verma, MSc
Technical Assistant, Department of Basic Sciences, College of Forestry, University of Horticulture and Forestry, Nauni, Solan – 173230, Himachal Pradesh, India. Mobile: +91-9418553489; E-mail: seema121274@gmail.com

Baljeet Singh Yadav, PhD
Associate Professor and Head, Department of Food Technology, Maharishi Dayanand University, Rohtak, Haryana – 124001, India, Mobile: +91-9896360766; E-mail: baljeetsingh.y@gmail.com

Nighat Ziaudin, MSc
Research Scholar, Department of Biochemistry, University of Agriculture, Faisalabad, 38040 (Punjab), Pakistan. Mobile: +92 3027060966; E-mail: nighatzia72@gmail.com

LIST OF ABBREVIATIONS

1D	one dimensional
2D	two dimensional
A-CoA	acetyl-CoA
AC	alternative current
AC	amylose content
AD	Alzheimer's disease
AMD	acid mine drainage
ANT	adenine nucleotide translocase
AOS	active oxygen species
APP	amyloid precursor protein
APPJ	atmospheric pressure plasma jet
Ar	argon
ATP	adenosine triphosphate
Aβ fibrils	amyloid-β fibril formation
B	blue
BT	black tea
C-CoA	coumaroyl-coa
CAT	catalase
CCD	charged coupled device
Cd	cadmium
Cg	catechin gallate
CHI	chalcone flavanone isomerase
CHS	chalcone synthase
CIE	Commission Internationale de L´eclairage (International Commission on Illumination)
CMOS	complementary metal oxide semiconductor
COMT	catechol-o-methyltransferase
COX-2	cyclo-oxygenase
CVD	cardiovascular disease
DA	dopamine
DBD	dielectric barrier discharge

DC	direct current
DNA	deoxyribonucleic acid
DS	degree of substitution
DTT	dithiothreitol
EC	epicatechin
ECg	epicatechin gallate
ECG	epicatechin-3-gallate
EDTA	ethylenediaminetetraacetic acid
EGC	epigallocatechin
EGCG	epigallocatechin-3-gallate
EPS	exopolymeric substance
EPS	exopolysaccharide substance
ERK	extracellular signal-regulated kinase
FAK	focal adhesion kinase
FAO	food and agriculture organization:
GA	gibblaric acid
GC	gallocatechin
GCg	gallocatechin gallate
GHz	giga hertz
GJIC	gap-junctional intercellular communication
GLUT-1	glucosetransporter 1
GR	glutathione reductase
GSK-3β	glycogen synthase kinase-3β
GT	green tea
GTC	green tea catachin
GTP	green tea polyphenol
HCAs	heterocyclic amines
HSI	hue saturation intensity
HSI	hyperspectral imaging
HSV	hue saturation value
IGF-1	insulin-like growth factor-1
IGFBP-3	IGF-binding protein-3
IP	intra peritoneal
ISL	*in-situ* leaching
K	potassium
KCL	potassium chloride

List of Abbreviations

LGC	least gelation concentration
lpm	liter per minute
LPS	lipopolysaccharides
LTE	local thermodynamic equilibrium
LWIR	long wavelength infrared
M-CoA	malonyl-CoA
MAPK	mitogen-activated protein kinase
MC	moisture content
MEs	micro-emulsions
MHCD	micro hollow cathode discharge
MLR	multiple linear regression
MMP	mitochondrial membrane permeabilization
MPa	mega Pascal
mRNA	messenger RNA
MSC	multiplicative scatter correction
MSCs	mesenchymal stem cells
MSNs	meso-porous silica nano-particles
MVS	machine vision system
MWIR	mid wavelength infrared
N	nitrogen
NaCl	sodium chloride
NaClO	sodium hypochlorite
NDEA	nitrosodiethylamine
NEs	nano-emulsions
NF-κB	nuclear factor-κb
NIR	near infrared
NLCs	nano-structure lipid carriers
NMR	nuclear magnetic resonance
NNK	4-(methylnitrosamino)-1-(3-pyridyl)-1-butanone
NTP	non-thermal plasma
OAUGDP	one atmospheric uniform glow discharge plasma
ODN	endozepineoctdecaneuropeptide
OH	ohmic heating
OH$^{\cdot}$	hydroxyl radical
ONPG	ortho-nitrophenyl-β-galactoside
OT	oolong tea

PCR	polymerase chain reaction
PD	Parkinson's disease
PEG	poly-ethoxy glycol
PG	polyethylene glycol
PGE2	prostaglandin E2
PGRs	plant growth regulators
PI3K	phosphoinositide 3-kinase
PKC	protein kinase C
PLS	partial least square
PNPG	4-nitrophenyl-beta-d-glucopyranoside
PNs	polymeric nano-particles
$POCl_3$	phosphorus oxychloride
$POCl_3$	phosphoryl chloride
PPARα	peroxisome proliferator-activated receptors
PV	peak viscosity
rDNA	ribosomal DNA
RF	radio frequency
RGB	red green blue
RNA	ribonucleic acid
ROI	region of interest
ROS	reactive oxygen species
rRNA	ribosomal RNA
sAPPα	soluble amyloid precursor protein α
SCLC	small cell lung carcinoma
SDBD	surface dielectric barrier discharge
SDS	sodium dodecyl sulfate
SLNs	solid lipid nano-particles
SMP	solid matrix priming
SOD	super oxide dismutase
sRGB	adobe red green blue
STMP	sodium trimetaphosphate
STPP	sodium tripolyphosphate
STZ	streptozotocin
SV	swelling volume
SWIR	short wavelength infrared
TA100/TA98	typhimurium *salmanella* strains

List of Abbreviations

TCA	tricarboxylic acid
TEAC	troloxequivalent antioxidant activity assay
TIFF	tagged image file format
TLC	thin layer chromatography
TNFα	tumor necrosis factor
TRs	thearubigins
USA	United States of America
USA-FDA	United States of America, Food and Drug Administration
UV	ultra violet
VDAC	voltage-dependant anion channel
VEGF	vascular endothelial growth factor
VIS	visible spectrum
VSMCs	vascular smooth muscle cells
W	watt
WBC	water binding capacity
WHO	World Health Organization
WT	white tea
XRD	x-ray diffraction
Zn	zinc

LIST OF SYMBOLS

a (−)	greenness value in hunter color meter, dimensionless
a (+)	redness value in hunter color meter, dimensionless
a_w	water activity
b (−)	blueness in hunter color meter, dimensionless
b (+)	yellowness in hunter color meter, dimensionless
Br_2	bromine
C_3H_5ClO	epichlorohydrin
$Ca(NO_3)_2$	calcium nitrate
$CaCl_2$	calcium chloride
$CaSO_4$	calcium sulfate
CO_2	carbon dioxide
Cr	chromium
Cu	copper
Cu_2S	copper sulfide
$CuFeS_2$	chalcopyrite
$CuSO_4$	copper sulfate
Ea	activition energy
Fe	iron
FeS_2	pyrite
Free Enz.	free enzyme
G	green
H_2	hydrogen
H_2O_2	hydrogen peroxide
He	helium
Hg	mercury
HK_2O_4P	dipotassium hydrogen phosphate
Hz	hertz
I	current, amperes
IKK-β	IκB kinase-β
Imm. Enz.	immobilized enzyme

IO_4	iodate
keV	kilo electron volt
kGy	kilogray
k_m	Michaelis constant
$KMnO_4$	permanganate
KNO_3	potassium nitrate
kV	kilo voltage
'L'	bright/dark color, dimensionless (hunter color meter)
mA	Milli Amphere
$MgSO_4$	magnesium sulfate
Mn	manganese
MoS_2	molybdenite
n_e	density of electrons
NH_3	ammonia
$(NH_4)_2SO_4$	ammonium sulfate
Ni	nickel
O_2	oxygen
$O_2^{\cdot-}$	superoxide anion
P	phosphorus
Pa	Pascal
Pb	lead
PbS	galena
Pgp	p-glycoprotein
R	red
R^2	coefficient of determination
R_m	relative mobility
T	neutral gas temperature
T_e	temperature of electrons
TF1	theaflavin
TF2a	theaflavin-3-gallate
TF2b	theaflavin-3'-gallate
TF3	theaflavin 3,3"-gallate
T_g	overall temperature of the gas
T_h	temperature of heavy particles
T_i	temperature of ions
T_p	temperature of plasma

List of Symbols

v	voltage
V-I	voltage current characteristics
V_e	elution volume
V_{max}	maximal velocity
V_{min}	minimum breakdown voltage
V_O	void volume
YCbCr	luminance, chroma blue, chroma red
ZnS	sphalerite

FOREWORD 1 BY R. K. SIVANAPPAN

Agricultural processing constitutes a significant sector of the economy of both developed and developing nations around the world. In the wake of rapid globalization, about 25% of food produced crosses national boundaries, which makes it imperative to evolve innovative methods of addressing food handling, food safety, processing and other related issues. Viewed against this background, engineering interventions or practices for agricultural processing continue to evolve, and novel technologies appear on the food industrial landscape.

Engineering Interventions in Agricultural Processing is a timely book covering the latest technologies in agricultural process engineering. This vital area of agricultural engineering has to address a whole range of post-production challenges throughout the world. Technological innovation has enabled crop production to achieve phenomenal growth, and global agricultural and food processing technologies are demanded at a faster rate than ever before.

Better handling of harvested produce using the latest technologies so as to minimize postharvest losses is a national agenda in many countries. Novel technologies in handling, storing and processing of food crops are vigorously sought by farmers, traders and food processors. Ever since agricultural engineering was introduced in agricultural universities 100 years ago, the engineers have played a commendable role in designing improved processing and postharvest equipment particularly suitable for marginal and small farmers. We are now at the threshold of making a foray into latest food processing technologies.

This book covers the important novel technologies that are likely to occupy the center stage of food processing in the near future. This book is organized in three parts. In Part 1, an attempt has been made to cover the latest developments in food manufacturing and processing such as non-thermal processing, Ohmic heating of meat products, fermentation of dairy products, machine vision based automation technologies and nanotechnology. A comprehensive overview of clean and economic alternative technologies has also been attempted in Part 2. Processing of medicinal plants and herbs is also gaining popularity due to the perceived health benefits. A brief coverage of potential health benefits of tea and lemon grass has also been made in Part 3.

This book has been edited by Dr. Megh Raj Goyal, a renowned agricultural engineer with many years of field experience in all branches of agricultural engineering. I have acquaintance with Dr. Goyal for more than 30 years. He is known for his excellence, hard work and dedication in his profession. Also joining him is Deepak Kumar Verma as Coeditor, who is a budding engineer. This book will be handy tool for students, postgraduate teachers and researchers in the area of agricultural processing.

R. K. Sivanappan, PhD, DSc (Hon)
Founder and Former Director of Water Technology, Centre; and Former Dean of College of Agricultural Engineering, Tamil Nadu Agricultural University, Coimbatore – 641043, TN, India
Mobile: +91-9443716255
E-mail: sivanappanrk@hotmail.com

FOREWORD 2 BY MURLIDHAR MEGHWAL

Agricultural processing, auto-mechanization, and its economic values are very important factors for the healthy growth of any country because this sector is the backbone of any country. Food grains can only be grown in the earth and that need lots of efforts. In the current modern situation more and more stress is given on mechanization of the auto-mechanization of agricultural practices because people's life styles have, and fewer changed a lot people are available to hande the manual laborious work. It means much food and agricultural processing has become mechanized and that leads to fast processing and production. It is a matter of great pleasure for me to write a foreword for the book, *Engineering Interventions in Agricultural Processing* by Megh R. Goyal and Deepak Kumar Verma, under book series *Innovations in Agricultural and Biological Engineering*. This reference book can bring a significant and useful message to readers.

Agricultural processing is one of the major segments of agro-base industry. Its significant contribution to the National GDP depends on auto-mechanization and introduction of innovative technologies feasible for transformation of raw agricultural material into new products exhibiting higher value addition in terms of nutrition, safety, security, consistent supply and money. Agricultural production and processing has become an important primary source of income for millions of people all over the world. The accelerated socio-economic development during the 21st century is associated with challenging issues like food security, food safety, quality and their linkages with national and international markets to compensate with increased demand.

This book volume encompasses a good number of chapters on various aspects of engineering interventions in agricultural processing and technology. The chapters are authored by scientists actively engaged in their respective area of specialization. The predominant chapters in this book provide interesting insights for budding entrepreneurs, policymakers, investors and students. It is an excellent resource on the subject.

Murlidhar Meghwal, PhD (IIT Kharagpur)
Assistant Professor, Department of Food Science and Technology, National Institute of Food Technology Entrepreneurship & Management, Kundli - 131028, Sonepat, Haryana, India; Mobile: +91 9739204027; Email: murli.murthi@gmail.com

Graduate students in engineering interventions in foods under Dr. Meghwal's supervision

FOREWORD 3 BY KHURSHEED ALAM KHAN

We are living in the twenty-first century, an era of new technologies. *Engineering Interventions in Agricultural Processing* covers new technology such as machine vision, cold plasma, and ohmic heating design that are the future of food processing world, making human life simple because these technologies are compatible with the nature. We have achieved a lot with the help of engineering technologies. For example, Machine Vision Technology can help in programmed harvesting of crops such as in apple picking, monitoring the quality of food products during processing (color of baked products and analyzing the cell structure for breads), thereby reducing the cost of production with less fatigue to manpower.

Cold plasma technology emerged as a powerful tool that has significant benefits. The most noticeable application is the disinfection of surfaces in particular equipment, such as packaging, food contact surfaces or even food itself. Compatibility with food products could allow shelf-life extension or online disinfection of processing equipment to reduce cross-contamination and the establishment of biofilms on equipment. Plasma treatment requires no liquids, making this technology an ideal disinfection tool for manufacturers of low water activity products. Ohmic heating technology has the potential to cook meat in a much shorter time than conventional cooking methods, thereby reducing the energy consumption and number of man-hours used while processing.

Ingestion of certain plants such lemon grass and tea polyphenols has been proved beneficial to the health of people. Nutritionally, lemon grass is a good source of vitamins A and C, folic acid, magnesium, zinc, copper, iron, potassium, phosphorus, calcium and manganese. It has many beneficial medicinal properties, including analgesic, anti-inflammatory, antidepressant, antipyretic, antiseptic, antibacterial, antifungal, astringent, carminative, diuretic, febrifuge, insecticidal, sedative, and anti-cancer properties. The leaves, stems and bulb of lemongrass are used in various treatments. Polyphenolic compounds present in tea may reduce the risk of a variety of illnesses, including cancer and coronary heart disease.

Finally, I would like to applaud the authors, who have exerted enormous effort and sincere work to publish this book.

Khursheed Alam Khan, PhD,
Assistant Professor (Agricultural Engineering),
College of Horticulture, Rajmata Vijayaraje
Scindia Agriculture University (RVSKVV),
Mandsaur-45800, Gwalior, India
Mobile: +91-9425942903
E-mail: khan_undp@yahoo.ca

PREFACE 1 BY MEGH R. GOYAL

According to http://asabe.org: *"Agricultural and Biological Engineering (ABE) is an engineering that applies engineering principles and the fundamental concepts of biology to agricultural and biological systems and tools, for the safe, efficient and environmentally sensitive production, processing, and management of agricultural, biological, food, and natural resources systems. Process engineers combine design expertise with manufacturing methods to develop economical and responsible processing solutions for agricultural industry. Also food and process engineers look for ways to reduce waste by devising alternatives for treatment, disposal and utilization."*

In 1955, S. M. Henderson and R. L. Perry published their first classical book on *Agricultural Process Engineering*. In their book on pages *vii–viii*, they define agricultural processing as, *"any processing activity that is or can be done on the farm or by local enterprises in which the farmer has an active interest—any farm or local activity that maintains or raises the quality or changes the form of a farm product may be considered as processing. Agricultural processing activities may include cleaning/sorting/grading/treating/drying/grinding or mixing/milling/canning/packing/dressing/freezing/conditioning/and transportation, etc. Specific farm product may involve a specific activity(ies)."* This focus has evolved over the years as new technologies have become available.

Each one of us eats a processed food daily. In today's era of engineering interventions in agriculture, processing of agricultural produce has become a necessity of our daily living. I am not an exception, as I have seen and tasted all kinds of processed food (except meat and fish as I am a vegetarian).

The Mango Festival was held at the Agricultural Experiment Station of the University of Puerto Rico, Juana Diaz (http://agris.fao.org/agris-search/search.do?recordID=US201300631580) during July 11–12 of 1987 (http://www.worldcat.org/title/2do-festival-del-mango-sab-11-dom-12-julio-87-fortuna-subestacion-experimental-agricola-juana-diaz-puerto-rico/

oclc/19783599), under my leadership as Program Chairman. I felt proud of my culinary skills as I was able develop, cook, display and sell about 20 processed products from mango fruits (first time outside my profession). I am not a professional cook, however I love to cook. My wife has prohibited me to enter into her kitchen as it has been too expensive for her (kitchen has been on fire three times, as I leave the food unattended most of the times: Of course my bad habit). Of course, our children love my food.

Apple Academic Press Inc. published my first book on *Management of Drip/Trickle or Micro Irrigation*, a 10-volume set under the book series *Research Advances in Sustainable Micro Irrigation*, in addition to other books in the focus areas of agricultural and biological engineering. The mission of this book volume is to introduce the profession of agricultural and biological engineering.

At 49th annual meeting of the Indian Society of Agricultural Engineers at Punjab Agricultural University during February 22–25 of 2015, a group of ABEs convinced me that there is a dire need to publish book volumes on the focus areas of agricultural and biological engineering (ABE). This is how the idea was born on new book series, *Innovations in Agricultural and Biological Engineering*.

The contributions by cooperating authors to this book volume have been most valuable in the compilation. Their names are mentioned in each chapter and in the list of contributors. This book would not have been written without the valuable cooperation of these investigators, many of whom are renowned scientists who have worked in the field of ABE throughout their professional careers.

Deepak Kumar Verma is pursuing a PhD degree as a DST INSPIRE Fellow at Agricultural and Food Engineering Department, Indian Institute of Technology, Kharagpur, India. He joins me as coeditor of this book volume. He is a frequent contributor to my book series and a staunch supporter of my profession. His contribution to the contents and quality of this book has been invaluable.

I thank editorial staff, Sandy Jones Sickels, Vice President, and Ashish Kumar, Publisher and President at Apple Academic Press, Inc., for making every effort to publish the book when the diminishing water resources are a major issue worldwide. Special thanks are due to the AAP Production

Staff for typesetting the entire manuscript and for the quality production of this book.

I request the reader to offer your constructive suggestions that may help to improve the next edition.

I express my deep admiration to my family for understanding and collaboration during the preparation of this book. As an educator, there is a piece of advice to one and all in the world: *"Permit that our almighty God, our Creator and excellent Teacher, allow us to process agricultural products wisely without contaminating our planet. I invite my community in agricultural engineering to contribute book chapters to the book series by getting married to my profession"*. I am in total love with our profession by length, width, height and depth. Are you?

—Megh R. Goyal, PhD, PE
Senior Editor-in-Chief

PREFACE 2 BY DEEPAK KUMAR VERMA

Agriculture is a gateway that contributes to the livelihood of communities and does not degrade the soil environment during production of food. Agriculture means keeping in existence; keep up; maintain or prolong, etc.: many goals are encompassed in its meaning.

In agriculture, a fundamental knowledge of biology has been a gateway to solutions for many practical problems. Today, we have found that there are many breakthroughs in agriculture processing, which have contributed spectacularly to human welfare over the past few decades, and scientists and researchers have continued to use biological principles and engineering tools in agriculture to develop usable, tangible, economically viable products. In addition, scientists and researchers also apply their scientific knowledge and expertise of pure and applied sciences such as biocatalysts, bioinformatics, biomechanics, bioreactor design, surface science, fluid mechanics, heat and mass transfer, kinetics, separation and purification processes, thermodynamics and polymer science, etc.

In agricultural education and research, the processing discipline has undergone tremendous changes in the past few decades. This covers human endeavors in the broad sense, such as transmission, absorption, and acquisition of knowledge for the better of understanding the processes that lead to scientific solutions for real-world problems in a cost-effective manner. Throughout the world, many organizations, institutions, and universities are playing a pivotal key role in the improvement of the present agricultural education system and research to provide graduates, postgraduates, research scholars, and world-class scientists with the best knowledge in the field of agricultural processing. The quality of research and education alone would compete in the international scenario worldwide. It is imperative that education should provide important skills as well as being commercially oriented in order to satisfycurrent market demands.

This volume, *Engineering Interventions in Agricultural Processing,* will be an asset to improve and to provide knowledge on agricultural processing, an important area in agricultural and biological engineering.

The volume is divided into in three main parts. Part I: Agricultural Processing: Interventions in Engineering Technologies consists of six chapters devoted to research opportunities in food and agricultural processing. In one chapter machine vision technology as a novel technological method described along with its principles and applications for assessment of the quality of an agricultural products. The major advantages of installed automatic software for quality control and inspection in food processing plants(replacing the visual quality of food products assessed by human inspection with a naked eye) are discussed.

Chapter 2, on cold plasma known as non-thermal technology,discusses a different state-of-the-art technology and covers flexible tools for sterilizing surfaces of food and products, whereas in Chapter 3, ohmic heating, known as thermal technology, describes research progress on meat cooking with ohmic heating as a novel technology in food processing stated better than traditional method.

Chapter 4, on solid-state fermentation associated with dairy products, describes the technology and application for β-galactosidase production from *Aspergillusoryzœ*. Two chapters on nanotechnology explore knowledge: (a) on application of nano-particle based delivery systems as a new approach in sustainable agriculture and eco-friendly environments; and (b) synthesis of silver nanoparticles via biological route.

Part II: Novel Practices in Agricultural Processing contains three chapters dealing with research opportunities and describing novel practices for bioleaching, seed priming, and future prospects of modified pearl millet starch. Part III: Agricultural Processing: Health Benefits of Medicinal Plants consists of two chapters devoted to research and developments in human health benefits from tea and lemon grass.

With contributions from a broad range of leading researchers and teachers, this book focuses on areas of processing technologies in agriculture as discussed above. It will provide a guide to students, instructors, and researchers in agriculture. In addition, agricultural professionals who are seeking recent advanced and innovative knowledge in processing will find this book helpful. It is envisaged that this book will also serve as a referencesource for individuals engaged in research, processing, and product development in agricultural processing.

Preface 2 by Deepak Kumar Verma

With great pleasure, I would like to extend my sincere thanks to all the learned contributors for the magnificent work. Their timely response, excellent devoted contributions to detail, accuracy of information, and consistent support and cooperation have made our task as editors a pleasure.

It is hoped that this edition will stimulate discussion and generate helpful comments to improve upon future editions.

I like to thank opportunity offered by Dr. Megh R. Goyal, who is a staunch supporter of the agricultural and biological engineering profession and also is the prime mover of this book series.

Finally, we acknowledge Almighty God, who provided all the inspirations, insights, positive thoughts, and channels to complete this book project.

—Deepak Kumar Verma, PhD
Co-Editor

WARNING/DISCLAIMER

PLEASE READ CAREFULLY

The goal of this compendium, *Engineering Interventions in Agricultural Processing*, is to guide the world engineering community on how to efficiently process agricultural products. The reader must be aware that the dedication, commitment, honesty, and sincerity are most important factors in a dynamic manner for a complete success.

The editors, the contributing authors, the publisher and the printer have made every effort to make this book as complete and as accurate as possible. However, there still may be grammatical errors or mistakes in the content or typography. Therefore, the contents in this book should be considered as a general guide and not a complete solution to address any specific situation in irrigation. For example, fruit or vegetable or meat or grain, etc. requires a different type of engineering intervention to process such produce.

The editors, the contributing authors, the publisher and the printer shall have neither liability nor responsibility to any person, any organization or entity with respect to any loss or damage caused, or alleged to have caused, directly or indirectly, by information or advice contained in this book. Therefore, the purchaser/reader must assume full responsibility for the use of the book or the information therein.

The mention of commercial brands and trade names are only for technical purposes. We do not endorse a particular products or equipment mentioned.

All web-links that are mentioned in this book were active on December 31, 2016. The editors, the contributing authors, the publisher and the printing company shall have neither liability nor responsibility, if any of the web-links is inactive at the time of reading of this book.

ABOUT SENIOR EDITOR-IN-CHIEF

Megh R. Goyal, PhD, PE
*Retired Professor in Agricultural and Biomedical
Engineering, University of Puerto Rico,
Mayaguez Campus Senior Acquisitions Editor,
Biomedical Engineering and Agricultural Science,
Apple Academic Press, Inc.*

Megh R. Goyal, PhD, PE, is a Retired Professor in Agricultural and Biomedical Engineering from the General Engineering Department in the College of Engineering at University of Puerto Rico–Mayaguez Campus; and Senior Acquisitions Editor and Senior Technical Editor-in-Chief in Agriculture and Biomedical Engineering for Apple Academic Press Inc. He has worked as a Soil Conservation Inspector and as a Research Assistant at Haryana Agricultural University and Ohio State University.

He was the first agricultural engineer to receive the professional license in Agricultural Engineering in 1986 from the College of Engineers and Surveyors of Puerto Rico. On September 16, 2005, he was proclaimed as "Father of Irrigation Engineering in Puerto Rico for the twentieth century" by the ASABE, Puerto Rico Section, for his pioneering work on micro irrigation, evapotranspiration, agroclimatology, and soil and water engineering. During his professional career of 50 years, he has received many prestigious awards such as Scientist of the Year, Blue Ribbon Extension Award, Research Paper Award, Nolan Mitchell Young Extension Worker Award, Agricultural Engineer of the Year, Citations by Mayors of Juana Diaz and Ponce, Membership Grand Prize for ASAE Campaign, Felix Castro Rodriguez Academic Excellence, Rashtrya Ratan Award and Bharat Excellence Award and Gold Medal, Domingo Marrero Navarro Prize, Adopted Son of Moca, Irrigation Protagonist of UPRM, Man of Drip Irrigation by Mayor of Municipalities of Mayaguez/Caguas/Ponce and Senate/Secretary of Agriculture of ELA, Puerto Rico.

The Water Technology Centre of Tamil Nadu Agricultural University in Coimbatore, India recognized Dr. Goyal as one of the experts "who rendered meritorious service for the development of micro irrigation sector in India" by bestowing "*Award of Outstanding Contribution in Micro Irrigation.*" This award was presented to Dr. Goyal during the inaugural session of the National Congress on "New Challenges and Advances in Sustainable Micro Irrigation on March 1, 2017, held at Tamil Nadu Agricultural University.

A prolific author and editor, he has written more than 200 journal articles and has edited over 45 reference and textbooks. He received his BSc degree in engineering from Punjab Agricultural University, Ludhiana, India; his MSc and PhD degrees from Ohio State University, Columbus; and his Master of Divinity degree from Puerto Rico Evangelical Seminary, Hato Rey, Puerto Rico, USA. Readers may contact him at: goyalmegh@gmail.com.

ABOUT CO-EDITOR

Deepak Kumar Verma
Research Scholar, Department of Agricultural and Food Engineering, Indian Institute of Technology, Kharagpur, West Bengal, India

Deepak Kumar Verma is an agricultural science professional and is currently a PhD Research Scholar in the specialization of food processing engineering in the Agricultural and Food Engineering Department, Indian Institute of Technology, Kharagpur (WB), India. In 2012, he received a DST-INSPIRE Fellowship for PhD study by the Department of Science and Technology (DST), Ministry of Science and Technology, Government of India.

Mr. Verma is currently working on the research project "Isolation and Characterization of Aroma Volatile and Flavoring Compounds from Aromatic and Non-Aromatic Rice Cultivars of India." His previous research work included "Physico-Chemical and Cooking Characteristics of Azad Basmati (CSAR 839-3): A Newly Evolved Variety of Basmati Rice (Oryza sativa L.)". He earned his BSc degree in agricultural science from the Faculty of Agriculture at Gorakhpur University, Gorakhpur, and his MSc (Agriculture) in Agricultural Biochemistry in 2011. He also received an award from the Department of Agricultural Biochemistry, Chandra Shekhar Azad University of Agricultural and Technology, Kanpur, India.

Apart of his area of specialization as *plant biochemistry*, he has also built a sound background in plant physiology, microbiology, plant pathology, genetics and plant breeding, plant biotechnology and genetic engineering, seed science and technology, food science and technology, etc. In addition, he is member of different professional bodies, and his activities and accomplishments include conferences, seminar, workshop, training, and also the publication of research articles, books, and book chapters.

BOOK ENDORSEMENTS

Processing of agricultural commodities plays a key role in improving quality or adding value to the produce and resulting into the increase in their market values along with the significant raising of the income of the growers. Appropriate technologies through engineering interventions are the effective means in achieving transformation, preservation and preparation of agricultural production for intermediary or final consumption. When agricultural productions are presently at the saturation level and their wastage are very high, it is the right time to think of the preservation and processing of the produces for feeding the ever-increasing population of the world. This goal can only be achieved through the development of appropriate technology and progress of sustainable food processing industry across the globe. We the educational community appreciate the efforts by Apple Academic Press Inc., for its efforts and inspiration to publish books in Agricultural and Biological Engineering.

—M. K. Ghosal, PhD
Professor,
Department of Farm Machinery and Power,
College of Agricultural Engineering and Technology,
Orissa University of Agriculture and Technology,
Bhubaneswar–751003, Odisha
E-mail: mkghosal1@rdiffmail.com

BOOKS ON AGRICULTURAL AND BIOLOGICAL ENGINEERING FROM APPLE ACADEMIC PRESS, INC.

Management of Drip/Trickle or Micro Irrigation
Megh R. Goyal, PhD, PE, Senior Editor-in-Chief

Evapotranspiration: Principles and Applications for Water Management
Megh R. Goyal, PhD, PE, and Eric W. Harmsen, Editors

Book Series: Research Advances in Sustainable Micro Irrigation
Senior Editor-in-Chief: Megh R. Goyal, PhD, PE
Volume 1: Sustainable Micro Irrigation: Principles and Practices
Volume 2: Sustainable Practices in Surface and Subsurface Micro Irrigation
Volume 3: Sustainable Micro Irrigation Management for Trees and Vines
Volume 4: Management, Performance, and Applications of Micro Irrigation Systems
Volume 5: Applications of Furrow and Micro Irrigation in Arid and Semi-Arid Regions
Volume 6: Best Management Practices for Drip Irrigated Crops
Volume 7: Closed Circuit Micro Irrigation Design: Theory and Applications
Volume 8: Wastewater Management for Irrigation: Principles and Practices
Volume 9: Water and Fertigation Management in Micro Irrigation
Volume 10: Innovation in Micro Irrigation Technology

Book Series: Innovations and Challenges in Micro Irrigation
Senior Editor-in-Chief: Megh R. Goyal, PhD, PE
Volume 1: Principles and Management of Clogging in Micro Irrigation
Volume 2: Sustainable Micro Irrigation Design Systems for Agricultural Crops: Methods and Practices
Volume 3: Performance Evaluation of Micro Irrigation Management: Principles and Practices
Volume 4: Potential Use of Solar Energy and Emerging Technologies in Micro Irrigation

Volume 5: Micro Irrigation Management: Technological Advances and
Their Applications
Volume 6: Micro Irrigation Engineering for Horticultural Crops: Policy
Options, Scheduling, and Design
Volume 7: Micro Irrigation Scheduling and Practices
Volume 8: Engineering Interventions in Sustainable Trickle Irrigation: Water
Requirements, Uniformity, Fertigation, and Crop Performance

Book Series: Innovations in Agricultural and Biological Engineering
Senior Editor-in-Chief: Megh R. Goyal, PhD, PE
- Dairy Engineering: Advanced Technologies and their Applications
- Developing Technologies in Food Science: Status, Applications, and
  Challenges
- Emerging Technologies in Agricultural Engineering
- Engineering Interventions in Agricultural Processing
- Engineering Interventions in Foods and Plants
- Engineering Practices for Agricultural Production and Water
  Conservation: An Interdisciplinary Approach
- Flood Assessment: Modeling and Parameterization
- Food Engineering: Modeling, Emerging Issues and Applications.
- Food Process Engineering: Emerging Trends in Research and Their
  Applications
- Food Technology: Applied Research and Production Techniques
- Modeling Methods and Practices in Soil and Water Engineering
- Novel Dairy Processing Technologies: Techniques, Management, and
  Energy Conservation
- Processing Technologies for Milk and Milk Products: Methods,
  Applications, and Energy Usage
- Soil and Water Engineering: Principles and Applications of Modeling
- Soil Salinity Management in Agriculture: Technological Advances
  and Applications
- State-of-the-Art Technologies in Food Science: Human Health,
  Emerging Issues and Specialty Topics
- Sustainable Biological Systems for Agriculture: Emerging Issues in
  Nanotechnology, Biofertilizers, Wastewater, and Farm Machines
- Technological Interventions in Dairy Science: Innovative Approaches
  in Processing, Preservation, and Analysis of Milk Products
- Technological Interventions in Management of Irrigated Agriculture
- Technological Interventions in the Processing of Fruits and Vegetables

EDITORIAL

Under the book series titled *Innovations in Agricultural and Biological Engineering*, Apple Academic Press, Inc., (AAP) is publishing book volumes in the specialty areas, over a span of 8 to 10 years. These specialty areas have been defined by the American Society of Agricultural and Biological Engineers (http://asabe.org). AAP wants to be the principal source of books in the field of agricultural and biological engineering. We seek book proposals from the readers in area of their expertise.

The mission of this series is to provide knowledge and techniques for agricultural and biological engineers (ABEs). The series aims to offer high-quality reference and academic content in agricultural and biological engineering (ABE) that is accessible to academicians, researchers, scientists, university faculty, and university-level students and professionals around the world.

The following material has been edited/modified and reproduced below from *Goyal, Megh R., 2006. Agricultural and biomedical engineering: Scope and opportunities. Paper Edu_47 at the Fourth LACCEI International Latin American and Caribbean Conference for Engineering and Technology (LACCEI' 2006): Breaking Frontiers and Barriers in Engineering: Education and Research by LACCEI University of Puerto Rico – Mayaguez Campus, Mayaguez, Puerto Rico, June 21–23.*

WHAT IS AGRICULTURAL AND BIOLOGICAL ENGINEERING (ABE)?

"Agricultural Engineering (AE) involves application of engineering to production, processing, preservation and handling of food, fiber, and shelter. It also includes transfer of technology for the development and welfare of rural communities," according to http://isae.in." *ABE is the discipline of engineering that applies engineering principles and the fundamental concepts of biology to agricultural and biological systems and tools, for the safe, efficient and environmentally sensitive production, processing,*

and management of agricultural, biological, food, and natural resources systems," according to http://asabe.org.

"AE is the branch of engineering involved with the design of farm machinery, with soil management, land development, and mechanization and automation of livestock farming, and with the efficient planting, harvesting, storage, and processing of farm commodities," definition by: http://dictionary.reference.com/browse/agricultural+engineering.

"AE incorporates many science disciplines and technology practices to the efficient production and processing of food, feed, fiber and fuels. It involves disciplines like mechanical engineering (agricultural machinery and automated machine systems), soil science (crop nutrient and fertilization, etc.), environmental sciences (drainage and irrigation), plant biology (seeding and plant growth management), animal science (farm animals and housing) etc.," by: http://www.ABE.ncsu.edu/academic/agricultural-engineering.php.

"According to https://en.wikipedia.org/wiki/Biological_engineering: *"BE (Biological engineering) is a science-based discipline that applies concepts and methods of biology to solve real-world problems related to the life sciences or the application thereof. In this context, while traditional engineering applies physical and mathematical sciences to analyze, design and manufacture inanimate tools, structures and processes, biological engineering uses biology to study and advance applications of living systems."*

SPECIALTY AREAS OF ABE

Agricultural and Biological Engineers (ABEs) ensure that the world has the necessities of life including safe and plentiful food, clean air and water, renewable fuel and energy, safe working conditions, and a healthy environment by employing knowledge and expertise of sciences, both pure and applied, and engineering principles. Biological engineering applies engineering practices to problems and opportunities presented by living things and the natural environment in agriculture. BA engineers understand the interrelationships between technology and living systems, have available a wide variety of employment options. *"ABE embraces a variety of following specialty areas,"* http://asabe.org. As new technology and information emerge, specialty areas are created, and many overlap with one or more other areas.

Editorial

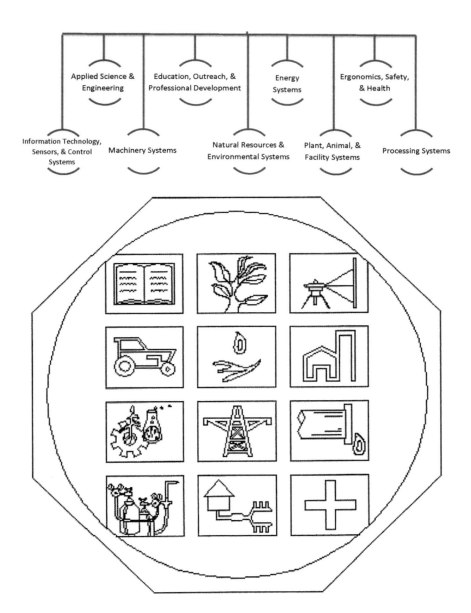

1. **Aquacultural Engineering**: ABEs help design farm systems for raising fish and shellfish, as well as ornamental and bait fish. They specialize in water quality, biotechnology, machinery, natural resources, feeding and ventilation systems, and sanitation. They

seek ways to reduce pollution from aquacultural discharges, to reduce excess water use, and to improve farm systems. They also work with aquatic animal harvesting, sorting, and processing.

2. **Biological Engineering** applies engineering practices to problems and opportunities presented by living things and the natural environment.

3. **Energy:** ABEs identify and develop viable energy sources – biomass, methane, and vegetable oil, to name a few – and to make these and other systems cleaner and more efficient. These specialists also develop energy conservation strategies to reduce costs and protect the environment, and they design traditional and alternative energy systems to meet the needs of agricultural operations.

4. **Farm Machinery and Power Engineering**: ABEs in this specialty focus on designing advanced equipment, making it more efficient and less demanding of our natural resources. They develop equipment for food processing, highly precise crop spraying, agricultural commodity and waste transport, and turf and landscape maintenance, as well as equipment for such specialized tasks as removing seaweed from beaches. This is in addition to the tractors, tillage equipment, irrigation equipment, and harvest equipment that have done so much to reduce the drudgery of farming.

5. **Food and Process Engineering:** Food and process engineers combine design expertise with manufacturing methods to develop economical and responsible processing solutions for industry. Also food and process engineers look for ways to reduce waste by devising alternatives for treatment, disposal and utilization.

6. **Forest Engineering**: ABEs apply engineering to solve natural resource and environment problems in forest production systems and related manufacturing industries. Engineering skills and expertise are needed to address problems related to equipment design and manufacturing, forest access systems design and construction; machine-soil interaction and erosion control; forest operations analysis and improvement; decision modeling; and wood product design and manufacturing.

7. **Information and Electrical Technologies Engineering** is one of the most versatile areas of the ABE specialty areas, because it is

applied to virtually all the others, from machinery design to soil testing to food quality and safety control. Geographic information systems, global positioning systems, machine instrumentation and controls, electromagnetics, bioinformatics, biorobotics, machine vision, sensors, spectroscopy: These are some of the exciting information and electrical technologies being used today and being developed for the future.

8. **Natural Resources:** ABEs with environmental expertise work to better understand the complex mechanics of these resources, so that they can be used efficiently and without degradation. ABEs determine crop water requirements and design irrigation systems. They are experts in agricultural hydrology principles, such as controlling drainage, and they implement ways to control soil erosion and study the environmental effects of sediment on stream quality. Natural resources engineers design, build, operate and maintain water control structures for reservoirs, floodways and channels. They also work on water treatment systems, wetlands protection, and other water issues.

9. **Nursery and Greenhouse Engineering**: In many ways, nursery and greenhouse operations are microcosms of large-scale production agriculture, with many similar needs – irrigation, mechanization, disease and pest control, and nutrient application. However, other engineering needs also present themselves in nursery and greenhouse operations: equipment for transplantation; control systems for temperature, humidity, and ventilation; and plant biology issues, such as hydroponics, tissue culture, and seedling propagation methods. And sometimes the challenges are extraterrestrial: ABEs at NASA are designing greenhouse systems to support a manned expedition to Mars!

10. **Safety and Health:** ABEs analyze health and injury data, the use and possible misuse of machines, and equipment compliance with standards and regulation. They constantly look for ways in which the safety of equipment, materials and agricultural practices can be improved and for ways in which safety and health issues can be communicated to the public.

11. **Structures and Environment:** ABEs with expertise in structures and environment design animal housing, storage structures, and greenhouses, with ventilation systems, temperature and humidity controls, and structural strength appropriate for their climate and purpose. They also devise better practices and systems for storing, recovering, reusing, and transporting waste products.

CAREER IN AGRICULTURAL AND BIOLOGICAL ENGINEERING

One will find that university ABE programs have many names, such as biological systems engineering, bioresource engineering, environmental engineering, forest engineering, or food and process engineering. Whatever the title, the typical curriculum begins with courses in writing, social sciences, and economics, along with mathematics (calculus and statistics), chemistry, physics, and biology. Student gains a fundamental knowledge of the life sciences and how biological systems interact with their environment. One also takes engineering courses, such as thermodynamics, mechanics, instrumentation and controls, electronics and electrical circuits, and engineering design. Then student adds courses related to particular interests, perhaps including mechanization, soil and water resource management, food and process engineering, industrial microbiology, biological engineering or pest management. As seniors, engineering students team up to design, build, and test new processes or products.

For more information on this series, readers may contact:

Ashish Kumar, Publisher and President	Megh R. Goyal, PhD, PE
Sandy Sickels, Vice President	Book Series Senior
Apple Academic Press, Inc.	Editor-in-Chief
Fax: 866-222-9549	*Innovations in Agricultural*
E-mail: ashish@appleacademicpress.com	*and Biological Engineering*
http://www.appleacademicpress.com/	E-mail: goyalmegh@gmail.com
publishwithus.php	

PART I

AGRICULTURAL PROCESSING: INTERVENTIONS IN ENGINEERING TECHNOLOGIES

CHAPTER 1

MACHINE VISION TECHNOLOGY IN FOOD PROCESSING INDUSTRY: PRINCIPLES AND APPLICATIONS—A REVIEW

P. S. MINZ, C. SINGH, and I. K. SAWHNEY

CONTENTS

1.1 Introduction 4

1.2 Machine Vision System (MVS) 4

1.3 Components of Machine Vision System 6

1.4 Classification of Machine Vision System Based on
Spectral Range 11

1.5 Classification of Machine Vision System on the Basis of
Number of Spectral Band 15

1.6 Application of Machine Vision System in Food Industry 16

1.7 Color Measurement Techniques Using Machine Vision System ... 22

1.8 Application of Open Source Software: MVS 24

1.9 Summary 25

Keywords 26

References 26

1.1 INTRODUCTION

Machine vision is an engineering technology that combines mechanics, optical instrumentation, and digital image processing technology. As an integrated mechanical-optical-electronic-software system, machine vision has been widely used for examining, monitoring, and controlling a very broad range of applications [66]. Recent advances in hardware and software have aided in this expansion by providing powerful solutions, leading to more studies on the development of computer vision systems in the food industry [52, 60, 80]. Research in this focus area started during 1950s. This concept did not become industrialized until early 1980s.

Machine vision technology was initially used in various industrial applications to read and verify letters, symbols, and numbers. The 1990s brought a boom of growth to the machine vision industry. Electronics and automobile industries were first to implement this technology. The advancement of computer technology was the main driver behind this expansion [46]. Looking at its vast potential, this technology was quickly picked up by number of industries. Machine vision technology is now one of the fastest adopted technologies in food processing industry.

This chapter focuses on principles and applications of machine vision technology in the food processing technology.

1.2 MACHINE VISION SYSTEM (MVS)

Machine vision is the construction of explicit informative and meaningful descriptions of a physical object via image analysis. MV is related to, though distinct from, computer vision (Figure 1.1). Appendix – A gives a detailed list of *Glossary of Terms Related to Machine Vision Technology* (https://en.wikipedia.org/wiki/Glossary_of_machine_vision). The machine vision image, obtained by a physical sensor, is analyzed by appropriate computing hardware and software utilities to perform a predefined visual task in order to improve the quality of human vision by electronically supported image perception and investigation [15]. The main steps in image-processing analysis are [26]:

Machine Vision Technology in Food Processing Industry

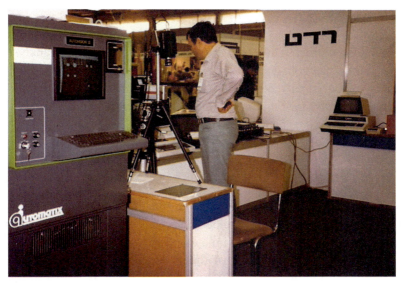

FIGURE 1.1 Early Automatix (now part of Microscan) machine vision system "Autovision II" at "Technology 83" trade show in Israel demonstrating blob analysis using backlighting (Source: ArnoldReinhold. https://en.wikipedia.org/wiki/Machine_visionhttps://creativecommons.org/licenses/by-sa/3.0/)

a. Image acquisition and conversion of the obtained images to digital form.
b. Image improvement for pre-processing.
c. Partitioning the digital image to separate non-overlapping regions by a segmentation process.
d. Obtaining the characteristics of objects by object measurement operations.
e. Classification to identify objects class groups.

In MVS, image capturing devices or sensors are used to view and generate images of the sample. Some of the devices or sensors used in generating images include: charged coupled device (CCD), scanners, X-ray and near infrared spectroscopy [48]. Its use in the field of agriculture has been the area of active research during the last decade. Several systems have been reported for inspecting a variety of agricultural produce ranging from fruits [6, 70, 83] and vegetables [3], grains [22, 82, 91], and seeds [75]. There is a rapid rise in machine vision research for application in food processing industry.

1.3 COMPONENTS OF MACHINE VISION SYSTEM

Machine vision process involves the capturing, processing and analysis of images, with others observing that it aims to duplicate the effect of human vision by electronically perceiving and understanding an image. A MVS consists of four basic components: image illumination, image acquisition, camera-hardware interface and image analysis (Figure 1.2).

1.3.1 IMAGE ILLUMINATION

It is similar to the human eye, the level and the quality of illumination that affect the vision system. The performance of the illumination system greatly influences the quality of image and plays an important role in the overall efficiency and accuracy of the system. Illumination systems are light sources. An appropriate light system can reduce the effect of reflection, shadow and some noise-making parameters, thereby reducing the

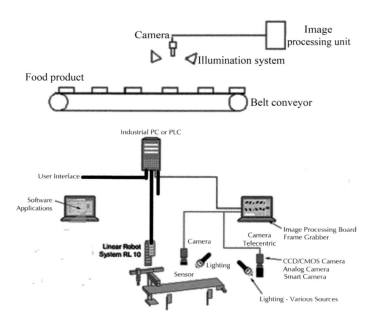

FIGURE 1.2 Components of a machine vision system.

required time of image processing [25]. Type of lighting, location and color quality play an important role in bringing out a clear image of the object [61]. The light range can be: the UV (200–400 nm), VIS (400–700 nm), or NIR (700–2500 nm).

1.3.2 DEVICES FOR IMAGE ACQUISITION

1.3.2.1 Vision Camera

An image can be generated by many different sensors. Cameras supported by silicon, complementary metal oxide semiconductor (CMOS) and charge-coupled-device (CCD) sensors are most common visible imaging systems. The sensor can be a one-dimensional (1D), linear array or a two-dimensional (2D) array [55]. Image sensors used in machine vision are usually based on solid-state CCD camera technology of the array type or the line-scan type using thermionic tube devices. Both monochrome and polychrome cameras have been used [15]. There can be three light detection sensors in the camera, dedicated to each primary color (Red, Green, Blue), or one sensor can be selectively used to handle all the three primary colors. The software can control the camera settings, timing of image acquisition, light source and can analyze the image to extract desired features to make decisions, which may include non-contact sensing, measuring object shape and dimensions, detecting product defects; providing process control feedback alerting production line operators for in-process system failures; and providing product quality statistics [7, 8, 74, 81].

1.3.2.2 Digital Camera

With advances in digital cameras, the camera and the image capture system generally merge into a single device. This device communicates with the computer via cables (e.g., USB or Fire-wire), or by Wi-Fi means. In a study, digital camera has been used to acquire image to predict immobilized yeast-biomass. Digital images were taken with different biomass concentration, and the RGB-analysis showed significant differences in the blue field. The histogram of the blue channel was used to develop a PLS multivariate calibration to predict biomass concentration [1]. A commercially available digital camera was used in a low-cost automatic observation system for monitoring crop growth change in open-air fields [73].

1.3.2.3 Scanner

Color scanners with improved functionality and quality are becoming increasingly cheaper and are now commonly available (Figure 1.3). Document scanners are being used. It would be beneficial if such scanners could be used as an inexpensive alternative for colorimetric measurements [49]. A MVS for color grading of lentils was developed using

FIGURE 1.3 3D-laser-scanner (FARO Laser Scanner LS) mounted on a tripod (Source: Dr. Schorsch. https://en.wikipedia.org/wiki/Glossary_of_machine_vision. https://creativecommons.org/licenses/by-sa/3.0/)

a flatbed scanner as the image gathering device. Grain samples belonging to different grades of large green lentils were scanned and analyzed for image color, color distribution, and textural features [75]. In another study, a flatbed scanner was used to capture the color image of the samples. The obtained images were saved in TIFF format. The color was quantitatively analyzed (in terms of L, a, b) using Photoshop software [24].

1.3.3 CAMERA HARDWARE INTERFACE

Camera (CMOS/CCD) can be connected to compatible hardware/computer for data transfer and control of camera. Image acquisition device can be interfaced using any of the five methods: Gigabit Ethernet; Camera link; Firewire; USB 2.0; and USB 3.0

1.3.4 IMAGE ANALYSIS

Image analysis can provide a wide range of information about a product from a single image in a fraction of a second, making it possible to analyze products as they pass on a conveyor belt [78]. Image acquisition, when carried out in the visible light range, will yield three grayscales in red, green and blue. Within the image, there is a region of interest (ROI) that requires removing the non-interesting parts of the image i.e.: segmenting the image into ROI and non-ROI [26]. The task of segmentation might be straightforward or so difficult that manual intervention is required to finish the extraction [38, 39] when trying to isolate the ribeye muscle in beef cuts. Irrespective of the level of difficulty, algorithms are available to successfully derive the ROI even if it is highly inefficient [41] and even if manual intervention is required. Some of the important segmentation algorithms are [26]:

- Graceful degradation algorithm which built its thresholds for isolating beef carcasses with multiple criteria ensuring that failure of one criterion did not cause the thresholding process to collapse [13].
- Active contour algorithm that monitored the ripening of ham muscle by creating contours with Variational Calculus, Dynamic Programming and Greedy Algorithms [5].

- B-spline algorithm which used a lofting technique to generate the muscle boundaries [33].
- Region growing algorithm that grew the defected and non-defected regions of oranges to identify the flawed fruits [12].
- Shadow elimination algorithm which solved the problem of shadows confounding the segmentation of confocal microscopy images by combining partial differential equation based diffusion and thresholding segmentation [28].
- Gradient and spline algorithm which identified the boundaries of a cooked ham by obtaining radii, applying a wavelet transform, locating protrusion and applying spline interpolation [27].
- Clustering and thresholding algorithm that combined simple thresholding with a crisp K-means clustering algorithm and a clipping step to isolate the muscle of a beef cut [40].

If it is absolutely required, segmentation can be sidestepped by keeping only the ROI inside the field of view. After the image has been successfully segmented into the ROI and non-ROI, the useful parts should be characterized in meaningful ways that reflect the food grading rules such as those issued by the USDA or another competent authority and also condense predictive power into few variables to avoid the curse of dimensionality. Once these variables are found, they should be linked by a trained and tested model to independent experimental verification of the food quality [41]. This will allow the qualities of any future sample to be safely estimated.

The kind of grading procedures that are envisaged by the USDA rules can be divided into four essential categories of features: color, texture, shape and size [42]. All of these have been proven to be highly suitable to a computer vision system. The types of predictive models typically applied in computer vision processes are classical statistical methods and neural networks. This points to two principal avenues for improving the image processing: (1) Extracting better color, texture, size and shape features or create stronger predictive models that can better condense the useful image information as it is by exploring these avenues that additional hardware cost will be avoided; and (2) The area of predictive modeling that has already been widely explored in the mathematical and statistical sciences [41]. Image processing/analysis can be broadly divided into three levels [34, 80]: low level processing, intermediate level processing, and high level processing.

1.3.4.1 Low Level Processing

Low level techniques involve primitive preprocessing operations. Image pre-processing refers to the initial processing of the raw image data for correction of geometric distortions, removal of noise, gray level correction and correction for blurring. Pre-processing aims to improve image quality by suppressing undesired distortions or by the enhancement of important features of interest. This type of processing is characterized by both inputs and outputs being images.

1.3.4.2 Intermediate Level Processing

Mid-level processing techniques involve segmentation (i.e., image partitioning into objects or regions, description for reduction in a suitable form for computer processing, and classification recognition of individual objects). In a mid-level process, inputs are generally images, while outputs (e.g., edges, contours, and the identity of individual objects) are attributes extracted from inputs.

1.3.4.3 High Level Processing

High-level processing is the far end of the continuum trying to "make sense" of an ensemble of recognized objects, and performing the cognitive functions normally associated with vision.

1.4 CLASSIFICATION OF MACHINE VISION SYSTEM BASED ON SPECTRAL RANGE

1.4.1 VISIBLE SPECTRAL (VIS)

The visible light (VIS) covers the wavelengths of approx. 380 to 770 nm (Table 1.1 and Figure 1.4). Many products are assigned to commercial grades according to their color. For color description, there are different color coordinate systems. In the well-known three-dimensional (3D) RBG system, the three basic colors, red, blue, and green are represented as the

TABLE 1.1 Range of Wavelengths Corresponding to Different Colors

Color	Wavelength	Frequency	Photon energy
Violet	380–450 nm	668–789 THz	2.75–3.26 eV
Blue	450–495 nm	606–668 THz	2.50–2.75 eV
Green	495–570 nm	526–606 THz	2.17–2.50 eV
Yellow	570–590 nm	508–526 THz	2.10–2.17 eV
Orange	590–620 nm	484–508 THz	2.00–2.10 eV
Red	620–750 nm	400–484 THz	1.65–2.00 eV

*Basic colors are red, blue and green (RBG).

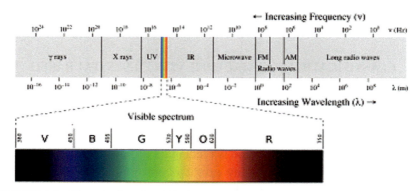

FIGURE 1.4 Three basic colors (RBG) and electromagnetic spectrum.

3 axles of a cube. However, other geometries are also used, for example, the spherical L*C*h * system, different *Commission Intl. de l'Eclairage* (CIE) systems, or the Hunter L, a, b system, which additionally describes brightness and saturation. They are aligned to adapt color distinctions to human seeing. Because color is a sensory perception, it is reasonable to adapt also the functionality of instrumental measurement to the physiology of the human eye. Sensors in colorimeters are provided with filters to simulate the three-color receptors of the eye [16].

1.4.2 ULTRAVIOLET (UV)

Ultraviolet light (UV) covers the optical spectrum from about 380 nm to 220 nm. UV luminescent sensors use the region from 320 nm to 380 nm, known as near UV or black light. Freeze damage detection in oranges by using machine vision and UV fluorescence was conducted by Slaughter et al. [77]. UV sensors are used to detect UV-enhanced inks used to print date codes on packages, and to detect glue on packages. UV based system can detect the presence of straws attached to juice box containers, when the orientation of the straw is too variable for a vision system [30].

1.4.3 INFRARED

- **Near Infrared (NIR):** The near-infrared range covers the wavelengths from 750 to 2500 nm, equal to the wave number range of 4000/cm to 12500/cm. Near Infrared System (NIRS) had been examined as a non-destructive method for the determination of firmness, freshness, Brix value, acidity, color, and other characteristics of many fruits; and from the results achieved, NIRS could be judged as an appropriate method for these applications [19, 45]. Today, NIRS is increasingly examined, tested, and used within all fields of food science and technology. Applications range from grains to potatoes, potato chips to meat, prepared meals, fruit juice and wine production where the contents of materials such as fat, water, protein, total nitrogen, sugar, and alcohol are assayed using this technique. NIRS is a method well suitable for the determination of chemical compounds containing OH-, CH-, and NH-groups. A hyperspectral imaging system in the near infrared (NIR) region

(900–1700 nm) was developed to predict the moisture content, pH and color in cooked, pre-sliced turkey hams [37]. NIR can again be classified as short-wave near infrared (SW-NIR), spectral range (400–1000 nm) and long-wave near infrared spectral range (LW-NIR, 900–2500 nm).

- **Short Wavelength Infrared (SWIR):** SWIR ranges from 1400 to 3000 nm. A shortwave infrared hyperspectral imaging system was explored to detect sour skin disease for onions [89].
- **Mid Wavelength Infrared (MWIR):** For mid-infrared (MIR), the spectral range is from 3000 to 8000 nm. Thermal MWIR imaging (3000–5000 nm) is useful for bruise recognition in apples when an active approach (lock-in or pulsed-phase) is applied [9].
- **Long Wavelength Infrared (LWIR):** LWIR ranges from 8000 to 15000 nm. Long waves (8000–12000 nm) have good transmission. Dual-wave band (long and short wave) cameras with two types of detectors are also available in the market.

1.4.4 X-RAY

X-rays cover the spectral range of 0.01–10 nm, which falls between gamma rays and ultra-violet rays. X-rays can penetrate through most horticultural products and the level of X-ray energy transmitted through the product depends on the incident energy and absorption coefficient, density and thickness of the product. X-ray imaging is thus useful for evaluating quality/maturity and internal defect of horticultural products. X-ray imaging technologies include: X-ray radiography which scans layers of the product to create 2D images; and computed tomography or CT scanning which creates 3D images. X-ray imaging showed potential for evaluating the maturity of peach, mango, and lettuce [10, 14,72].

Nuclear Magnetic Resonance (NMR) Spectroscopy

Since the discovery of the magnetic resonance phenomenon in 1946 and subsequent achievements, nuclear magnetic resonance (NMR) has become one of the most significant non-invasive techniques for internal inspection of biological objects. Derived from NMR are NMR spectroscopy, NMR relaxometry and magnetic resonance imaging (MRI). For NMR spectroscopy, resonance frequency encodes chemically equivalent nuclei populations at different electronic and chemical environments so that the outcome is an NMR spectrum where intensity is plotted versus frequency. MRI devoted spatial codification of the signal intensity produces a 2D- or 3D image. NMR technique offers information about the inside of an object, thus making it feasible for quality classification of fruits and vegetables. NMR is a useful non-destructive monitoring technique for a wide range of applications because it is sensitive to the concentration, chemical environment, mobility, and diffusion among other phenomena, related to certain nuclei [72].

1.5 CLASSIFICATION OF MACHINE VISION SYSTEM ON THE BASIS OF NUMBER OF SPECTRAL BAND

1.5.1 HYPERSPECTRAL IMAGING (HSI)

Hyperspectral deals with imaging narrow spectral bands over a continuous spectral range, and produce the spectra of all pixels in the scene. Generally, HSI ($n > 10$) is a useful tool to identify the optimal bands for developing MSI systems [65]. A main drawback of applying HSI is that the large amount data from the hyperspectral images increases the complexity of data analysis and slows the speed for processing [35]. Therefore, a sensor with only 20 bands can also be hyperspectral when it covers the range from 500 to 700 nm with 20 bands each 10 nm wide.

1.5.2 MULTISPECTRAL IMAGING (MSI)

MSI only uses a small number of key wavebands ($n < 10$) and thus it can obtain images that are much smaller than hyperspectral images. In addition, some MSI systems can acquire spatially-coherent band images simultaneously to reduce the image acquisition time [47]. MSI deals with several images at discrete and somewhat narrow bands. A multispectral sensor may have many bands covering the spectrum from the visible to the long-wave infrared. A sensor with 20 discrete bands covering the VIS, NIR, SWIR, MWIR, and LWIR would be considered multispectral.

1.5.3 ULTRASPECTRAL

It has interferometer type imaging sensors with a very fine spectral resolution. These sensors often have (but not necessarily) a low spatial resolution of several pixels only, a restriction imposed by the high data rate.

1.6 APPLICATION OF MACHINE VISION SYSTEM IN FOOD INDUSTRY

The primary uses for machine vision are automatic inspection and industrial robot guidance. Functions of MVS are shown in Figure 1.5. MVS encompasses different focus area (Figure 1.6). The https://en.wikipedia.org/wiki/Machine_vision indicates following applications of MVS:

- Automated Train Examiner (ATEx) Systems
- Automatic PCB inspection
- Checking medical devices for defects
- Classification of Non-Woven Fabrics
- Engine part inspection
- Final inspection cells
- Final inspection of sub-assemblies
- Food pack checks
- Food products
- Inspection of Ball Grid Arrays (BGAs)
- Inspection of Punched Sheets
- Inspection of Saw Blades
- Label inspection on products
- Measuring of Spark Plugs
- Medical vial inspection
- Molding Flash Detection
- Packaging Inspection
- Pose Verification of Resistors
- Reading of Serial Numbers
- Robot guidance and checking orientation of components
- Surface Inspection
- 3D Plane Reconstruction with Stereo
- Verifying engineered components
- Wafer Dicing
- Wood quality inspection

Machine Vision Technology in Food Processing Industry 17

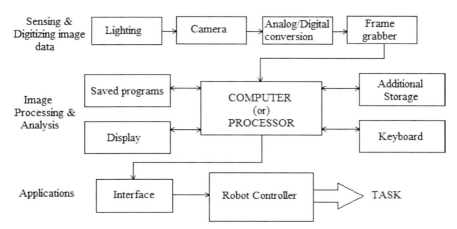

FIGURE 1.5 Flow chart indicating different functions of a typical MVS.

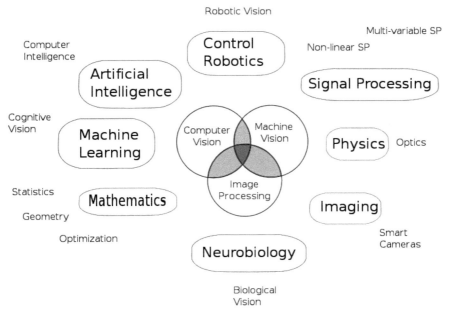

FIGURE 1.6 Relation between computer vision and various other fields [https://en.wikipedia.org/wiki/Glossary_of_machine_vision].

1.6.1 HORTICULTURAL PRODUCE AND PRODUCTS

Shape, size, color, blemishes and diseases are important aspects, which need to be considered when grading fruits and vegetables. Color provides valuable information to estimate the maturity and to examine the freshness of fruits and vegetables. To meet the quality requirements of customer, computer vision is being implemented for the automated inspection and grading of horticulture produce to increase product throughput and to improve objectivity of the industry [15]. Image acquisition of apples using CCD cameras has been studied [50]. Acquired images were segmented and the position and diameter of the fruits were measured. Blobs were found, characterized by 16 features which included: color, position, shape and texture features. Machine vision systems have been successfully used for quality inspection of number of horticulture produce viz.: citrus [53], mushroom [85], strawberry [51], potato [29], lettuce [79], etc.

An intelligent virtual grader was developed for automatic grading of red delicious apples based on their surface color using machine vision. The heart of the proposed virtual grader is executed in the form of k-nearest neighbor classifier. k-nearest neighbor classifier is chosen for this particular application since it is more robust to noise as compared to other classification algorithms. The performance of the implemented virtual grader was examined experimentally with an industrial grade camera connected to an image grabber of a computer based machine vision system. Results of this study are quite promising. In fact, efficiency achieved using proposed virtual grader was 95.12% compared to reference value of 100% in manual grading [17].

The potential of near-infrared (NIR) spectroscopy in the wavelength range of 1200–2200 nm was evaluated to determine total soluble solids and pH for seven major cultivars of mangos from seven states of India. NIR models were developed based on multiple-linear regression (MLR) and partial least square (PLS) regression employing pre-processing technologies (baseline correction, smoothening, multiplicative scatter correction (MSC) and second order derivatisation). The multiple correlation coefficients for calibration and validation were 0.782 and 0.762 for total soluble solids; and 0.715 and 0.703 for pH, respectively [43]. In another study, formula for prediction of maturity index (IM) was proposed using physico-chemical characteristics and overall acceptability (OA) of a sensory panel for mangos from orchards of nine Indian states [44]. Computed IM values were in agreement with both OA scores and the perceptions of experienced farmers. NIR spectra of 1180 mangos were acquired. MLR and PLS models were developed in the wavelength range of 1200–2200 nm to predict IM. The best prediction was achieved using PLS model after MSC data treatment in the wavelength range of 1600–1800 nm. Multiple correlation coefficients (R) for calibration and validation of PLS model were 0.74 and 0.68, respectively.

1.6.2 DAIRY

Machine vision application is limited to dairy products like cheese. Computer vision algorithms have been used to:

- Evaluate the oiling off property of cheese and the results are correlated well with the fat ring test traditionally used in the industry [88].
- Determine functional properties of cheese like melting and browning [87].
- Recognize individual cheese shred and automatically measure the shred length [63].

Computer vision system was used for color measurement of *Kunda* an Indian dairy product. A digital camera and flatbed scanner was used to measure R, G, B color values of *Kunda*. Product image was analyzed using Photoshop software and color parameters R, G, B was measured using graphical editor. Overall results showed that computer vision system can be used for color measurement of *kunda* [86].

1.6.3 BEVERAGES

Machine vision can determine bubble size distribution and can predict rates of bubble nucleation, growth, and motion in beer [36]. Color in coffee, tea, juices, etc., can be quantified using such automated systems. Such data can be used to identify better beverages and to grade the product. It can also be used for checking fill levels in bottling lines.

1.6.4 MEAT, POULTRY AND FISH

Machine vision application in meat industry can be grouped as: determination of composition, fat/muscle ratio, marbling, measurement and evaluation of size and volume, measurement of shape parameters, quantification of the outside or meat color, and detection of defects during quality evaluation. Moisture and fat content of meat has been correlated with the color. Online poultry inspection by a multi-camera system can be employed to accurately detect and identify carcasses unfit for human consumption [18]. Another market within the food industry for machine vision involves applications with automatic portioners in the seafood, poultry and meat industries. Several companies have developed water-jet cutters that employ three-dimensional machine vision systems to calculate the volume. The volume dictates where the cutting should take place in order to obtain the optimal yield from a piece. This is appropriate in products where the density is relatively consistent.

1.6.5 EGG SORTING

Manual inspection of eggs is an extremely laborious process and is prone to error. By automating this process, the level of accuracy in identifying defective eggs increases and the rate of sorting is higher. In a study, 1.4 Megapixel cameras were positioned in such a fashion as to capture images from every angle as the eggs roll down a conveyor belt. The camera monitored the quality of eggs passing through the system and the images were analyzed digitally, with complex algorithms identifying any hairline cracks or detritus on the egg's surface [23].

1.6.6 BAKERY PRODUCTS, SNACKS AND CONFECTIONARIES

Quality maintenance of snacks, chips and bakery products is a challenging problem in the food industry. At present, it is done by a panel of experts. Vision systems have been used in inspection applications such as checking the presence and quality of wafers in chocolate biscuits, or ensuring that molds used in the production of confectionery are empty and properly cleaned. An intelligent system for color inspection of biscuits was designed [62] with classifiers like support vector machine and Wilk's (λ) analysis to classify biscuits into four classes: under baked, moderately baked, over baked and substantially over baked. Machine vision can be used to:

- classify potato chips according to their color in different categories
- identify broken crackers
- determine edge
- ensure uniform baking/cooking and color development
- detect defects in color, shape, topping and packaging
- find pizza topping percent and distribution [80, 81]
- visually inspect chocolate chip cookies and muffins

1.6.7 GRAINS AND CEREALS

Quality inspection of cereal grains and pulses (like rice, corn, wheat, gram, beans, etc.) can be performed based on size (length/width) and color quantification of samples [20, 56]. Vision systems are being used to sort grains falling off the end of a conveyor belt. The cameras capture images as the beans are in mid-air, identify the produce that do not meet the quality standards and direct air nozzles to pick these out. The air nozzles remove black grains and bits of grit and stones. Few more applications include: disease infection, weed identification, size (whole and broken kernel), whiteness and grading, etc.

Color based recognition system has been proposed [4] to estimate the temperature level of boiled grains. The effect of boiling and automatic recognition of images of boiled food grains was observed. The boiling temperatures chosen were 40°C, 50°C, 60°C, 80°C and 100°C. A color feature centered knowledge based classifier was used. The classification accuracy was high at lower and higher temperatures and low at medium temperatures.

A comparative study was carried out among HSV and YCbCr color models in the classification of food grains by combining color and texture features without performing pre-processing. The non-uniformity of RGB color space is eliminated by HSV and YCbCr color space. The good classification accuracy was achieved using both the color models [67]. In another study, food grains were classified using different color models (such as L*a*b, HSV, HSI and YCbCr) by combining color and texture features. The classification results for different color models were quite good [68].

1.7 COLOR MEASUREMENT TECHNIQUES USING MACHINE VISION SYSTEM

CCD (charge-coupled device) image acquisition systems have been effectively used for color measurement of food products like potato chips, fish, orange juice, wine, etc. [32, 54, 69, 90]. The major advantage of these systems comes from its ability to determine L*a*b* values for each pixel of the sample's image. The entire surface of the food is analyzed and average values are determined for the object [64].

In general, a computer vision camera (CVC) employs a single array of light-sensitive elements on a CCD chip, with a filter array that allows some elements to see red (R), green (G) and blue (B). 'White balance' is conducted to measure relative intensities manually or automatically [57]. A digital color image is then generated by combining three intensity images (R, G, and B) in the range 0–255. As being device-dependent, RGB signals produced by different cameras are different for the same scene. These signals will also change over time as they are dependent on the camera settings and scenes [84]. Therefore, measurements of color and color differences cannot be conducted on RGB images directly. On the other hand, different light sources present different emission spectra dominated by diverse wavelengths that affect those reflected by the object under analysis [21]. Therefore, in order to minimize the effects of illuminants and camera settings, color calibration prior to photo/image interpretation is required in food processing to quantitatively compare sample's color during workflow with many devices [59].

The sRGB is a device-independent color space that has relationship with the CIE colorimetric color spaces. Most of the variability introduced by the camera and illumination conditions can be eliminated by finding the relationship between the varying and unknown camera RGB and the sRGB color space [84]. Different calibration algorithms defining the relationship between the input RGB color space of the camera and the sRGB color space have been published using various methods [84]. Several softwares are available to perform color calibration using a color profile assignable to the image that deals with different devices (e.g., ProfileMaker, Monaco Profiler, EZcolor, i1Extreme). However, they are often too imprecise for scientific purpose. Therefore, polynomial algorithms, multivariate statistics, neural networks, and their combinations are proposed for the color calibration [59].

A computer vision system was developed to analyze the effect of drying on shrinkage, color and image texture of apple discs. A standardized image acquisition system consisting of a digital camera, illumination, computer hardware and software was developed to capture and process the images. All parameters related to shape (area, perimeter, Fourier energy, etc.) were decreased with drying time. With regard to sample color, lightness (L*) remained almost constant while the chromatic co-ordinates (a* and b*) increased steadily as drying proceeded. Parameters related to the texture of the image and calculated from the color co-ordinates represented well the complexity and non-homogeneity of the visual appearance of samples [31].

Hand-held Minolta colorimeter and machine vision technique were compared for measuring the color of irradiated Atlantic salmon filets. The L*, a*, b* values of Atlantic salmon filets subjected to different electron beam doses (0, 1, 1.5, 2 and 3 kGy) were measured using a Minolta CR-200 Chroma Meter and a machine vision system. For both Minolta and machine vision, L* value increased and a* and b* values decreased with increasing irradiation dose. However, the machine vision system showed significantly higher readings for L*, a*, b* values than the Minolta colorimeter. Because of this difference, colors that were actually measured by the two instruments, were illustrated for visual comparison. Minolta readings resulted in a purplish color based on average L*, a*, b* values, while machine vision readings resulted in an orange color, which was expected for Atlantic salmon filets [90].

A computer vision system has been implemented [57] to identify the ripening stages of bananas based on color, development of brown spots, and image texture information. Nine simple features of appearance (L*, a*, b* values; brown area percentage; number of brown spots per cm^2, homogeneity, contrast, correlation, and entropy of image texture) extracted from images of bananas were used for classification purposes.

The results showed that in spite of variations in data for color and appearance, a simple classification technique is as good to identify the ripening stages of bananas as professional visual perception.

A computerized image analysis technique (with a flatbed scanner for image acquisition) has been developed [79] in order to measure the amount and distribution of the most important visual aspects of potato chips: color components (L*, a* and b*) and brown and oily areas on the surface. The potato slices were fried at three temperatures, i.e., 170, 180 and 190°C for 2, 3 and 4 min. Pre-processing, segmentation and color analysis were carried out by software programmed in Matlab v 6.5. Results showed a high linear relationship ($R^2 > 0.962$) between image RGB values and those measured by conventional colorimeter. The applied image analysis technique was able to differentiate with high sensitivity among potato chip colors after the frying processes.

A simple method has been proposed [2] to show that digital imaging and software analysis can be combined for color measurement. The results showed that L*, a*, and b* values from Hunter colorimeter and the digital imaging method had appropriate correlation with R^2 of greater than 0.98. However, color values obtained from digital imaging method can be used only to monitor the trend of color changes and relative comparison and there is a noticeable difference between L*, a*, and b* from digital imaging and values of Hunterlab colorimeter.

1.8 APPLICATION OF OPEN SOURCE SOFTWARE: MVS

There are three important components of a MVS: (i) Image acquisition device or camera; (ii) Hardware (computer or embedded system, camera-hardware interface, etc.); and (iii) Software for image processing. The performance of MVS depends on all these three components.

Software is the backbone of MVS and one of the major implications in development of MVS is the software cost. There can be three approaches in MVS software development. First option is to develop the software or program in-house. However, it requires a strong project team and could be a time consuming process. Second way is to get the program developed by a software company. The last alternative is to customize the software provided by MVS manufacturing companies. One of the unexplored areas is the application of open source software for MVS. It can result in significant reduction in development cost of MVS. Authors have successfully evaluated Scilab software for image analysis and food color measurement. The features and limitations of Scilab software are as follows:

1.8.1 FEATURES

- Scilab is free and open source software for numerical computation providing a powerful computing environment for engineering and scientific applications.
- Scilab includes hundreds of mathematical functions and the programming syntax is similar to Matlab.
- Image processing design tool box and Scilab Image and Video Processing toolbox enables advanced image analysis.
- Graphic user interface toolbox helps to develop input/output menu. It becomes easy for the users with low programming skills.

1.8.2 LIMITATIONS

- As Scilab software is still evolving with many tools being developed, it lacks many build-in-functions which are available in proprietary software. However, it is not difficult to write program for these functions.

1.9 SUMMARY

In the past, the visual quality of food products was assessed by human inspection with a naked eye. But with high speed processing lines, it is not possible. Therefore, it becomes necessary to install automatic software for quality control and inspection. MVS provides rapid, economic, constant and objective assessment of the quality of an agricultural product. It allows increasing the throughput of the plant without compromising accuracy. The major advantage of using such system is that the system can be programmed as per the inspection requirement in food processing plant.

KEYWORDS

- computer vision system
- hyperspectral imaging
- image acquisition
- machine vision system
- nuclear magnetic resonance (NMR) spec
- short wavelength infrared
- ultraspectral

REFERENCES

1. Acevedo, C. A., Skurtys, O., Young, M. E., Enrione, J., Pedreschi, F., & Osorio, F. (2009). A non-destructive digital imaging method to predict immobilized yeast-biomass. *LWT: Food Science and Technology, 42*(8), 1444–1449.
2. Afshari-Jouybari, H., & Farahnaky, A. (2011). Evaluation of Photoshop software potential for food colorimetry. *J. Food Engineering, 106*, 170–175.
3. Al-Mallahi, A., Kataoka, H., Okamoto, T., & Shibata, Y. (2010). An image process-ing algorithm for detecting in-line potato tubers without singulation. *Computers and Electronics in Agriculture, 70*(1), 239–244.
4. Anami, B. S., & Burkpalli, V. C. (2010). Color based recognition and estimation of temperature levels of images of boiled food grains. *International Journal of Computer Applications, 1*(14), 98–103.
5. Antequera, T., Caro, A., Rodriguez, P. G., & Perez, T. (2007). Monitoring the ripen-ing process of Iberian ham by computer vision on magnetic resonance imaging. *Meat Science, 76*(3), 561–557.
6. Arivazhagan, S., Shebiah, R. N., Nidhyanandhan, S. S., & Ganesan, L. (2010). Fruit Recognition using Color and Texture Features. *Journal of Emerging Trends in Computing and Information Sciences, 1*(2), 90–94.
7. Balaban, M. O., & Odabasi, A. Z. (2006). Measuring color with machine vision. *Food Technology, 60*, 32–36.
8. Balaban, M. O., Kristinsson, H. G., & Otwell, W. S. (2005). Evaluation of color parameters in a machine vision analysis of carbon monoxide-treated fish. Part 1: Fresh tuna. *Journal of Aquatic Food Product Technology, 14*, 5–24.
9. Baranowski, P., Mazurek, W., Wozniak, J., Majewska, U. (2012). Detection of early bruises in apples using hyperspectral data and thermal imaging. *J. Food Engineering, 110*(3), 345–355.

Machine Vision Technology in Food Processing Industry   27

10. Barcelon, E. G., Tojo, S., & Watanabe, K. (1999a). Relating X-ray absorption and some quality characteristics of mango fruit (*Mangifera indica* L.). *Journal of Agricultural and Food Chemistry, 47*(9), 3822–3825.

11. Barcelon, E. G., Tojo, S., & Watanabe, K. (1999b). X-ray CT imaging and quality detection of peach at different physiological maturity. *Transactions of the ASAE, 42*(2), 435–441.

12. Blasco, J., Aleixos, N., & Molto, E. (2007). Computer vision detection of peel defects in citrus by means of a region oriented segmentation algorithm. *J. Food Engineering, 81*(3), 535–543.

13. Borggaard, C., Madsen, N. T., & Thodberg, H. H. (1996). In-line image analysis in the slaughter industry, illustrated by beef carcass classification. *Meat Science, 43*(S1), 151–163.

14. Brecht, J. K., Shewfelt, R. L., Garner, J. C., & Tollner, E. W. (1991). Using X-ray-computed tomography to nondestructively determine maturity of green tomatoes. *Horticultural Science, 26*(1), 45–47.

15. Brosnan, T., & Sun, D. W. (2004). Improving quality inspection of food products by computer vision—a review. *J. Food Engineering, 61*, 3–16.

16. Butz, P., Hofmann, C., & Tauscher, B. (2005). Recent Developments in Noninvasive Techniques for Fresh Fruit and Vegetable Internal Quality Analysis. *J. Food Science, 70*, R131–R141.

17. Chauhan, A. P. S., & Singh, A. (2013). Development of intelligent virtual grader for estimation of fruit quality. *International Journal of Computer Applications, 62*(17), 25–29.

18. Chen, Y., Chao, K., & Kim, M. S. (2002). Machine vision technology for agricultural applications. *Computers and Electronics in Agriculture, 36*, 173–191.

19. Cho, R. W. (1996). Non-destructive quality evaluation of intact fruits and vegetables by near infrared spectroscopy. *Proc. International Symposium on Non destructive Quality Evaluation of Horticultural Crops*, Kyoto, Japan, 26 Aug. pp. 8–14.

20. Choudhary, R., Paliwal, J., & Jayas, D. S. (2008). Classification of cereal grains using wavelet, morphological, color, and textural features of non-touching kernel images. *Biosystems Engineering, 99*(3), 330–337.

21. Costa, C., Pallottino, F., Angelini, C., Proietti, P., Capoccioni, F., & Aguzzi, J. (2009). Color calibration for quantitative biological analysis: a novel automated multivariate approach. *Instrumentation Viewpoint, 8*, 70–71.

22. Courtois, F., Faessel, M., & Bonazzi, C. (2010). Assessing breakage and cracks of parboiled rice kernels by image analysis techniques. *Food Control, 21*(4), 567–572.

23. Deng, X., Wang, Q., Chen, H., & Xie, H. (2010). Eggshell crack detection using a wavelet-based support vector machine. *Computers and Electronics in Agriculture, 70*(1), 135–143.

24. Dogan, I. S., Javidipour, I., & Akan, T. (2007). Effects of interesterified palm and cottonseed oil blends on cake quality. *International Journal of Food Science and Technology, 42*(2), 157–164.

25. Dowlati, M., Mohtasebi, S. S., & Guardia, M. D. L. (2012). Application of machine vision techniques to fish quality assessment. *Trends in Analytical Chemistry, 40*, 168–179.

26. Du, C. J., & Sun, D. W. (2004). Recent developments in the applications of image processing techniques for food quality evaluation. *Trends Food Science Technology, 15*, 230–249.

27. Du, C. J., & Sun, D. W. (2006). Learning techniques used in computer vision for food quality evaluation: a review. *J. Food Engineering, 72*(1), 39–55.
28. Du, C. J., & Sun, D. W. (2009). Retrospective shading correction of confocal laser scanning microscopy beef images for three-dimensional visualization. *Food Bioprocess Technology, 2*(2), 167–176.
29. ElMasry, G., Cubero, S., Molto, E., & Blasco, J. (2012). In-line sorting of irregular potatoes by using automated computer-based machine vision system. *J. Food Engineering, 112*(1), 60–68.
30. EMXIN. (2007). UV Luminescence Sensor Application Handbook. http://www.omniray.se/userData/omniray/docs/EMX-UVX-Handbook.pdf.
31. Fernandez, L., Castillero, C., & Aguilera, J. M. (2005). An application of image analysis to dehydration of apple discs. *J. Food Engineering, 67*, 185–193.
32. Fernandez-Vazquez, R., Stinco, C. M., Melendez-Martinez, A. J., Heredia, F. J., & Vicario, I. M. (2011). Visual and instrumental evaluation of orange juice color: a consumers' preference study. *J. Sensory Studies, 26*, 436–444.
33. Goni, S. M., Purlis, E., & Salvadori, V. O. (2008). Geometry modeling of food materials from magnetic resonance imaging. *J. Food Engineering, 88*(4), 561–567.
34. Gonzalez, R. C., & Woods, R. E. (2008). *Digital Image Processing*, 3rd ed., Prentice Hall, Upper Saddle River, NJ.
35. Gowen, A. A., O'Donnell, C. P., Cullen, P. J., Downey, G., & Frias, J. M. (2007). Hyperspectral imaging – an emerging process analytical tool for food quality and safety control. *Trends Food Science Technology 18*(12), 590–598.
36. Hepworth, N. J., Hammond, J. R. M., & Varley, J. (2004). Novel application of computer vision to determine bubble size distributions in beer. *J. Food Engineering, 61*(1), 119–124.
37. Iqbal, A., Sun, D. W, & Allen, P. (2013). Prediction of moisture, color and pH in cooked, pre-sliced turkey hams by NIR hyperspectral imaging system. *J. Food Engineering, 11*, 742–751.
38. Jackman, P., Sun, D. W., Du, C. J., Allen, P., & Downey, G. (2008). Prediction of beef eating quality from color, marbling and wavelet texture features. *Meat Science, 80*(4), 1273–1281.
39. Jackman, P., Sun, D. W., Du, C. J., & Allen, P. (2009a). Prediction of beef eating qualities from color, marbling and wavelet surface texture features using homogenous carcass treatment. *Pattern Recognition, 42*(5), 751–763.
40. Jackman, P., Sun, D. W., & Allen, P. (2009b). Automatic segmentation of beef longissimus dorsi muscle and marbling by an adaptable algorithm. *Meat Science, 83*(2), 187–194.
41. Jackman, P., & Sun, D. W. (2011a). Recent advances in the use of computer vision technology in the quality assessment of fresh meats. *Trends Food Science and Technology, 22*(4), 185–197.
42. Jackman, P., & Sun, D. W. (2011b). Application of computer vision systems for objective assessment of food qualities. In: Y. J. Cho (ed.). *Emerging Technologies for Evaluating Food Quality and Food Safety*, Taylor and Francis, Boca Raton, Florida, USA.
43. Jha, S. N., Jaiswal, P., Narsaiah, K., Gupta, M., Bhardwaj, R., & Singh, A. K. (2013). Non-destructive prediction of sweetness of intact mango using near infrared spectroscopy. *Scientia Horticulturae, 138*, 171–175.

44. Jha, S. N., Narsaiah, K., Jaiswal, P., Bhardwaj, R., Gupta, M., Kumar, R., & Sharma, R. (2014). Nondestructive prediction of maturity of mango using near infrared spectroscopy. *J. Food Engineering. 124*, 152–157.
45. Kawano, S. (1999). Non-destructive methods for quality analysis-especially for fruits and vegetables. *Proc. of XXXIV Lecture Session on Non-Destructive Quality Analysis,* Freising – Weihenstephan , Germany, 22–23 Mar. pp. 5–12.
46. Kirsch, K. (2009). *A Vision of the Future: The Role of Machine Vision Technology in Packaging and Quality Assurance.* http://www.iopp.org/files/public/MSUKathleen-Kirsch.pdf.
47. Kise, M., Park, B., Heitschmidt, G. W., Lawrence, K. C., & Windham, W. R. (2010). Multispectral imaging system with interchangeable filter design. *Computers and Electronics in Agriculture, 72*(2), 61–68.
48. Kodagali, J. A., & Balaji, S. (2012). Computer Vision and Image Analysis based Techniques for Automatic Characterization of Fruits – a Review. *International Journal of Computer Applications, 50*(6), 6–12.
49. Lee, B. S., Bala, R., & Sharma, G. (2007). Scanner characterization for color measurement and diagnostics. *Journal of Electronic Imaging, 16*(4), 43009–43009.
50. Leemans, V., & Destains, M. F. (2004). A real time grading method of apples based on features extracted from defects. *J. Food Engineering, 61*, 83–89.
51. Liming, X., & Yanchao, Z. (2010). Automated strawberry grading system based on image processing. *Computers and Electronics in Agriculture, 71*, S32–S39.
52. Locht, P., Thomsen, K., & Mikkelsen, P. (1997). Full color image analysis as a tool for quality control and process development in the food industry. Paper No. 973006. St. Joseph, MI: ASAE.
53. Lopez-Garcia, F., Andreu-Garcia, G., Blasco, J., Aleixos, N., & Valiente, J. M. (2010). Automatic detection of skin defects in citrus fruits using a multivariate image analysis approach. *Computers and Electronics in Agriculture, 71*(2), 189–197.
54. Martin, M. L. G. M., Ji, W., Luo, R., Hutchings, J., & Heredia, F. J. (2007). Measuring color appearance of red wines. *Food Quality and Preference, 18*, 862–871.
55. Mathiassen, J. R., Misimi, E., Bondo, M., Veliyulin, E., & Ostvik, S. O. (2011). Trends in application of imaging technologies to inspection of fish and fish products. *Trends in Food Science and Technology, 22*, 257–275.
56. Mebatsion, H. K., Paliwal, J., & Jayas, D. S. (2012). Evaluation of variations in the shape of grain types using principal components analysis of the elliptic Fourier descriptors. *Computers and Electronics in Agriculture, 80*, 63–70.
57. Mendoza, F., Dejmek, P., & Aguilera, J. M. (2006). Calibrated color measurements of agricultural foods using image analysis. *Postharvest Biology and Technology, 41*, 285–295.
58. Mendoza, F., & Aguilera, J. M. (2004). Application of image analysis for classification of ripening bananas. *J. Food Science, 69*(9), E471–E477.
59. Menesatti, P., Angelini, C., Pallottino, F., Antonucci, F., Aguzzi, J., & Costa, C. (2012). RGB color calibration for quantitative image analysis: the "3D thin-plate spline" warping approach. *Sensors, 12*, 7063–7079.
60. Minz, P. S., Sharma, A. K., & Raju, P. N. (2012). Automatic food quality evaluation using computer vision system-A framework. *Beverage and Food World, 39*(3), 23–26.

61. Narendra, V. G., & Hareesh, K. S. (2010). Prospects of computer vision automated grading and sorting systems in agricultural and food products for quality evaluation. *International Journal of Computer Applications, 1*(4), 1–9.

62. Nashat, S., Abdullah, A., Aramvith, S., & Abdullah, M. Z. (2011). Support vector machine approach to real-time inspection of biscuits on moving conveyor belt. *Computers and Electronics in Agriculture, 75*, 147–158.

63. Ni, H., & Gunasekaran, S. (1998). A computer vision method for determining length of cheese shreds. In: *Artificial Intelligence for Biology and Agriculture.* Springer Netherlands. pp. 27–37.

64. Oliveira, A. C. M., & Balaban, M. O. (2006). Comparison of a colorimeter with a machine vision system in measuring color of Gulf of Mexico sturgeon filets. *Applied Engineering in Agriculture, 22*(4), 583–587.

65. Park, B., Lawrence, K. C., Windham, W. R., & Smith, D. P. (2006). Performance of hyperspectral imaging system for poultry surface fecal contaminant detection. *J. Food Engineering, 75*, 340–348.

66. Patel, K. K., Kar, A., Jha, S. N., & Khan, M. A. (2012). Machine vision system: a tool for quality inspection of food and agricultural products. *J. Food Science Technology, 49*(2), 123–141.

67. Patil, N. K., Yadahalli, R. M., & Pujari, J. (2011a). Comparison between HSV and YCbCr Color Model Color-Texture based Classification of the Food Grains. *International Journal of Computer Applications, 34*(4), 975–987.

68. Patil, N. K., Malemath, V. S., & Yadahalli, R. M. (2011b). Color and texture based identification and classification of food grains using different color models. *International Journal of Computer Science and Engineering, 3*(12), 3669–3680.

69. Pedreschi, F., Leon, J., Mery, D., & Moyano, P. (2006). Development of a computer vision system to measure the color of potato chips. *Food Research International, 39*, 1092–1098.

70. Pydipati, R., Burks, T. F., & Lee, W. S. (2006). Identification of citrus disease using color texture features and discriminant analysis. *Computers and Electronics in Agriculture, 52*(1), 49–59.

71. Romani, S., Rocculi, P., Mendoza, F., & Rosa, M. D. (2009). Image characterization of potato chips during frying. *J. Food Engineering, 93*, 487–494.

72. Ruiz-Altisent, M., Ruiz-Garcia, L., Moreda, G. P., Lu, R., Hernandez-Sanchez, N., Correa, E. C., Diezma, B., Nicolai, B., & Garcia-Ramos, J. (2010). Sensors for product characterization and quality of specialty crops—a review. *Computers and Electronics in Agriculture, 74*(2), 176–194.

73. Sakamoto, T., Shibayama, M., Kimura, A., & Takada, E. (2011). Assessment of digital camera-derived vegetation indices in quantitative monitoring of seasonal rice growth. *ISPRS Journal of Photogramtry and Remote Sensing. 66*(6), 872–882.

74. Sarkar, N. R. (1991). Machine vision for quality control in the food industry. In: D. Y. C. Fung, & R. F. Mathews (eds.), *Instrumental Methods for Quality Assurance in Foods,* ASQC Quality Press, Marcel Dekker Inc., New York, pp. 166–188.

75. Shahin, M. A., & Symons, S. J. (2001). A machine vision system for grading lentils. *Canadian Biosystems Engineering, 43*, 1–7.

76. Shahin, M. A., Symons, S. J., & Poysa, V. W. (2006). Determining Soya Bean Seed Size Uniformity with Image Analysis. *Biosystems Engineering, 94*(2), 191–198.

77. Slaughter, D. C., Obenland, D. M., Thompsona, J. F., Arpaia, M. L., & Margosan, D. A. (2008). Non-destructive freeze damage detection in oranges using machine vision and ultraviolet fluorescence. *Postharvest Biology and Technology, 48*, 341–346.
78. Storbeck, F., & Daan, B. (2001). Fish species recognition using computer vision and a neural network. *Fisheries Research, 51*, 11–15.
79. Story, D., Kacira, M., Kubota, C., Akoglu, A., & An, L. (2010). Lettuce calcium deficiency detection with machine vision computed plant features in controlled environments. *Computers and Electronics in Agriculture, 74*(2), 238–243.
80. Sun, D. W. (2000). Inspecting pizza topping percentage and distribution by a computer vision method. *J. Food Engineering, 44*, 245–249.
81. Sun, D. W. (2004). Computer vision - an objective, rapid and non-contact quality evaluation tool for the food industry. *J. Food Engineering, 61*, 1–2.
82. Szczypinski, P. M., & Zapotoczny, P. (2012). Computer vision algorithm for barley kernel identification, orientation estimation and surface structure assessment. *Computers and Electronics in Agriculture, 87*, 32–38.
83. Throop, J. A., Aneshansley, D. J., Anger, W. C., & Peterson, D. L. (2005). Quality evaluation of apples based on surface defects: development of an automated inspection system. *Postharvest Biology and Technology, 36*(3), 281–290.
84. Van Poucke, S., Haeghen, Y. V., Vissers, K., Meert, T., & Jorens, P. (2010). Automatic colorimetric calibration of human wounds. *BMC Medical Imaging, 10*, 7.
85. Vízhanyo, T., & Felfoldi, J. (2000). Enhancing color differences in images of diseased mushrooms. *Computers and Electronics in Agriculture, 26*, 187–198.
86. Vyawahre, A. S., & Rao, K. J. (2011). Application of computer vision system in color evaluation of *Kunda*-A heat desiccated dairy product. *International Journal of Dairy Science, 6*(4), 253–266.
87. Wang, H. H., & Sun, D. W. (2001). Evaluation of the functional properties of cheddar cheese using a computer vision method. *J. Food Engineering. 49*(1), 49–53.
88. Wang, H. H., & Sun, D. W. (2004). Evaluation of the oiling off property of cheese with computer vision: Influence of cooking conditions and sample dimensions. *J. Food Engineering, 61*(1), 57–66.
89. Wang, W., Li, C., Tollner, E. W., Gitaitis, R. D., & Rains, G. C. (2012). Shortwave infrared hyperspectral imaging for detecting sour skin infected onions. *J. Food Engineering, 109*(1), 38–48.
90. Yagiz, Y., Balaban, M. O., Kristinsson, H. G., Welt, B. A., & Marshall, M. R. (2009). Comparison of Minolta colorimeter and machine vision system in measuring color of irradiated Atlantic salmon. *Journal of the Science of Food and Agriculture, 89*, 728–730.
91. Zapotoczny, P. (2011). Discrimination of wheat grain varieties using image analysis and neural networks. Part I. Single kernel texture. *Journal of Cereal Science, 54*(1), 60–68.

CHAPTER 2

COLD PLASMA TECHNOLOGY: AN EMERGING NON-THERMAL PROCESSING OF FOODS—A REVIEW

R. MAHENDRAN, C. V. KAVITHA ABIRAMI, and K. ALAGUSUNDARAM

CONTENTS

2.1 Introduction 34
2.2 Definition of Plasma 34
2.3 Plasma Generation 35
2.4 Classification of Plasma 35
2.5 Sources and Production of Plasmas 37
2.6 Electrode Configurations 39
2.7 Applications of Plasma in Food Products 43
2.8 Sterilization Mechanism Using Plasma 44
2.9 Potential Applications and Microbial Inactivation
     Mechanism for Different Agricultural and Processed Food
     Products by Plasma 44
2.10 Effect of Plasma on Quality of Food Products 45
2.11 Summary 49
Keywords 49
References 50

2.1 INTRODUCTION

Plasma is an ionized gaseous matter consisting of entirely or partially charged particles such as electrons and ions. When an energy applied on a matter increases, it transforms from a solid, to a liquid, to a gas and then to plasma [20]. When gas is given sufficient energy, the molecules in the gas dissociate to form gas of atoms. Further energy will cause atoms to break up into electrons and positive ions creating plasmas. Plasmas are classified into two categories: thermal and non-thermal based on the mean temperatures of their heavy particles like ions and neutral species (atoms, molecules). In the thermal plasma, all the particles (electrons, ions and atoms) are in thermodynamic equilibrium, while a significant difference in kinetic energy caused by the temperature of electrons and the ambient gas particles is observed in non-thermal plasma (NTP). Such plasmas are also commonly called as non-equilibrium plasmas or cold plasmas because of the temperature difference between heavy and light particles. The temperature of thermal plasmas may range from a few thousand Kelvin (e.g., Plasma torches) to a few million Kelvin (e.g., in fusion plasmas). In contrast, non-thermal plasmas typically range from room temperature to a few times the room temperature [4]. Sources of plasma, pressure, type of gases, temperature, type and geometry of the plasma electrodes and excitation power are important factors that influence the plasma intensity.

This chapter reviews the technology of cold plasma for non-thermal processing of foods and milk products.

2.2 DEFINITION OF PLASMA

Plasma is an ionized gas that consists of a large number of electrons, positive and negative ions, free radicals, and gas atoms, molecules in the ground or excited state; and quanta of electromagnetic radiation (photons) [73]. Irving Langmuir was introduced the term plasma in the first half of 20th century to describe the part of neutral charge of a gas discharge. David A. Frank-Kamenezki was identified plasma as the fourth state of matter [58]. Plasma can be generated in large ranges of temperatures and pressures by means of coupling energy to gaseous medium.

This energy can be mechanical, thermal, nuclear, radiation or carried by an electric current. These energies dissociate the gaseous molecules into collection of ions, electrons, charge-neutral gas molecules, and other species.

2.3 PLASMA GENERATION

Depending on the method of generation, the plasma may have a high or low density, high or low temperature, it may be steady or transient, stable or unstable, and so on. The most commonly used method of generating and sustaining a low-temperature plasma is by applying an electric field to a gas. A given volume of a gas always contains few electrons and ions. These free electrons and ions are accelerated by the electric field and new charged particles may be created when these charge carriers collide with atoms and molecules in the gas or with the surfaces of the electrodes. This leads to an avalanche of charged particles that is eventually balanced by charge carrier losses, so that steady-state plasma develops.

Low-pressure glow discharge plasmas are of great interest in fundamental research. In the last two decades, non-thermal plasmas (NTPs) have attracted more attention due to their significant industrial advantage. NTP may be obtained by applying a diversity of electrical discharges such as corona discharge, micro hollow cathode discharge, atmospheric pressure plasma jet, gliding arc discharge, one atmospheric uniform glow discharge, dielectric barrier discharge, and plasma needle, all of which have important technological applications.

2.4 CLASSIFICATION OF PLASMA

Depending on the amount of energy transferred and the type of energy supplied to the plasma, electron density and the temperature of the electrons are changed. These lead plasma to be distinguished into two groups (Table 2.1): high temperature plasma and low temperature plasma [52]. High temperature plasma implies that electron, ions and neutral species are in a thermal equilibrium state. Low temperature plasma is subdivided into: (i) thermal plasma, also called as Local Thermodynamic Equilibrium

TABLE 2.1 Classification of Plasma According to Their State and Sources [52]

Plasma	State	Example
High temperature plasma		
Equilibrium plasma	$T_e \approx T_i \approx T_g$, $T_p \approx 10^6\text{–}10^8$ K $n_e \geq 10^{20}$ m^{-3}	Laser fusion plasma
Low temperature plasma		
Thermal Plasma (Quasi equilibrium plasma)	$T_e \approx T_i \approx T_g \leq 2 \times 10^4$ K $n_e \geq 10^{20}$ m^{-3}	RF inductively coupled discharges, Plasma torches Arc plasma
Non thermal plasma (Non equilibrium plasma), NTP	$T_e \gg T_i \approx T_g = 300$ to 1000 K $n_e \approx 10^{10}$ m^{-3}	OAUGDP, Glow, APPJ, DBD, MHCD, Plasma needle, Crona.

T_e – temperature of electrons; T_h – temperature of heavy particles; T_g – overall temperature of the gas; n_e – density of electrons; T – neutral gas temperature; T_p – temperature of plasma; T_i – temperature of ions; OAUGDP – One Atmospheric Uniform Glow Discharge Plasma; APPJ – Atmospheric Pressure Plasma Jet; DBD – Dielectric Barrier Discharge; MHCD – Micro Hollow Cathode Discharge. (Adapted from Nehra, V., Kumar, A., & Dwivedi, H. K. (2008). Atmospheric non-thermal plasma sources. Int. J. Eng., 2, 53–68. https://creativecommons.org/licenses/by/4.0/)

(LTE) plasmas; and (ii) NTP, also called as non-Local Thermodynamic Equilibrium (non-LTE) plasmas [6, 39, 73]. An equilibrium or near equality between electrons, ions and neutrals is the main characterization of thermal plasmas. In generation of cold plasma, most of the coupled electrical energy is channeled to electron component instead of heating entire gas stream so that the temperature of heavy particle remains near the room temperature. These characteristics make it suitable to be used in processes where high temperature is not desirable [52].

Thermal plasmas are plasmas in which the plasma is said to approach a state of LTE. LTE occurs when the temperatures of the electrons and the relatively heavier particles (ions and neutrals) are equal implying that the particles are in thermal equilibrium with each other. These thermal plasmas are generally produced by atmospheric arcs, sparks and flames. In thermal plasma, the temperature of electrons (T_e), temperature of heavy particles (T_h) and the overall temperature of the gas (T_g) are almost the same and generally equivalent to 10,000 K (i.e., $T_e \approx T_h \approx T_g \approx 10,000$ K) [73].

However, with non- thermal plasmas, the thermal motion of the ions can be ignored. As a result, there is no pressure force, the magnetic force can be ignored, and only the electric force is considered to act on the particles. Furthermore, *the electrons are not in thermal equilibrium with the heavier particles*. The temperature of the ions and neutrals are generally at much lower levels and sometimes near the room temperature, whereas the electrons are at much higher temperatures. This is sometimes referred to as non-LTE (nLTE). Similar types of nLTE or NTP production system have been used by authors for studying the surface sterilization of bread slices.

In LTE plasma, the temperature of the gas, heavy particles and electrons are the same but in non-LTE plasmas the temperature of electrons (T_e) is much higher than the temperature of heavy particles (T_h). Because of huge mass difference between heavy particles and electrons, the temperature of plasma or the temperature of the gas (T_g) is governed by the temperature of heavy particles (i.e., $T_e >> T_h\, T_g$). The deviation of non-LTE plasmas from Boltzman distribution for the density of electrons could be explained by the fact that the electron induced de-excitation rate of atoms is lower than the corresponding electron induced excitation rate because of significant radiative de-excitation rate [45].

Electrons move very fast and the heavy particles are nearly static in comparison to the electrons. Therefore, unlike LTE plasmas, non-LTE plasmas have local gradients of plasma properties such as temperature, electron density; and thermal conductivity should be high enough and diffusion time should be less than the time, the particles need to reach the equilibrium. In this scenario, non-equilibrium plasma is formed. Inelastic collisions between electrons and heavy particles are responsible for plasma chemistry and the *only few elastic collisions* heat up the heavy particles slightly ($T_h \approx 300–1000$ K). Therefore, overall temperature of the plasma remains low (cold plasma) and electrons remain highly energetic ($T_e \approx 10{,}000–100{,}000$ K), as seen in glow discharges [73]. The density of electrons, n_e, degree of ionization (ratio of electron to neutral particle density, n_e/N), and the average energy of the electrons (electron temperature, T_e) can also be used to characterize the plasmas as thermal and non-thermal [20, 36].

2.5   SOURCES AND PRODUCTION OF PLASMAS

There are many different sources and methods of producing plasma. The most commonly used method of generating and sustaining a low-temperature plasma, nLTE for technological and technical applications is by applying an electric energy to a gas.

The most commonly used method of generating and sustaining low temperature plasma is by applying an electric field to a neutral gas. A given volume of a gas always contains few electrons and ions that are formed, for example, as a result of collision. These free electrons and ions are accelerated by the applied electric field and new charged particles may be created, when these charged carriers collide with atoms and molecules in the gas or with the surfaces of the electrodes. This leads to an avalanche of charged particles that is eventually balanced by charge carrier losses, so that steady-state plasma develops [10].

One characteristic of this process is that the applied electric field transfers energy much more efficiently to the light electrons than to the relatively heavy electrons. The electron temperature in gas discharges is therefore usually higher than the ion temperature, since the transfer of electrons to the heavier particles is slow. When the ionizing source is turned off, the ionization decreases gradually because of recombination until it reaches an equilibrium value consistent with the temperature of the medium.

Many approaches have been proposed in recent years to overcome the problems of generating and sustaining stable, uniform and homogeneous non-thermal atmospheric pressure plasma discharges [64]. One novel approach in generating atmospheric pressure non-thermal plasma discharge is based on Paschen's law, which describes the product of pressure and inter electrode distance as follows:

$$p \times d = \text{constant} \tag{1}$$

where, p = pressure; d = distance between electrodes.

Paschen's Law is an equation that gives the breakdown voltage, which is necessary to start a discharge or electric arc, between two electrodes in a gas; and this is as a function of pressure and gap length. It is named after Friedrich Paschen who discovered it empirically in 1889. Figure 2.1 shows the Paschen curves [60] for air, nitrogen, hydrogen, helium and argon. The parameter '$p \times d$' in the Paschen's curve is simply proportional to the ratio of the discharge gap length to the mean free path or the inverse of the Knudsen number. There is a minimum breakdown voltage on the

Cold Plasma Technology

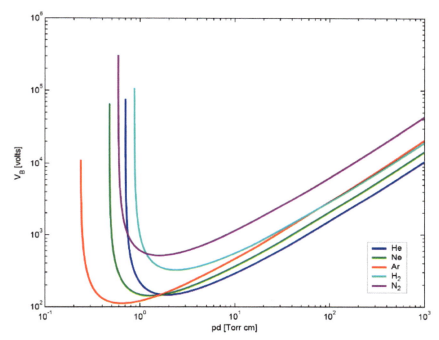

FIGURE 2.1 Paschen's curves for helium, neon, argon, hydrogen and nitrogen, using the expression for the breakdown voltage as a function of the parameters A, B that interpolate the first Townsend coefficient, (Source: Krishnavedala. https://en.wikipedia.org/wiki/Paschen%27s_law. https://creativecommons.org/licenses/by-sa/4.0/)

Paschen curve (V_{min}) and a corresponding pd_{min}. To the left of the minimum point, the discharge gap is so small that the mean free path in the discharge is insufficient to provide ignition. On the right, the required voltage is higher since the gap is longer. Paschen's law therefore suggests that the inter-electrode separation has to be scaled down to the micrometer range, in order to ignite plasma at atmospheric pressure with a minimum breakdown voltage.

2.6 ELECTRODE CONFIGURATIONS

Recent findings on different plasma sources, methods of plasma generation, design features and electrode configuration are given in Table 2.2. Typical electrode configurations such as planar and cylindrical electrodes

40 Engineering Interventions in Agricultural Processing

TABLE 2.2 Recent Findings on Plasma Sources, Methods of Generation of Plasma and Electrode Configurations

Plasma sources/ methods of generation	Design features and electrode configurations	Ref.
Atmospheric pressure needle plasma	Voltage in the range of 3.95 kV up to 12.83 kV at 60 Hz was applied on 12 steel electrode needles (Nickel coated) with radius of 50 μm.	[5]
Atmospheric pressure plasma	Two different sized electrode (400 × 500 × 3 mm and 200 × 200 × 3 mm) used with a gap of 6 mm. Power applied for plasma generation was 4–20 kV at 10–20 kHz and He and O_2 used for treatment.	[34]
Atmospheric pressure plasma jet	RF power (13.56 MHz) applied between two planar square aluminum electrodes of area 100 cm² each with gap between electrodes in the range of 0.16 to 0.32 cm. Helium gas was used at pressure of 3×10^{-3} torr.	[56]
Cold atmospheric pressure plasma	Ar gas used on the 1.88 liter capacity of reactor vessel made of pyrex and electrodes used were of copper.	[70]
Cold jet plasma	The device consisted of the plasma jet itself (length: 170 mm; diameter: 20 mm; weight, 170 g), a gas flow controller and a DC power supply. The process gases were argon as well as mixtures of argon and oxygen, and the plasma exposure time was varied between 0 and 480 s.	[71]
Cold plasma torch	4 kV voltage was applied on a tungsten electrode (3.2 mm diameter, 30 mm long) placed in center of discharge tube and copper was ground electrode.	[9]
DBD cold atmospheric pressure treatment	27 kV voltage at frequency of 27.8 kHz was applied between two aluminum electrodes of size 100 × 100 mm with electrode gap of 10 mm.	[61]
Dielectric barrier discharge	Helium gas (4 lpm) passed between Aluminum electrodes with a gap of 5 mm and power applied was 3.5 kV at 50 kHz.	[33]
Dielectric barrier discharge cold plasma	120 kV voltage at 50 Hz was applied between two aluminum circular (15 cm diameter) disk electrode with electrode gap of 40 mm	[79]
Dielectric barrier discharge powered atmospheric cold plasma	Two aluminum electrodes of circular geometry (outer diameter 158 mm), resting over two polypropylene (PP) dielectric layers of 2 mm thick.	[44]
	Plasma source comprised of two aluminum disc electrodes of 15 cm diameter, over two polypropylene dielectrics (2 mm thick) between which the PET package with samples were placed.	[65]

Cold Plasma Technology

TABLE 2.2 (Continued)

Plasma sources/ methods of generation	Design features and electrode configurations	Ref.
Gliding arc cold plasma system	15 kV power at 60 mA and 60 Hz was applied between copper electrodes of 2 mm thickness.	[53]
Low pressure cold plasma prototype processor/unit	Air and sulfur hexafluoride was used as gas on quartz sterilization tube (12×40 cm) with thickness of 0.8 cm and operating pressure of 500 m Torr. Energy applied in the form of sinusoidal voltage 20 kV at 1 kHz.	[3]
	Sinusoidal voltage of 20 kV with 1 kHz was applied in a chamber size of 5 mm diameter with length of 40 mm. Operating pressure of the chamber was 500 mTorr and exposure time used between 30 sec to 30 min.	[66]
Low pressure microwave plasma	Cylindrical stainless steel vacuum chamber was used with a size 35 cm diameter and 40 cm height. Microwave power applied at frequency of 2.45 Ghz on quartz tube. Pressure maintained was 10 to 50 Pa and Ar, H_2, N_2, O_2 and NH_3 gases were used.	[18]
Low temperature atmospheric pressure plasma system	9 kV Ac voltage was applied between two tungsten electrodes of 0.8 mm radius.	[24]
Microwave cool plasma	Chamber used for developing microwave plasma was of diameter of 120 mm and height 100 mm. Pressure maintained in the chamber was 0.1 to 0.2 mbar and microwave generator 2.45 GHz used for power generation.	[26]
N_2 gas plasma	N_2 gas plasma was generated by applying a short high-voltage pulse to N_2 using a static induction thyristor power supply (1.5 kilo pulse per second) and a Cathode electrode (earth electrode) was placed between the anode electrodes (high voltage electrodes).	[38]
Plasma corona discharge	System consisted of a 9 kV AC power supply and two tungsten electrodes (0.8 mm radius).	[29]
Plasma jet	The plasma jet consists of a tungsten rod with a sharp tip, inserted in a quartz capillary with 1.3 mm inner diameter. The tungsten rod and quartz capillary together are centered inside a grounded aluminum tube (ground electrode).	[74]

TABLE 2.2 (Continued)

Plasma sources/ methods of generation	Design features and electrode configurations	Ref.
Pulsed plasma reactor	Diameter and height of the pulsed plasma chamber used were 250 and 500 mm resp. Pressure maintained in the reactor was 0.8 MPa with exposure time in the range of 5 to 10 min.	[19]
Resistive Barrier Discharge prototype	Brass electrode was energized with 15 kV transformer and 5 mm thick glass was used as dielectric barrier. The capacity of treatment chamber used was 70 dm^3 maximum volumes.	[60]
RF powered cold atmospheric pressure plasma	Input power (75–150 W) given to electrode (110 × 15 mm) through RF (13.56 MHz) and He gas was used at flow rate of 4 lpm with exposure time of 30 to 120 seconds.	[78]
	RF power source was provided to generate high voltage (10 kV) and high frequency for the plasma jet discharge. The powers required to sustain the discharges ranged from 0 to 40 W and 50 to 600 kHz. The plasma jet was produced at atmospheric pressure using argon at a gas flow rate of 10 L/min.	[69]
	Rod electrode with a size of 110 × 15 mm was powered by RF 13.56 MHz, which was covered by dielectric material. Helium gas (10 lpm) was used in the electrode gap of 0.6 mm.	[68]
Surface Dielectric Barrier Discharge (SDBD)	The SDBD plasma source consisted of an array of 7 concentric ring-shaped electrodes (85 mm outer diameter) embedded in a 1.5 mm thick epoxy–glass bulk material and mounted into the upper shell of a Petri dish (90 mm diameter)	[7]

with dielectric-barriers (DBDs) are shown in Figure 2.2. DBDs are characterized by the presence of one or more insulating layers in the current path between the metal electrodes in addition to the discharge gap(s) [52].

The measurement of basic plasma characteristics, such as gas temperature, electron temperature, electron density, species concentration and electric field is important for understanding the discharge behavior. Although extensive investigations of micro discharge modes and voltage current (V-I) characteristics have been reported, yet direct plasma diagnostics of micro discharges are still limited [63, 76].

Cold Plasma Technology

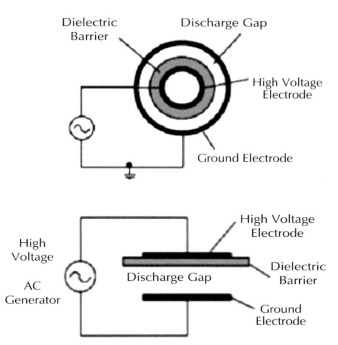

FIGURE 2.2 Typical configurations of planar and cylindrical electrodes with dielectric barriers.

2.7 APPLICATIONS OF PLASMA IN FOOD PRODUCTS

The combination of highly energetic plasma species with a non-thermal treatment mode makes NTP particularly suited for surface sterilization in food processing applications [77]. NTP has been used in the food industry for sterilization of raw agricultural products (e.g., golden delicious apple, lettuce, almond, mangoes, and melon), dry disinfection of food surfaces (like meat, poultry, fish and freshly harvested horticultural produce), granular and particulate foods (like dried milk, herbs and spices) and sprouted seeds. There is a significant scope for application of NTP in sterilizing particulate foods, particularly after the ban of ethylene oxide gases. This technology has also been successfully applied for the surface sterilization of packaging material [15] and for their functional modification for imparting desired properties [23, 55]. A considerable body of data has already accumulated in recent years addressing the efficacy of NTP in inactivating microorganisms on the surfaces of abiotic materials such as glass and synthetic membranes.

2.8 STERILIZATION MECHANISM USING PLASMA

During plasma treatment, killing microorganisms on the food products are result of direct contact to antimicrobial active species. Oxidation of the lipids, amino acids and nucleic acids by reactive oxygen and nitrogen species cause changes leading to microbial death or injury. In addition to reactive species, UV photons can modify DNA of microorganisms and as a result disturb cell replication. Contribution of mentioned mechanisms depends on plasma characteristics and to the type of microorganisms [41].

2.9 POTENTIAL APPLICATIONS AND MICROBIAL INACTIVATION MECHANISM FOR DIFFERENT AGRICULTURAL AND PROCESSED FOOD PRODUCTS BY PLASMA

Several mechanisms are considered to be responsible for microbial inactivation. During plasma treatment, killing microorganisms are result of direct contact of antimicrobial active species to the treatment samples and by different mechanisms. Contribution of mentioned mechanisms depends on plasma characteristics and type of microorganisms. The former includes voltage, working gas, water content in the gas, distance of the microorganism from the discharge glow, etc.; whereas the latter takes account of Gram-positive, Gram-negative, spores and other types [31, 32, 40, 46, 47].

Applications of NTP for decontaminating raw and processed foods have been the subject of research for the past one decade. Several attempts have been made to apply NTP to decontaminate and to control microorganisms:

- Raw agricultural products such as: Bell pepper [75], almonds [14], grains and legumes [66], apple [53], raw almonds [13], tomato, lettuce, carrots [1], apples, cantaloupe and lettuce [12], whole black pepper [25], red pepper [28], peas [7] and wheat seeds [17];
- Processed foods such as: Apple juice [48, 71], nut surfaces [3], sliced cheese and ham [68], egg shells [20], meat surface [19], sliced bacon

Cold Plasma Technology 45

[27], chicken meat and chicken skin [54], pork [49], cheese slices [33], milk [24, 29], chicken breast and chicken leg [16], cold smoked salmon [8], ready-to-eat bresanla ham [61], orange juice [67], brown rice cereal bars [69], white flour [44], whey protein [65], and fish oil [74].

The methods of generation of the NTP and the microbes that were treated upon for controlling vary among researchers. A comprehensive summary of the past research to control various microorganisms, the methods of creation of NTP and results obtained are summarized in Table 2.3, which also shows the type of microorganisms that have been exposed under different plasma sources and indicates the levels of log reduction of microorganisms after the treatments.

2.10 EFFECT OF PLASMA ON QUALITY OF FOOD PRODUCTS

2.10.1 EFFECT OF NTP ON THE COLOR OF THE TREATED FOOD PRODUCTS

The potential use of dielectric barrier discharge (DBD) plasma system was evaluated to improve microbial safety of sliced cheese and no visible change was observed in the color of plasma treated cheese slices, even though the instrumental analysis showed a significant decrease in L^* value and increase in the b^* value [33]. The effect of large area type atmospheric pressure plasma on surface color of bacon was investigated and there was a significant decrease in the L^* value after treatment [27]. When low-pressure cold plasma is exposed on surface micro flora, there was no significant difference among color parameters of meat [19].

A researcher recorded that there were no significant changes in color of pork surface after atmospheric pressure plasma treatment for one minute at an input power of 150 W [49]. On the other hand, another researcher found that increase of a^* values and decrease of b^* values of pork meat after plasma treatment in comparison to untreated meat samples [22]. There were no important changes in color of carrots and tomato, when evaluated at different voltages and processing times under atmospheric pressure cold plasma [1]. Similarly, it was also found that there was no significant difference between the L^*, a^* and b^* values of fresh, control and in-package atmospheric pressure cold plasma treated tomatoes [42]. The

46 Engineering Interventions in Agricultural Processing

TABLE 2.3 Recent Findings on Potential Applications of Plasma for Various Food Products And Level of Inactivation

Method of plasma generation	Treated microbes under plasma	Application in type of food product	Log reduction	Ref.
Cold atmospheric pressure plasma	*Aspergillus flavus*	Brown rice cereal bars	3	[69]
Atmospheric pressure cold plasma, He & O_2	*E. Coli, Staphylococcus, Saccharamyces Cerevisiae*	Deposited on nitro cellulose	5	[34]
Atmospheric pressure glow discharge	*E. Coli*	Raw almonds	5	[13]
Atmospheric pressure plasma	*E. Coli*	Pork and human skin	Comparable to UV	[49]
	Listeria Monocytogenes	Disposable plastic trays, aluminum foil and paper cup	Satisfactory inactivation	[78]
Atmospheric pressure plasma, He and He & O_2	*Listeria monocytogenes, E. Coli, Salmonella Typhimurium*	Sliced bacon	1 to 3	[27]
Atmospheric pressure RF-plasma jet	*E. Coli*	Fresh produce	3 to 5	[2]
Barrier discharge	*Salmonella enteritidis*	Egg shell	2.2 to 2.5	[60]
Cascaded dielectric barrier discharge	*Aspergillus Niger, Bacillus atrophaeus, Bacillus pumilus, E. Coli, Staphyloccus aureus* etc.	PET foils	5	[51]
Cold air plasma with DBD	*Geobacillus sterothermophillus Bacillus cereus*	Direct and indirect exposure on microbes	Significant	[50]
Cold atmospheric gas plasma	*Listeria innoua*	Chicken meat and chicken skin	1 in skin	[54]
Cold atmospheric plasma	*S. enterica, B. subtilis spores* and *B. atrophaeus*	Whole black pepper	4.1, 2.4 and 2.8 resp.	[25]
Cold atmospheric pressure air plasma micro jet	*Staphylococcus aureus*	Aqueous suspension	Complete inactivation	[37]

Cold Plasma Technology

TABLE 2.3 (Continued)

Method of plasma generation	Treated microbes under plasma	Application in type of food product	Log reduction	Ref.
Cold atmospheric pressure plasma	*E. Coli*	Fresh produce: tomato, lettuce, carrots	1.6	[1]
	Yeast/mold	Blueberries	0.8 to 1.6	[30]
Cold plasma	*Aspergillus* Spp. & *Pencillium* Spp	Grains and legumes	3	[66]
DBD	*E. Coli*	Almonds	4	[14]
	Deinococcus radiadurans	Suspended in distilled water	4	[11]
DBD air	*Staphylococcus aureus, E. Coli, Candida Albicans*	Orange juice	5	[67]
	Camphlobacter jejumi, Salmonella entrica	Chicken breast and chicken leg	0.5 to 3	[16]
DBD Ar, CO_2	*Lactobacillus Sakei, Photobacterium Phosphoreum*	Cold smoked salmon	1 to 5	[8]
DBD atmospheric cold plasma	*E. Coli, Salmonella enterica serovar Typhimurium* and *Listeria monocytogenes*	Cherry tomatoes and strawberries	3.5, 3.8 & 4.2 resp.	[79]
DBD O_2, Ar	*Listeria Innocula*	Ready-to-eat bresanla ham	0.4 to 1.6	[61]
DBD plasma	*E. coli* and *Staphylococcus aureus*	Cheese slices	0.9 to 1.98 (He/O_2) 0.45 to 1.47 (He)	[33]
Gliding arc plasma	*E. Coli* and *Salmonella*	Apple	3	[53]
He-O_2 plasma (two plate electrode)	*Pantoea agglomerans*	Bell pepper	2	[75]
Low pressure cold plasma	*Psychrotrophus, total bacteria, yeast* and *mold*	Meat surface	2 to 3	[19]
Low pressure glow discharge plasma	*Escherichia coli O157:H7* and *Staphylococcus aureus*	Packaging material's surface	4	[35]

TABLE 2.3 (Continued)

Method of plasma generation	Treated microbes under plasma	Application in type of food product	Log reduction	Ref.
Low pressure microwave plasma	*Bacillus subtilis, Aspergillus niger, Bacillus Stearothermophillus* and *Sacchromyces Cerevisiae*	Microbes sprayed on PET foils	Significant reduction	[18]
Low pressure plasma	*Aspergillus parasiticus*	Nut surfaces	5	[3]
Low temperature plasma	*E. Coli*	Milk	3	[24]
Non thermal cold plasma	*E. Coli*	Milk	Significant	[29]
One atmosphere uniform glow discharge	*E. Coli, Salmonella* and *Listeria monocytogens*	Apples, cantaloupe and lettuce	2	[12]
Plasma jet	*Migration microorganisms*	Mango, melon	3	[57]
	Citrobacter freundii	Apple juice	5	[71]
Plasma jet with He	*Listeria monocytogenes*	Sliced cheese and ham	5.8 & 1.7 resp	[68]
Plasma needle	*E. Coli*	Apple juice	7	[48]
Pulsed plasma gas discharge	*E. Coli and Bacillus Cereus* etc.	Chilled poultry wash water	≤ 8	[62]
Resistive Barrier Discharge	*Salmonella enteritidis* and *Salmonella typhimurium*	Egg shells	2.5 to 4.5	[20]

overall change in hue and chroma value (indicating color saturation) of the control and treated samples was also insignificant.

In another study, a change in the L^*, a^* and b^* color parameters of atmospheric cold plasma treated strawberries was observed. However, the changes in individual color parameters like, lightness, redness or greenness were statistically insignificant ($p > 0.05$) in comparison to the untreated control stored under same conditions [43].

2.10.2 EFFECT OF NTP ON THE TEXTURAL PROPERTIES OF TREATED FOOD PRODUCTS

It has been reported that gas plasma treatments cause loss of firmness in fresh-cut apples with treatment time of 10 to 30 min [72]. It was also summarized that the plasma treatment of cherry tomatoes does not adversely affect critical quality parameters of color, firmness, pH and weight loss. An insignificant difference ($p > 0.05$) between the firmness values of control and treated tomatoes was recorded at the end of storage period, meaning that the tissue structure of the tomato remained intact [42]. In another study, it was reported that a significant ($p \leq 0.05$) decrease in firmness of atmospheric cold plasma treated strawberries was recorded within 24 hours. The difference in firmness among untreated control and treated group was statistically insignificant ($p > 0.05$) [43].

2.11 SUMMARY

Cold plasma is an emerging non-thermal food preservation technology. It represents a different state-of-the-art and flexible tools for sterilizing surfaces of food and products. Cold plasma technology will continue to be explored for potential use in food processing, to address issues like: food sensory, antimicrobial activity and the technical issues of integration of NTP into food processing systems.

KEYWORDS

- **atmospheric pressure plasma jet**
- **cold jet plasma**
- **dielectric barrier discharge**
- **fourth state of matter**
- **gliding arc cold plasma system**
- **Knudsen number**
- **micro hallow cathode discharge**
- **non thermal plasma**

- **non-local thermodynamic equilibrium plasma**
- **one atmospheric uniform glow discharge**
- **Paschen's law**
- **plasma corona discharge**
- **rf powered atmospheric pressure plasma**

REFERENCES

1. Aguirre Daniela Bermudez, Erik Wemlingerb, Patrick Pedrowb, Gustavo Barbosa-Canovasa, & Manuel Garcia-Perez. (2013). Effect of atmospheric pressure cold plasma (APCP) on the inactivation of Escherichia coli in fresh produce. *Food Control. 34*(1), 149–157.
2. Baier, M., Gorgen, M., Ehlbeck, J., Knorr, D., Herppich, W. B., & Schluter, O. (2014). Non-thermal atmospheric pressure plasma: screening for gentle process conditions and antibacterial efficiency on perishable fresh produce. *Innov. Food Sci. Emerg. Technol. 22*, 147–157.
3. Basaran, P., Basaran-Akgul, N., & Oksuz, L. (2008). Elimination of *Aspergillus parasiticus* from nut surface with low pressure cold plasma (LPCP) treatment. *Food Microbiology. 25*(4), 626–632. doi: 10.1016/j.fm.2007.12.005.
4. Becker, K. H., Kogelschatz, U., Schoenbach, K. H., Barker, R. J. (2005). Non-Equilibrium air plasmas at atmospheric pressure. The Institute of Physics, London, CRC Press.
5. Bermudez-Aguirre, D., Wemlinger, E., Pedrow, P., Barbosa-Canovas, G., & Garcia-Perez, M. (2013). Effect of atmospheric pressure cold plasma (APCP) on the inactivation of *Escherichia coli* in fresh produce, *Food Control, 34*(1), 149–157.
6. Boulos, M. I., Fauchais, P., & Pfender, E. (1994). Thermal plasmas: Fundamental and Applications. Volume 1, plenum Press, New York, ISBN: 0-306-44607-3, 452.
7. Bubler Sara, Werner, B. Herppich, Susanne Neugart, Monika Schreiner, Jorg Ehlbeck, Sascha Rohn, & Oliver Schluter. 2015. Impact of cold atmospheric pressure plasma on physiology and flavonol glycoside profile of peas (*Pisum sativum* 'Salamanca'). *Food Research International.* http://dx.doi.org/10.1016/j.foodres. 2015.03.045.
8. Chipper, A.S, Chen, W, Mejlholm, O, Dalgaard, P., & Stamate, E. (2011). Atmospheric pressure plasma produced inside a closed package by a dielectric barrier discharge in Ar/CO(2) for bacterial inactivation of biological samples. *Plasma Sources Sci Technol, 20*, 10.
9. Choi Yoon-Ho, Ji-Hun Kima, Kwang-Hyun Paekb, Won-Tae Jub, & Hwanga, Y. S. (2005). Characteristics of atmospheric pressure N_2 cold plasma torch using 60-Hz AC power and its application to polymer surface modification, *Surface & Coatings Technology, 193*, 319–324.

10. Conrads, H., & Schmidt, M. (2000). Plasma generation and plasma source. *Plasma Sources Sci Technol 9*, 441–454.
11. Cooper, M., Fridman, G., Staack, D., Gutsol, A. F., Vasilets, V. N., Anandan, S., Cho, Y. I., Fridman, A., Tsapin, A. (2009). Decontamination of surfaces from extremophile organisms using non-thermal atmospheric-pressure plasmas. *Plasma Science, IEEE Transactions. 37*(6), 866–871.
12. Critzer, F. J., Kelly-Wintenberg, K., South, S. L., & Golden, D. A. (2007). Atmospheric plasma inactivation of food borne pathogens on fresh produce surface. *Journal of Food Protection 70*(10), 2290–2296.
13. Deng, X. J., & Kong, M. (2009). Protein destruction by a helium atmospheric pressure glow discharge: capability and mechanisms. *J Appl Phys. 101*(7), 074701.
14. Deng, S., Ruan, R., Mok, C. K., Huang, G., Lin, X., & Chen, P. (2007). Inactivation of *Escherichia coli* on Almonds Using Non thermal Plasma. *Journal of Food Science, 72*(2), M62–M66.
15. Deilmann, M., Halfmann, H., Bibinov, N., Wunderlich, J., & Awakowicz, P. (2008). Low pressure microwave plasma sterilization of polyethylene terephthalate bottles. *Journal of Food Protection. 71*(10), 2119–2123.
16. Dirks, Brian, P. Dobrynin, Danil Fridman, Gregory Mukhin, Yuri Fridman, Alexander Quinlan, & Jennifer, J. (2012). Treatment of raw poultry with nonthermal dielectric barrier discharge plasma to reduce *Campylobacter jejuni* and *Salmonella enterica. Journal of Food Protect. 75*(1), 4–206.
17. Dobrin Daniela, Monica Magureanu, Nicolae Bogdan Mandache, & Maria-Daniela Ionita (2015). The effect of non-thermal plasma treatment on wheat germination and early growth. *Innovative Food Science and Emerging Technologies.* http://dx.doi.org/10.1016/j.ifset.2015.02.006.
18. Feichtinger, J., Schulz, A., Walker, M., & Schuhmacher, U. (2003). Sterilization with low-pressure microwave plasmas. *Surf Coat Technol. 174–175*, 564–569.
19. Figlewicz-Natalia Ulbin, Andrzej Jarmoluk, & Krzysztof Marycz (2013). Antimicrobial activity of low-pressure plasma treatment against selected food borne bacteria and meat microbiota. *Ann Microbiol*, doi: 10.1007/s13213-014-0992-y.
20. Fridman, A., & Kennedy, L. A. (2004). *Plasma Physics and Engineering.* Taylor & Francis Books Inc., New York.
21. Fridman, G., Friedman, G., Gutsol, A., Shekhter, A. B., Vasilets, V. N., & Fridman, A. (2008). Applied plasma medicine. *Plasma Process Polym. 5*, 503–533.
22. Frohling, A., Durek, J., Schnabel, U., Ehlbeck, J., Bolling, J., & Schluter, O. (2012). Indirect plasma treatment of fresh pork: Decontamination efficiency and effects on quality attributes. *Innovative Food Science and Emerging Technologies, 16*, 381–390.
23. Gulec, H. A., Sarloglu, K., & Mutlu, M. (2006). Modification of food contacting surfaces by plasma polymerisation technique. Part I: Determination of hydrophilicity, hydrophobicity and surface free energy by contact angle method. *Journal of Food Engineering. 75*(2), 187–195. doi: 10.1016/j.jfoodeng.2005.04.007.
24. Gurol, C., Ekinci, F. Y., Aslan, N., & Korachi, M. (2012). Low temperature plasma for decontamination of, E. coli in milk. *International Journal of Food Microbiology. 157*(1), 1–5.

25. Hertwig Christian, Kai Reineke, Jorg Ehlbeck, Dietrich Knorr, & Oliver Schluter. (2015). Decontamination of whole black pepper using different cold atmospheric pressure plasma applications. *Food Control, 55*, 221–229.
26. Kai Knoerzer, Anthony, B. Murphy, Mark Fresewinkel, Peerasak Sanguansri, & John Coventry. (2012). Evaluation of methods for determining food surface temperature in the presence of low-pressure cool plasma. *Innovative Food Science and Emerging Technologies, 15*, 23–30.
27. Kim, B., Yun, H., Jung, Y., Jung, H., Choe, W., & Jo, C. (2011). Effect of atmospheric pressure plasma on inactivation of pathogens inoculated onto bacon using two different gas compositions. *Food Microbiol. 28*, 9–13.
28. Kim Jung Eun, Dong-Un Lee, & Sea, C. Min. (2014). Microbial decontamination of red pepper powder by cold plasma. *Food Microbiology, 38*, 128–136.
29. Korachi May, Fatma Ozen, Necdet Aslan, Lucia Vannini, Maria Elisabetta Guerzoni, Davide Gottardi, & Fatma Yesim Ekinci. (2015). Biochemical changes to milk following treatment by a novel, cold atmospheric plasma system. *International Dairy Journal. 42*, 64–69.
30. Lacombe Alison, Brendan, A. Niemira, Joshua, B. Gurtler, Xuetong Fan, Joseph Sites, Glenn Boyd, & Haiqiang Chen. (2014). Atmospheric cold plasma inactivation of aerobic microorganisms on blueberries and effects on quality attributes. *Food Microbiology, 46*, 479–484.
31. Laroussi, M., Lu, X., & Malott, C. M. (2003). A non-equilibrium diffuse discharge in atmospheric pressure air. *Plasma Sources Science and Technology, 12*, 53.
32. Laroussi, M., & Leipold, F. (2004). Evaluation of the roles of reactive species, heat, and UV radiation in the inactivation of bacterial cells by air plasmas at atmospheric pressure. *International Journal of Mass Spectrometry, 233*(1–3), 81–86.
33. Lee Hyun Jung, Jung Samooel, Jung Heesoo, Park Sanghoo, Choe Wonho, Ham Jun Sang, & Jo Cheorun (2012). Evaluation of a Dielectric Barrier Discharge Plasma System for Inactivating Pathogens on Cheese Slices. *Journal of Animal Science and Technology. 54*(3), 191–198.
34. Lee Kyenam, Kwang-hyun Paek, Won-Tae Ju, & Yeonhee Lee (2006). Sterilization of bacteria, yeast, and bacterial endospores by Atmospheric-Pressure Cold Plasma using Helium and Oxygen. *The Journal of Microbiology, 44*, 3.
35. Lee Taehoon, Pradeep Puligundla, & Chulkyoon Mok (2015). Inactivation of foodborne pathogens on the surfaces of different packaging materials using low-pressure air plasma. *Food Control, 51*, 149–155.
36. Lieberman, M. A., & Lichtenberg. A. J. (1994). Principles of Plasma Discharges and Materials Processing. John Wiley & Sons. Inc., New York.
37. Liu Fuxiang, Peng Sun, Na Bai, Ye Tian, Haixia Zhou, Shicheng Wei, Yanheng Zhou, Jue Zhang, Weidong Zhu, Kurt Becker, & Jing Fang (2010). Inactivation of Bacteria in an Aqueous Environment by a Direct Current, Cold Atmospheric Pressure Air Plasma Micro jet. *Plasma Processes and Polymers. 7*(3–4), 231–236.
38. Maeda Kojiro, Yoichi Toyokawa, Naohiro Shimizu, Yuichiro Imanishi, & Akikazu Sakudo (2015). Inactivation of Salmonella by nitrogen gas plasma generated by a static induction thyristor as a pulsed power supply. *Food Control, 52*, 54–59.

39. Massines, F., Segur, P., Gherardi, N., Khamphan, C., & Ricard, A. (2003). *Surf. and Coat Techn.*, 174–175:8.
40. Mendis, D., Rosenberg, M., & Azam, F. (2000). A note on the possible electrostatic disruption of bacteria. *Plasma Science, IEEE* Transactions. 1304–1306.
41. Misra, N. N., Tiwari, B. K., Raghavarao, K. S. M. S., & Cullen, P. J. (2011). Nonthermal plasma inactivation of food-borne pathogens. *Food Engineering Reviews*, 3(3–4), 59–170.
42. Misra, N. N., Keener, K. M., Mosnier, J. P., Bourke, P., & Cullen, P J. (2014a). Effect of in-package atmospheric pressure cold plasma treatment on quality of cherry tomatoes. *Journal of Bioscience and Bioengineering. 118*(2), 177–182.
43. Misra, N. N., Patil, S., Moiseev, T., Bourke, P., Mosnier, J. P., Keener, K. M., & Cullen, P. J. (2014b). In-package atmospheric pressure cold plasma treatment of strawberries, *Journal of Food Engineering, 125*, 131–138.
44. Misra. N. N., Seeratpreet Kaur, Brijesh, K. Tiwari, Amritpal Kaur, Narpinder Singh, & Cullen, P. J. (2015). Atmospheric pressure cold plasma (ACP) treatment of wheat flour. Food Hydrocolloids, *44*, 115–121.
45. Moisan, M., Calzada, M. D., Gamero, A., & Sola, A. (1996). Experimental investigation and characterization of the departure from local thermodynamic equilibrium along a surface-wave-sustained discharge at atmospheric pressure, *J. Appl. Phys. 80*(1), 46–55.
46. Moisan, M., Barbeau, J., Crevier, M. C., Pelletier, J., Philip, N., & Saoudi, B. (2002). Plasma sterilization. Methods and mechanisms. *Pure and Applied Chemistry, 74*(3), 349–358.
47. Moisan, M., Barbeau, J., Moreau, S., Pelletier, J., Tabrizian, M., & Yahia, L. H. (2001). Low-temperature sterilization using gas plasmas: a review of the experiments and an analysis of the inactivation mechanisms. *International Journal of Pharmaceutics. 226*, 1–21.
48. Montenegro, J., Ruan, R., Ma, H., & Chen, P. (2002). Inactivation of, E. coli O157:H7 Using a Pulsed Nonthermal Plasma System. *Journal of Food Science, 67*(2), 646–648.
49. Moon, S. Y., Kim, D. B., Gweon, B., Choe, W., Song, H. P., & Yo, C. (2009). Feasibility study of the sterilization of pork and human skin surfaces by atmospheric pressure plasmas. *Thin Solid Films, 14*, 4272–4275.
50. Morris, A., Akan, T., McCombs. G. B, Hynes. W. L., & Laroussi (2007). Bactericidal effects of non-equilibrium cold plasma on *Geobacillus stearothermophilis* and *Bacillus cerus*. 28th ICPIG, Prague, Czech Republic.
51. Muranyi, P., Wunderlich J and Heise, M. (2007). Sterilization efficiency of a cascaded dielectric barrier discharge. *Journal of Applied Microbiology. 103*(5), 1535–1544.
52. Nehra, V., Kumar, A., & Dwivedi, H. K. (2008). Atmospheric non-thermal plasma sources. *Int. J. Eng., 2*, 53–68.
53. Niemira, B. A., & Sites, J. (2008). Cold Plasma Inactivates Salmonella Stanley and Escherichia coli O157:H7 Inoculated on Golden Delicious Apples. *Journal of Food Protection, 174, 71*(7), 1357–1365.
54. Noriega, E Sharma, G., Laca, A., Diaz M and Kong MG. (2011). Cold atmospheric gas plasma disinfection of chicken meat and chicken skin contaminated with *Listeria innocua*. *Food Microbiology, 28*(7), 1293–1300.

55. Ozdemir, M., Yurteri, C. U., Sadikoglu, H. (1999). Physical polymer surface modification methods and applications in food packaging polymers. *Critical Reviews in Food Science and Nutrition. 39*(5), 457–477.
56. Park Jaeyoung, Henins, I., Herrmann. H. W, Selwyn. G. S., & Hicks. R. F. (2001). Discharge phenomena of an atmospheric pressure radio-frequency capacitive plasma source. *Journal of Applied Physics, 89*(1), 20–28.
57. Perni, S., Liu, D. W., Shama, G., & Kong, M. G. (2008). Cold Atmospheric Plasma Decontamination of the Pericarps of Fruit. *Journal of Food Protection. 174, 71*(2), 302–308.
58. Piel, Alexander. (2010). Plasma Physics. *An Introduction to Laboratory, Space, and Fusion Plasmas.* Springer: Berlin, 344.
59. Ragni, L., Berardinelli, A., Vannini, L., Montanari, C., Sirri, F., Guerzoni, M. E., & Guarnieri, A. (2010). Non-thermal atmospheric gas plasma device for surface decontamination of shell eggs. *Journal of Food Engineering, 100,* 125–132.
60. Raizer. Y. P. (1997). Gas Discharge Physics. Springer, Berlin.
61. Rod Sara Katrine, Flemming Hansenb, Frank Leipoldc, & Susanne Knochela. (2012). Cold atmospheric pressure plasma treatment of ready-to-eat meat: Inactivation of *Listeria innocua* and changes in product quality. *Journal of Food Microbiology. 30*(1), 233–238.
62. Rowan, N., Espie, S., Harrower, J., Anderson, J., Marsili, L., & MacGregor, S. (2007). Pulsed plasma gas-discharge inactivation of microbial pathogens in chilled poultry wash water. *Journal of Food Protection, 70*(12), 2805–2810.
63. Sakai, O., Kishimoto, Y., & Tachibana. K. (2005). Integrated coaxial-hollow micro dielectric barrier discharges for a large-area plasma source operating at around atmospheric pressure. *Journal of Physics D: Applied Physics 38,* 431–441.
64. Schutze, A., Jeong. J. Y, Babayan. S. E, Park, J., Selwyn. G. S., & Hicks. R. F. (1998). The atmospheric- pressure plasma jet: A review and comparison to other plasma sources. *IEEE Transactions on Plasma Science, 26*(6), 1685–1694.
65. Segat Annalisa, N., Misra, N., Cullen, P. J., & Nadia Innocente. (2015). Atmospheric pressure cold plasma (ACP) treatment of whey protein isolate model solution. *Innovative Food Science and Emerging Technologies.* http://dx.doi.org/10.1016/j.ifset.2015.03.014.
66. Selcuk, M., Oksuz, L., & Basaran, P. (2008). Decontamination of grains and legumes infected with *Aspergillus spp.* and *Penicillum spp.* by cold plasma treatment, *Bioresource Technology, 99,* 5104–5109.
67. Shi, X. M., Zhang, G J., Wu, X. L., Li, Y. X., Ma, Y., & Shao, X. J. (2011). Effect of low temperature plasma on microorganism inactivation and quality of freshly squeezed orange juice. *IEEE Transactions on Plasma Science, 39,* 1591–1597.
68. Song, H. P., Kim, B., Choe, J. H., Jung, S., Moon, S. Y., Choe, W., & Jo, C. (2009). Evaluation of atmospheric pressure plasma to improve the safety of sliced cheese and ham inoculated by 3-strain cocktail Listeria monocytogenes. *Food Microbiology, 26*(4), 432–436.
69. Suhem Kitiya, Narumol Matan, Mudtorlep Nisoa, & Nirundorn Matan. 2013. Inhibition of *Aspergillus flavus* on agar media and brown rice cereal bars using cold atmospheric plasma treatment. *International Journal of Food Microbiology, 161,* 107–111.

70. Sulmer, A. Fernandez-Gutierrez, & Patrick, D. Pedrow. (2010). Cold Atmospheric-Pressure plasmas applied to active packaging of apples. *IEEE Transactions of Plasma Science. 38*(4), 957–965.

71. Surowsky Bjorn, Antje Frohling, Nathalie Gottschalk, Oliver Schlüter, & Dietrich Knorr. (2014). Impact of cold plasma on *Citrobacter freundii* in apple juice: Inactivation kinetics and mechanisms. *International Journal of Food Microbiology, 174*, 63–71.

72. Tappi, S., Berardinelli, A., Ragni, L., Rosa, M. D., Guarnieri, A., & Rocculi, P. (2013). Atmospheric gas plasma treatment of fresh-cut apples. *Innov Food Sci Emerg Technol.* doi:10.1016/j.ifset.2013.09.012.

73. Tendero, C., Tixier, C., Tristant, P., Desmaison, J., & Leprince, P. (2006). Atmospheric pressure plasmas: A review. *Spectrochimica Acta Part B: Atomic Spectroscopy, 61*(1), 2–30.

74. Vandamme Jeroen, Anton Nikiforov, Klaas Dujardin, Christophe Leys, Luc De Cooman, & JimVanDurme. (2015). Critical evaluation of non-thermal plasma as an innovative accelerated lipid oxidation technique in fish oil. *Food Research International, 72*, 115–125.

75. Vleugels, M., Shama, G., Deng, X. T., Greenace, E., Brocklehurst, T., & Kong, M. G. (2005). Atmospheric plasma inactivation of biofilm-forming bacteria for food safety control. *IEEE Trans Plasma Sci 33*, 824–828.

76. Wang, Q., Koleva, I., Donnelly, V., & Economou. D. (2005). Spatially resolved diagnostics of an atmospheric pressure direct current helium micro plasma. *Journal of Physics D: Applied Physics 38*, 1690–1697.

77. Yu, Q. S. (2006). Bacterical Inactivation using low-temperature atmospheric plasma brush sustained with Argon gas. *Journal of Biomedical Materials Research Part B: Applied Biomaterials, 211*–219.

78. Yun Hyejeong, Binna Kim, Samooel Jung, Zbigniew, Kruk, A., Dan Bee Kim, Wonho Choe, & Cheorun Jo, (2010). Inactivation of *Listeria monocytogenes* inoculated on disposable plastic tray, aluminum foil, and paper cup by atmospheric pressure plasma. *Food Control, 21*, 1182–1186.

79. Ziuzina, D., Patil, S., Cullen. P. J., Keener, K. M., & Bourke, P. (2014). Atmospheric cold plasma inactivation of *Escherichia coli, Salmonella enterica serovar Typhimurium* and *Listeria monocytogenes* inoculated on fresh produce. *Food Microbiology, 42*, 109–116.

CHAPTER 3

TECHNOLOGY OF OHMIC HEATING IN MEAT PROCESSING

ASAAD REHMAN SAEED AL-HILPHY

CONTENTS

3.1 Introduction 57
3.2 Principles of Ohmic Heating 59
3.3 Pasteurization Value of Meat 64
3.4 Ohmic Assisted Cooking Burger 65
3.5 Microorganisms' Inhibition in Meat and Meat Products
    Using Ohmic Heating 66
3.6 Future Research Opportunities 67
3.7 Conclusions 67
3.8 Summary 67
Keywords 68
References 68

3.1 INTRODUCTION

Finland used Ohmic heating (OH) for meat processing first time in 1970, however the difficulty in operation of OH led to failure of the enterprise. In 1990, cooking liver pate and hams by Ohmic heating were executed in France by the Meat Development Association (MDA) and Électricité de France. Although, the Ohmic heating is a successful technology for treatment of liquid foods such as milk, juice, soup, etc., yet it is still only an industrial application for processed meat and still lot of research is needed [6, 16].

Ohmic thawing gave higher quality of product compared with conventional method. Also Ohmic thawing technology in frozen meat resulted in uniform, better quality product and quicker operation. In addition, pH and color of thawed meat were not changed significantly [36]. As shown in Figure 3.1, Ohmic heating system consists of a voltage source, connecting wires, two electrodes, plastic container, water and food product such as meat. Piette et al. [23] designed Ohmic cooker for processed meat (sausage manufactured from pork meat) consisting of a nylon tube with 7.5 cm I.D., 30 cm length and 0.64 cm thick (Figure 3.2), electrodes of titanium, and a 7.5 cm flat circular plate between the electrodes to hold the meat [23]. Temperature of processed meat ranges from 70 to 80°C with heating rate from 3.9 to 10.3°C/min and holding time of 20 minutes.

This chapter presents technology of Ohmic heating to process the meat.

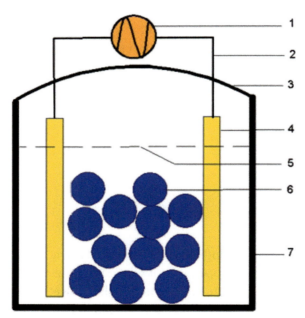

FIGURE 3.1 Layout of Ohmic heating system: (1) voltage source, (2) wire, (3) cover, (4) electrode, (5) water level, (6) food material, and (7) plastic container.

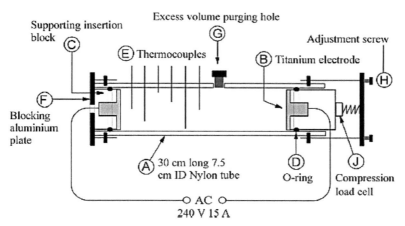

FIGURE 3.2 Schematic of Ohmic cooking unit.

3.2 PRINCIPLES OF OHMIC HEATING

Ohmic heating is also known as electric resistance, joule heating, direct electric resistance heating, electro-conductive heating or electro-heating. Ohmic heating is a novel thermal process where alternative electric current is passed within the food and the food is heated to a desired temperature [25]. Due to an electric resistance in the electric circuit, the heat is generated instantaneously and volumetrically within the food because of ionic movement. The amount of heat generated in the food depends on many factors such as: current, voltage, electric field and electric conductivity of the food [8, 9]. In Ohmic heating, heat transfers volumetrically from the interior of the food towards outside periphery contrary to traditional heating where heat transfer from hot surface to the inside surface of food is via conduction and convection. In addition, heat transfer causes a thermal lag in the mixture of food [13, 38].

3.2.1 ADVANTAGES AND DISADVANTAGES OF OHMIC HEATING

3.2.1.1 Advantages [11, 29]

a. Does not heat surfaces,
b. Uniform and rapid heating,

c. Minimum heat damage to food,
d. Minimum nutrient loss,
e. Less heat loss,
f. Reducing fouling,
g. Maintenance costs are low,
h. Environmentally friendly,
i. Heating food by inner heat generation,
j. Temperature of particulates is higher than that of liquid and can be accomplished which is not possible in traditional heating, and
k. Energy efficiency is high, because of conversion of most electric energy into heat.

3.2.1.2 Disadvantages [6]

a. High initial cost,
b. Inadequate information about Ohmic heating,
c. Heat generation rate is affected by the electric heterogeneity of food particles, complicated coupling between electric field distribution and temperature, heat channeling and particles form and orientation, and
d. Ohmic heating application varies from one product to another, thus resulting in increase of development costs.

3.2.2 THERMAL POWER

Alternative current (AC) is used to heat the food because it creates an oscillatory movement for ions resulting in heat dissipation due to resistance or Ohmic heating phenomenon [6]. Thermal power dissipated into the food can be calculated as follows:

$$P = R I^2 K \tag{1}$$

$$R = \frac{V}{I} \tag{2}$$

where: P = thermal power dissipated into the food (W), I = electric current (A), K = shape factor of apparatus, R = food electric resistance (Ω), and

Technology of Ohmic Heating in Meat Processing 61

V = voltage (V). Units of resistivity are ohm/m. Also, reciprocal of conductivity (Siemens/m) is used to express the resistivity.

3.2.3 ELECTRIC FIELD INTENSITY

Electric field intensity is calculated as follows [31]:

$$E = \frac{V}{d} \tag{3}$$

$$E = \frac{V}{r \left[1n \dfrac{R_2}{R_1} \right]}, \text{ and} \tag{4}$$

$$r = \left(R_1 - \left(R_2 - R_1 \right) / 2 \right) \tag{5}$$

where: E = electric field intensity (V cm^{-1}), d = distance between electrodes (cm). For a cylindrical shape of the electrode, electric field intensity is calculated with Eq. (4) with R_1 = inner radius of large pipe (cm) and R_2 = outer radius of small pipe (cm).

3.2.4 ELECTRIC CONDUCTIVITY

Electric conductivity means how a food material transmits electric charge and is affected by the chemical composition of food. On the other hand, electric conductivity increased with increasing ionic (salts as NaCl) [2]. Electric conductivity is calculated from the following equation [10, 37]:

$$\sigma = \frac{I\,d}{V\,A} \tag{6}$$

$$\therefore R = \frac{V}{I} \tag{7}$$

where, σ = electric conductivity (Sm^{-1}) and A = sectional area (m^2).

Combining Eqs. (6) and (7), one gets:

$$\sigma = \frac{d}{R\,A} \tag{8}$$

From Eqs. (3) and (6), one gets

$$\sigma = \frac{I}{E\,A} \tag{9}$$

The heat generated (Q_G) in the meat during Ohmic heating can be calculated as follows:

$$Q_G = I^2\,R \tag{10}$$

Also, Q_G can be calculated using the voltage gradient (∇V) and electric conductivity as follows:

$$Q_G = |\nabla V|^2\,\sigma \tag{11}$$

where, σ is a function of temperature and position.

Also, Q_G is calculated from the following equation:

$$Q_G = E^2\sigma, \text{ or} \tag{12}$$

$$Q_G = \frac{J^2}{\sigma} \tag{13}$$

where, J = current intensity (Amperes m^{-2}).

The general equation for calculating Q_G is as follows:

$$Q_G = m\,C_p\,\Delta T \tag{14}$$

where, m = meat mass (kg), C_p = specific heat of meat (kJ kg$^{-1\circ C-1}$), and ΔT = the difference in the temperature between ambient and meat.

When current passes through the meat tissue, the Ohmic heating occurs. It causes the temperature to increase. Temperature can be estimated using the following equation [35]:

Technology of Ohmic Heating in Meat Processing

$$\frac{dT}{dt} = \frac{\sigma E^2}{\rho C_p} \tag{15}$$

where, ρ = the density.

The relationship between temperature and electric conductivity is a straight line as shown in the following equation [20]:

$$\sigma_T = \sigma_{ref.}\left[1 + m\left(T - T_{ref.}\right)\right] \tag{16}$$

where, σ_T = the electric conductivity at temperature T, m = the temperature coefficient, and σ_{ref} = the electric conductivity at reference temperature T_{ref}

De Alwis and Fryer [5] indicated that the quality of meat cooked by Ohmic heating was acceptable. They indicated several problems that hinder the fabrication of industrial Ohmic cookers, such as: the good contacts between meat and electrodes surfaces are difficult; there is a difference in the electric conductivity of meat because of heterogeneous texture of meat during Ohmic heating. Meat texture, fiber direction, fat content and meat type have effects on the electric conductivity [24]. Piette et al. [22] stated that for cooking meat (increasing temperature) by Ohmic heating, a very large current or voltage is needed. The animal fat electric conductivity is low and ranges from 0.001 to 0.1 Sm^{-1} [7]. The electric conductivity of a processed meat ranges from 1 to 3.5 Sm^{-1} [22]. Bozkurt and Icier [4] demonstrated that the meat with the low fat ratio had a highest electric conductivity while the electric conductivity was reduced when the meat fat was low. They also observed that fat content of meat determines the efficiency of Ohmic cooking process. Therefore, distribution of fat globules in the meat product and blend meat has a direct effect on the homogeneous heating and probability of occurrence of the cold and hot zones in the meat.

Piette et al. [23] found that the electric conductivity of the processed meat was increased significantly with increasing temperature. In addition, it was noticed that increasing salt concentration led to increasing electric conductivity, while it was decreased with increasing fat content. Also, the electric conductivity ranged from 1 to 6 Sm^{-1} during cooking, due to variation of electric conductivity of processed meat for salt concentration range of 1 to 4%. De Halleux et al. [6] found following empirical equation to predict electric conductivity of emulsion for temperature range of 15 to 80°C:

$$\sigma = 0.13 + 0.0039TR^2 = 0.996 \tag{17}$$

The following nonlinear model can be used to predict the electric conductivity of meat [4]:

$$\sigma = A + BT + C(fat\%)^N \tag{18}$$

where, A = Electric conductivity constant (Sm^{-1}), B = Electric conductivity constant (Sm^{-1} °C^{-1}), N = Dimensionless constant, and A, B, and C are nonlinear regression coefficients.

3.2.5 SYSTEM PERFORMANCE COEFFICIENT

The system performance coefficient (SPC) is defined in the following equation [9]:

$$SPC = \frac{Q_G}{E_g} \tag{19}$$

$$E_g = Q_G + E_{loss} = \sum \Delta VIT \tag{20}$$

where, E_g = amount of energy (J).

Heat loss (E_{loss} in J) to the ambient is calculated as follows [9]:

$$E_{loss} = \bar{h}\pi DL\left(\overline{T_w} - T_{amb.}\right) \tag{21}$$

where, $\bar{h}$ is the heat transfer coefficient (Wm^{-2}°C)

3.3 PASTEURIZATION VALUE OF MEAT

The purpose of Ohmic heating is to eliminate pathogenic bacteria from liquid and solid foods such as meat, milk, juice, etc. Pasteurization value is a common method, which is used to determine grade of microbiological security in the coolest zone inside food during Ohmic cooking. De Halleux et al. [6] stated that the Ohmic cooking has three phases: heating phase (increasing temperature), temperature holding phase and cooling phase. Pasteurization value is calculated from the following equation:

Technology of Ohmic Heating in Meat Processing

$$P_v = \int_{t_{T55-}}^{t_{T55+}} 10^{(T-T_P)/Z} \, dt \tag{22}$$

$$z = \frac{T_2 - T_1}{\log D_2 - \log D_1} \tag{23}$$

$$D = \frac{t_2 - t_1}{\log N_2 - \log N_1} \tag{24}$$

In Eqs. (22)–(24): P_v = pasteurization value, T = meat temperature at time t (°C), T_P = pasteurization temperature (°C), t_{T55+} = final pasteurization time (min.) at temperature below 55°C, t_{T55-} = initial pasteurization time (min.) at temperature above 55°C, Z = thermal resistance (°C) is the temperature rise [27] that is needed to reduce required heating time by 90% as shown in Eq. (23), D_1 = decimal reduction time at temperature of T_1 (min.), D_2 = decimal reduction time at temperature of T_2 (min.), $(T_2 - T_1)$ = the variation in the temperature, D = decimal reduction time that is required to destroy 90% of microorganisms [34] and is calculated by the Eq. (24), $(T_2 - t_1)$ = the variation in the pasteurization time, N_1 = initial count of microorganisms (CFU ml^{-1}) at t_1, and N_2 = last count of microorganisms (CFU ml^{-1}) at t_2. boughs

Inactivation of microorganisms [15] is defined as follows:

$$log\left[\frac{N_2}{N_1}\right] = -\int_0^t \frac{dt}{D_2} \tag{25}$$

$$D_2 = D_1 10^{(T_2 - T_1)/z} \tag{26}$$

3.4 OHMIC ASSISTED COOKING BURGER

Mohammed [18] developed a method and apparatus of cooking food by Ohmic heating. Ohmic assisted cooking means using both traditional heating and Ohmic heating together for cooking the burger. In this method, there is a conversion to electric resistance due to contacts with two electrodes from top and bottom. There are two types of heat emitted to burger: the first is the heat emitted from hot plate via contact which leads to increase in temperature to 180°C; and the second is heat generated inside the burger due to passage of electric current in the burger which kills all microorganisms that are present in the burger. Alternative current is used in this technology with voltage range from 30 to 100 V, however preferable voltage is 70 V. In addition, the burger resistance can approximately reach 600 Ohm. Mohammed used burger diameter of 90 mm and thickness of 6 mm that gave Ohmic resistance of 5.7 Ohm, the beak current of 12A, consumed peak power of 821 W with applied voltage of 70 V. This means that 0.19 Acm^{-2} and 13 Wm^{-2} are needed to produce burger that is safe for consumption.

Özkan and Farid [19] studied quality of hamburger cooked by combined Ohmic heating and traditional heating. They found that no significant differences between Ohmic heating and traditional heating in the quality of hamburger (oil content, moisture content and mechanical properties). The advantages of using Ohmic assisted cooking of hamburger are [19]: (1) quicker cooking, (2) less power consumption, and (3) more safe hamburger.

3.5 MICROORGANISMS' INHIBITION IN MEAT AND MEAT PRODUCTS USING OHMIC HEATING

Ohmic heating eliminates microorganisms from meat via mild electroporation and heating. Electroporation is a non-thermal treatment that causes cellular damage in the cell walls of microorganisms because of generation of electric field on the cells [21, 30]. Electroporation leads to enhance cell membrane permeability and increases material diffusion through the cell membrane via electro-osmosis [1, 14]. Many researchers have assumed that the electroporation or breakdown mechanism always occurs with non-thermal effect through Ohmic heating [12, 26]. On the other hand, USA-FDA [32] stated that the low frequency of 50–60 Hz allows cell walls to accrual charges and create pores.

Mitelut et al. [17] used Ohmic treatment device to inhibit microbes in meat and meat products with applied voltage range from 20 to 220 V, power of 0.5 kW at 50 Hz. They used different meat ball diameters from minced pork meat such as: 10, 15, 20 and 30 mm at 5 minutes of Ohmic treatment time and at temperature of Ohmic heating of 125°C. They found that the count of *Pseudomonas aeruginosa* ATCC 27853 and *Staphylococcus aureus* ATCC 25923 before Ohmic heating were 10^3 and 10^3 respectively. Then after Ohmic heating, these microorganisms were absent for all meat balls diameters.

3.6 FUTURE RESEARCH OPPORTUNITIES

Ohmic cooking is one of emerging technologies in food processing and lot of research is still needed. Moreover, there is a challenge to construct a large scale factory for Ohmic cooking units. For increasing productivity of Ohmic cookers, engineers should design a continuous Ohmic cookers. A vacuum Ohmic cookers can be designed to enhance quality of cooked meat.

3.7 CONCLUSIONS

Meat cooking by using Ohmic heating is a new technology and it gives a high quality product compared with traditional cooking. Meat cooking by Ohmic heating depends on fat content in the meat, electric conductivity, applied voltage, and distance between electrodes. Electric conductivity of meat and processed meat increases with increasing temperature, while it decreases with increasing fat content. Quality of meat cooked by Ohmic heating is more acceptable. All microorganisms in meat cooked by Ohmic heating are absence. Ohmic cooking method is faster than other traditional methods. Also, it needs less power consumption.

3.8 SUMMARY

Ohmic heating is a novel technology. There is a very little research on meat cooking with Ohmic heating. Quality of meat cooked by Ohmic heating is better than traditional methods. Also Ohmic cooking method is faster than other traditional methods. On the other hand, Ohmic cooking kills all microorganisms in the cooked meat and gives a safer product. Power consumption in Ohmic cooking is significantly lesser than in the traditional method.

KEYWORDS

- electro conductive heating
- inhibition of microbes
- joule heating
- meat processing
- ohmic heating
- pasteurization value
- performance coefficient
- safe burger

REFERENCES

1. An, H. J., & King, J. M. (2007). Thermal characteristics of Ohmically heated rice starch and rice flours. *Journal of Food Science, 72*(1), C84–C88.
2. Anderson, D. (2008). *Ohmic Heating as an Alternative Food Processing Technology.* MS thesis, Kansas state university, Food Science Institute, College of Agriculture. Manhattan, p. 45.
3. Barbosa-Cánovas, G. V., Juliano, P., & Peleg, M. (2004). Engineering properties of foods. In: Barbosa-Cánovas, G. V. (Ed.). *Encyclopedia of Life Support Systems (EOLSS),* Developed under the Auspices of the UNESCO, Eolss Publishers, and Oxford, UK.
4. Bozkurt, H., & Icier, F. (2010). Electrical conductivity changes of minced beef-fat blends during Ohmic cooking. *J. Food Eng., 96*, 86–92.
5. De Alwis, A. A. P., & Fryer, P. J. (1992). Operability of the Ohmic heating process: electrical conductivity effects. *J Food Eng., 15*, 21–48.
6. De Halleux, D., Piette, G., Buteau, M. L., & Dostie, M. (2005). Ohmic cooking of processed meat: energy evaluation and food safety considerations. *Canadian Biosystems Engineering, 47*, 341–347.
7. Ede, A. J., & Haddow, R. R. (1951). The electrical properties of food at high frequencies. *Food Manuf., 26.* 156–160.
8. Icier, F., & Ilicali, C. (2005a). The effects of concentration on electrical conductivity of orange juice concentrates during Ohmic heating. *European Food Research and Technology, 220*, 406–414.
9. Icier, F., & Ilicali, C. (2005b). The use of tylose as a food analog in Ohmic heating studies. *J. Food Eng., 59*, 67–77.
10. Icier, F., Yildiz, H., & Baysal, T. (2008). Polyphenoloxidase deactivation kinetics during Ohmic heating of grape juice. *Journal of Food Engineering, 85*, 410–417.

Technology of Ohmic Heating in Meat Processing

11. Kim, H. J., Choi, Y. M., Yang, T. C. S., Taub, I. A., Tempest, P., Skudder, P., Tucker, G., & Parrot, D. L. (1996). Validation of Ohmic heating for quality enhancement of food products. *Food Technol., 50*(5), 253–261.
12. Kulshrestha, S., & Sastry, S. K. (2003). Frequency and voltage effects on enhanced diffusion during moderate electric field (MEF) treatment. *Innovative Food Science and Emerging Technologies, 4*(2), 189–194.
13. Lima, M. (2007). Ohmic heating: Quality improvements, Encyclopedia of Agricultural, *Food and Biological Engineering, 1*, 1–3.
14. Lima, M., & Sastry, S. K. (1999). The effects of Ohmic heating frequency on hot-air drying rate and juice yield. *Journal of Food Engineering, 41*, 115–119.
15. Maroulis, Z. B., & Saravacos, G. D. (2003). *Food process Design.* Marcel Dekker Inc., USA.
16. Marra, F., Zell, M., Lyng, J. G., Morgan, D. J., & Cronin, D. A. (2009). Analysis of heat transfer during Ohmic processing of a solid food. *J. Food Eng., 1*, 56–63.
17. Mitelut. A., Popa, M., Geicu, M. A., Niculita, P., Vatuiu, D., Vatuiu, I., Gilea, B., Balin, R. & Cramariuc, R. (2011). Ohmic treatment for microbial inhibition in meat and meat products. *Romanian Biotechnological Letters, 16*(1).
18. Mohammed, M. F. (2004). *Method and Apparatus of Cooking Food.* Patent 0197451A1, USA.
19. Özkan, N., & Farid, I. H. M. (2004). Combined Ohmic and plate heating of hamburger patties: quality of cooked patties. *Journal of Food Engineering, 63*, 141–145.
20. Palaniappan, S., & Sastry, S. K. (1991). Electrical conductivity of selected juices: influences of temperature, solids content, applied voltage, and particle size. *Journal of Food Process Engineering, 14*, 247–260.
21. Pereira, R., Martins, J., Mateus, C., Teixeira, J. A., & Vicente, A. (2007). Death kinetics of Escherichia coli in goat milk and Bacillus licheniformis in cloudberry jam treated by Ohmic heating. *Chemical Papers, 61*(2), 121–126.
22. Piette, G., Buteau, M. L., de Halleux, D., Chiu, L., Raymond, Y., & Ramaswamy, H. S. (2004). Ohmic cooking of processed meats and its effects on product quality. *Journal of Food Science, 69*, E71–E78.
23. Piette, G., Buteau, M. L., De Halleux, D., Chiu, L., Raymond, Y., Ramaswamy, H. S., Dostie, M. (2013). Ohmic Cooking of Processed Meats and its Effects on Product Quality. *Journal of Food Science, 69*(2), 71–78.
24. Sarang, S., Sastry, S. K., & Knipe, L. (2008). Electrical conductivity of fruits and meats during Ohmic heating. *J. Food Eng., 87*, 351–356.
25. Sastry, S. K., & Barach, J. T. (2000). Ohmic and inductive heating. *Journal of Food Science Supplement, 65*(4), 42–46.
26. Sensoy, I., & Sastry, S. K. (2004). Extraction using moderate electric fields. *Journal of Food Science, 69*, 7–13.
27. Sieber, R., Eberhard, P., Fuchs, D., Gallmann, P. U., & Strahm, W. (1996). Effect of microwave heating on vitamins A, E, B1, B2 and B6 in milk. *J Dairy Res., 63*, 169–172.
28. Sieber, R., Eberhard, P., Fuchs, D., Gallmann, P. U., & Strahm, W. (1996). Effect of microwave heating on vitamins A, E, B1, B2 and B6 in milk. *J Dairy Res., 63*, 169–172.
29. Skudder, P. J. (1988). Ohmic heating: new alternative for aseptic processing of viscous foods, *Food Engineering, 60*, 99–101.

30. Sun, H. X., Kawamura, S., Himoto, J. I., Itoh, K., Wada, T., & Kimura, T. (2008). Effects of Ohmic heating on microbial counts and denaturation of proteins in milk. *Food Science and Technology Research, 14*, 117–123.

31. Toepfl, S., Heinz, V., & Knorr, D. (2001). Overview of pulsed electric field processing for food. In: Barbosa-Cánovas, G. V., Zhang, Q. H., & Tabilo-Munizaga, G. (Eds.). *Pulsed Electric Fields in Food Processing: Fundamental Aspects and Applications.* Lancaster, PA: Technomic Publishing Co. Inc., Chapter 10.

32. USA-FDA. (2009). Kinetics of microbial inactivation for alternative food processing technologies: Ohmic and inductive heating. http://www.cfsan.fda.gov/wcomm/ift-ohm.html.

33. Valentas, K. J., Rotstein, E., & Singh, R. P. (1997). *Handbook of Food Engineering Practice*. CRC Press Boca Raton, New York, USA.

34. Valentas, K. J., Rotstein, E., & Singh, R. P. (1998). *Handbook of Food Engineering*. CRC Press, pp. 736. ISBN 0-8493-8694-2.

35. Vorobiev, E., & Lebovka, N. (2008). *Electro Technologies for Extraction From Food Plants and Biomaterials*. Springer Science Business Media, LLC. New York. p. 280.

36. Wang, C. S., Kuo, S. Z., Kuo-Huang, L. L., & Wu, J. S. B. (2001). Effect of tissue infrastructure on electric conductance of vegetable stems. *Journal of Food Science: Food Engineering and Physical Properties, 66*(2), 284–288.

37. Wang, W. C., & Sastry, S. K. (1993). Salt diffusion into vegetable tissue as a pretreatment for Ohmic heating: determination of parameters and mathematical model verification. *Journal of Food Engineering, 20*, 311–323.

38. Zareifard, M. R., Ramaswamy, H. S., Trigui, M., & Marcotte, M. (2003). Ohmic heating behavior and electrical conductivity of two-phase food systems. *Innovative Food Science and Emerging Technologies, 4*, 45–55.

CHAPTER 4

TECHNOLOGY OF SOLID STATE FERMENTATION IN DAIRY PRODUCTS: PRODUCTION OF β-GALACTOSIDASE FROM MOLD ASPERGILLUS ORYZAE

ALAA JABBAR ABD AL-MANHAL, ALI KHUDHAIR JABER ALRIKABI, GHEYATH H. MAJEED, and ABDULLAH M. ALSALIM

CONTENTS

4.1 Introduction 71
4.2 Steps in Purification of the Enzyme 73
4.3 Methods to Immobilize β-galactosidase 90
4.4 Future Perspectives and Research Opportunities 101
4.5 Summary 102
Keywords 103
References 103

4.1 INTRODUCTION

The enzymes are biological catalysts of protein origin produced by living cells to catalyze biochemical reactions necessary for their metabolism. Given the importance of enzymes and their role, the global production of enzymes has reached 53,000 tons per year headed by Denmark with 24910 tons. During 2009, the sale of synthetic enzymes in the world has ranged from 650 to 750 million dollars compared to 150 million dollars in the mid-seventies. Although, the enzymes were extracted traditionally from plants and animals, yet its production from microorganisms is growing rapidly up to 90%. The filamentous fungi tops the list of microorganisms, which are considered inexhaustible for various types of enzymes. Their number is more than approximately 2000 to 3000 enzymes, however the number of species used in different areas does not exceed 25, where the food industries has a share of 45%. The 6% of the enzyme source come from animals compared to 4% from plants [4, 33].

The β-galactosidase is considered as one of the important enzymes in the dairy industry and resides in small intestines of the mammals. Its existence has also been noted in vegetables and microorganisms. It works on the decomposition of lactose to its primary units of monosaccharides (Glucose and Galactose). As a result, the solubility and sweetness is increased and problems arising are decreased because of crystallization in the ice cream. Also, the enzyme has an important role in addressing the phenomenon of lactose intolerance afflicting more than 70% of the world's population [27].

The β-galactosidase production of microorganisms, especially molds in the last decades, has attracted the attention of many researchers because of ease in extraction of mold enzymes (especially external ones: Extracellular Enzymes) as a result of secretion into the growth environment, which in turn is reflected in the reduction of production costs compared to other microorganisms that have their production limited to internal enzymes (intracellular enzymes) as well as small nutritional requirements of molds, achieving an opportunity to produce the enzyme at low cost through the use of cheap and appropriate raw materials [3].

The enzyme β-galactosidase produced from molds has higher purity than that produced using other organisms. The molds are more convenient for the fermentation system of solid state witnessing at present a great interest in the production of many biomaterials such as enzymes. As this technique is suitable for the production of many of the microbial enzymes that are characterized by high activity and stability and with less costs, compared to the submerged fermentation. In addition to the available possibilities of

Technology of Solid State Fermentation in Dairy Products 73

solid state fermentation, there is exploitation of agricultural and industrial waste in the production of commercially useful and important materials as well as their accumulation and resulted pollution of the environment [1]. The researchers also began to apply the immobilization technique on many enzymes, including the β-galactosidase enzyme because of their important role in overcoming the inhibition occurring to the enzyme during decomposition of lactose and the accumulation of galactose. In general, the immobilization can identify cells or enzymes movement or its immobilizing or enveloping activities by adsorption, covalent binding, entrapment, encapsulation, as well as other means such as ionic binding and chelation of metal binding [14]. The immobilization process can prolong the period of use of immobilized enzymes as they retain their activity during the transformation process and thereby reducing production costs, and increasing the stability of immobilized enzymes to heat due to sensitivity of the free enzymes to thermal changes. The immobilization process provides an opportunity to end the reaction process in a timely manner and reduces undesired reactions that occur after the end of fermentation. The restricted enzymes can be used in continuous or intermittent fermentation processes with the ease of purification and extraction of the fermentation process products, since they do not contain free enzyme and offer the diversity of the use of immobilized enzymes [9, 33].

In this study, isolation and screening processes were carried out to get to the most efficient isolation of fungal in the production of enzyme. The isolation was diagnosed in the light of the diagnostic keys described by Klich [13] and it turned out to be *Aspergillus oryzae*. The production of the enzyme was conducted by this isolation using solid state fermentation.

4.2 STEPS IN PURIFICATION OF THE ENZYME

4.2.1 PRECIPITATION WITH AMMONIUM SULFATE

The Figure 4.1 shows first step of enzyme progressive precipitation from the crude extract, then a clear gradual rise to the activity of quality of the enzyme is noted in the resulted sediment accompanied by a significant decrease in the quality of the enzyme activity in the resulted filtrate up to a saturation of 80%, where the activity of the enzyme quality amounted

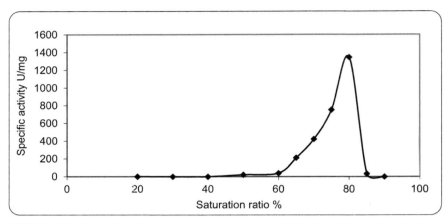

FIGURE 4.1 Procedure for enzyme progressive precipitation from the crude extract.

to its maximum in the sediment, followed by dialysis process of the resulted sediment (dissolved in buffer acetate) to get rid of salts, ammonium sulfate versus acetate buffer solution [2]. This step has given the qualitative activity of 1093.21 units/mg, purification number of 3.29 times and the enzymatic outcome of 55.16% when using saturations ranging from 65 to 80% (Table 4.1).

These results are in agreement with findings of Shaikh et al. [31] who obtained the qualitative activity of 3.07 units/mg to sediment the enzyme extracted from *Rhizomucor* sp. mold and the number of purification 1.5 times, with a saturation ratio of 90% of ammonium sulfate. However, the results were different with those of Chen et al. [6], who used a saturation ratio of 30–65% to sediment the enzyme produced from the *Bacillus stearothermophilus* bacteria with a qualitative activity of 80.3 units/mg, the number of purification 12.4 times and the enzymatic outcome of 73.6%. The cause of the discrepancy may be due to microbial source variation.

4.2.2 ION EXCHANGE

The passage of the enzymatic solution after concentration with ammonium sulfate on the ionic exchanger DEAE-Sephadex A-50 is shown in Figure 4.2, which shows the appearance of protein peaks in the washing zone free of enzymatic activity with a positive charge, and the emergence

Technology of Solid State Fermentation in Dairy Products

TABLE 4.1 Summary of Purification of β-galactosidase from *Aspergillus oryzae*

Purification Step	Volume (ml)	Activity (U/ml)	Total protein (mg/ml)	Specific activity (U/mg)	Total activity (U)	Fold purification	Yield (% recovery)
1 Crude enzyme	200	248.52	0.75	331.36	49704	1	100
2 Ammonium sulfate fraction 80–65 and dialysis %	38	721.52	0.66	1093.21	27417.76	3.29	55.16
3 Ion exchange by DEAE Sephadex A-50	65	294.86	0.17	1734.47	19165.9	5.23	38.56
4 Gel filtration by Sephadex G-100 fraction	50	257.29	0.08	3216.12	12864.5	9.70	25.88

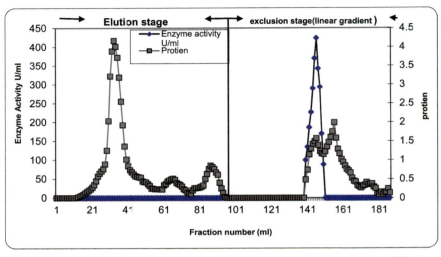

FIGURE 4.2 Anion-exchange chromatography for purification of β-galactosidase on DEAE Sephadex A-50(2.5 × 34 cm), washed with 0.01 M sodium acetate buffer (pH 5) at flow rate of 1 ml/min and fraction volume of 5 mL The β-galactosidase was eluted out of column against a gradient of NaCl of from 0 to 1 M.

of several protein peaks in the recovery stage and one peak of enzymatic activity almost coincided with one of the protein peaks. This implies that the enzymatic extract contains other proteins similar in the charge but different in the charge outcome and density, leading to a contrast in the binding power with exchanger substance and separation with it because of the effect of progressive salt power of sodium chloride concentrations ranged from 0 to 1 Molar, then qualitative activity of recovered part reached 1734.47 units/mg, the number of purification 5.23 times and the enzymatic outcome of 38.56% (Table 4.1).

4.2.3 GEL FILTRATION

The dialysis and concentration of the enzyme with freeze drying device (freeze dryer) are followed by the step of gel filtration using Sephadex G-100 column. Figure 4.3 shows that parts recovery included two protein peaks (one: the first containing the enzyme activity; the other: completely free of any activity) and the activity peak matching to a large extent the first protein peak. The curve conformity of activity and protein to this

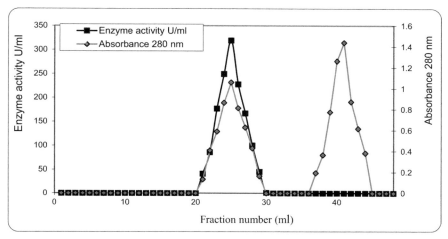

FIGURE 4.3 Gel filtration elution profile of *Aspergillus oryzae* β-galactosidase on Sephadex G-100 (2.5×86) cm, washed with 0.05 M sodium acetate buffer (pH 5) at flow rate 30 ml/h and fraction volume of 5 mL.

extent is one of the primary signs of purity [40]. The extraction activity of the enzyme was 3216.12 units/mg (Table 1), and the outcome amounted to 25.88%, while the number of purification was 9.70 times.

4.2.4 DETERMINATION OF ENZYME PURITY

Figure 4.4 shows steps in enzyme purification: gel A represents the raw enzymatic extract movement, which appeared as several protein bundles in a polyacrylamide gel; the appearance of two protein bundles in gel B that represents the part acquired after the ion exchange step; and gel C represents the part acquired after the gel filtration step. Then the appearance of protein single bundle is observed that this gives a clear idea on the extent of activity of the purification operations, which aim to eliminate all the proteins associated with the enzyme in the raw extract and obtains it individually indicating the high purity of the enzyme [30].

The results in this chapter agree with those reported by O'Connell et al. [23], because authors were able to acquire one bundle of protein, using electrophoresis with polyacrylamide gel, when confirmation of purity of the acquired enzyme from the *Aspergillus niger*.

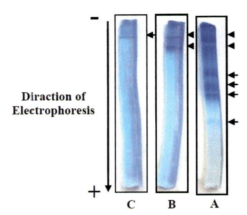

FIGURE 4.4 Polyacrylamide gel electrophoresis of purified *Aspergillus oryzae* β-galactosidase: (A) Croud extract, (B) Enzyme after ion exchange, and (C) Enzyme after gel filtration.

4.2.5 DETERMINATION OF THE OPTIMUM pH FOR THE ACTIVITY AND STABILITY OF THE ENZYME

The Figure 4.5 shows that the optimum pH of activity was 5, where the highest activity of the enzyme appeared (Figure 4.5A), and these results agree with those by Shaikh et al. [31] for the enzyme extracted from *Rhizomucor* sp mold. The optimum pH of the enzyme activity was 4.5.Saad[29] indicated that the best activity of the enzyme was at pH of 5.2 for *Aspergillus japonicus* mold. Further O'Connell et al. stated [23] that the best activity of the enzyme was at pH 4 due to decline in the activity of β-galactosidase in the highest basal-range and the highest acid-range is due to the effect of pH of the reaction medium on ionizable totals in the active site, or a change in the substrate of the reaction, or as a result of a change in the ionic state of the reaction substances, which include the enzyme complex with the substrate (ES) and the enzyme complex with the resultants (EP) [40].

The optimum pH for the enzyme stability is shown in Figure 4.5B, which indicates that the optimum pH for the enzyme stability ranges from 4 to 6.5, as the enzyme maintained approximately its full activity when incubated in this range of pH; and the enzyme lost more than 65% of its activity at the pH of 8. A decline in enzyme activity was observed at high acidic pH values as it retained 35.99% at pH of 2 and 56.2% of its activity at pH 3, respectively.

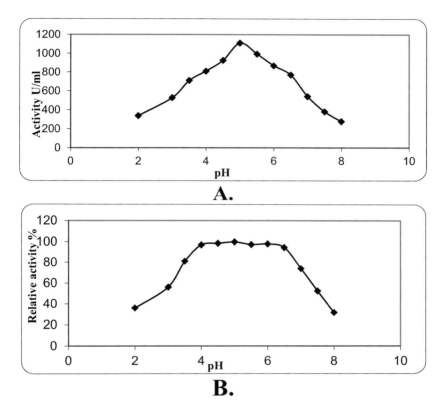

FIGURE 4.5 Effects of pH on enzyme activity (top: Figure A) and stability (bottom: Figure B) from *Aspergillus oryzae*.

The decline in activity at the acidic or basic pH values may be due to changes in the secondary and tertiary structure of the enzyme molecule in addition to the change in ion state of the enzyme active site [16]. The source of the enzyme and the chemical nature of the buffer solution are important factors that influence the determining of the optimum pH for the enzyme stability [30]. The results in this chapter did not agree in determining the optimum pH for the enzyme stability, with those of Shaikh et al. [31] who found that the extracted enzyme from a *Rhizomucor* spp. has a wide range of stability between 3 to 8 at 4°C for a period of 24 hours, and it lost 30% of its activity at pH of 9. However, the enzyme produced from *Aspergillus niger* mold was stable at pH 3–6 for a period of 5 days at a temperature of 25°C [12].

4.2.6 EFFECTS OF TEMPERATURE ON THE ACTIVITY AND STABILITY OF β-GALACTOSIDASE

Figure 4.6 indicates an obvious increase in the enzyme activity by an increase in the reaction temperature until it reaches its maximum at 50°C, then the activity was decreased gradually until it achieved 4.44% of the maximal enzyme activity at 80°C. The increase in temperature leads to obvious changes in the reaction components that include each of the enzyme and the substrate [39], which leads to an increase in the (Kinetic Energy) of the reactant molecules under the influence of temperature. However, the increase in temperature from the thermal optimum degree of the reaction leads to a decline in the activity due to the negative enzymatic impact of the reaction components [40]. However, the severe decrease in enzymatic activity at high temperature may be due to the absorption of high energy by the reactant molecules leading to change in the tertiary structure of the enzyme and then the deformation and loss of a part of its activity [30].

The results of this study are comparable with those of Saad [29], who has found that the optimal temperature of the enzyme extracted from the *Aspergillus japonicus* mold is 45°C. Haider et al. [10] indicated that the maximal activity of the purified enzyme from *Aspergillus oryzae* mold was at 50°C. Also Ustok et al. [38] indicated that the best enzymatic activity of β-galactosidase extracted from *Streptococcus thermophilus* bacteria and *Lactobacillus delbrueckii* ssp. bulgaricus bacteria was at

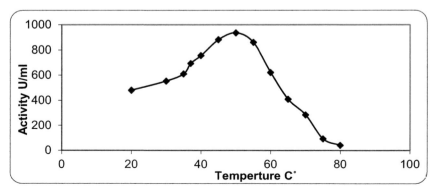

FIGURE 4.6 Effects of temperature on enzyme activity from *Aspergillus oryzae*.

Technology of Solid State Fermentation in Dairy Products 81

50, 50 and 45°C, respectively, in case of each bacteria alone as well as when using both sexes together.

4.2.7 DETERMINATION OF ACTIVATION ENERGY

Figure 4.7 shows the relationship between the logarithm of enzyme β-galactosidase reaction speed and the reciprocal of the absolute temperature [2]. The logarithm of reaction speed was reached the activation energy value (activation energy) that is needed to convert the substrate to resultant 6.19 (kcal/mol), that is considered within the range of activation energy (Ea) for the reactions to convert the substrate to the resultant between 6–15 kcal/mol [40].

These results were similar to the results of Neri et al. [22] who indicated that the activation energy value of the purified enzyme from *Kluyveromyces lactis* yeast was 25.5 kJ/mol (6.09 kcal/mol). It should be noted that the activation energy value gives an idea on the activity of the enzyme work in the conversion of the substrate to the resultant, the higher value implies greater activity in the conversion of the substrate and then accelerating to complete the reaction [40]. The enzyme denaturation energy was 52.31 kcal/mol. This value gives an idea on the stability of the enzyme at high temperature. Higher value implies that the enzyme is more stable towards heat and the enzyme denaturation energy value ranges for most enzymatic reactions between 40–150 kcal/mol [39]. Ustok et al. [38]

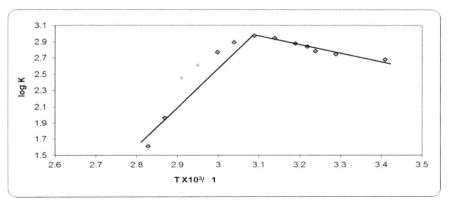

FIGURE 4.7 Arrhenius plot to determine the activation energy for β-galactosidase.

found that the denaturation energy of the raw enzyme of *Streptococcus thermophilus* bacteria and *Lactobacillus delbrueckii* sp bulgaricus and a mixture of the two cultures was 51.3, 48.3 and 44 kcal/mol, respectively.

Results of β-galactosidase incubation showed temperatures ranging from 20 to 90°C for 15 minutes; and the enzyme retained its full activity at temperatures ranging between 20–50°C; then enzymatic activity was decreased gradually and lost 25% of its activity at 60°C and lost 97% of its activity at 90°C. It was also noted that the purified enzyme retained full activity at the optimal temperature 50°C for 60 minutes [2] while it has lost 42.98% of its activity for 6 hours at optimal temperature (Figure 4.8).

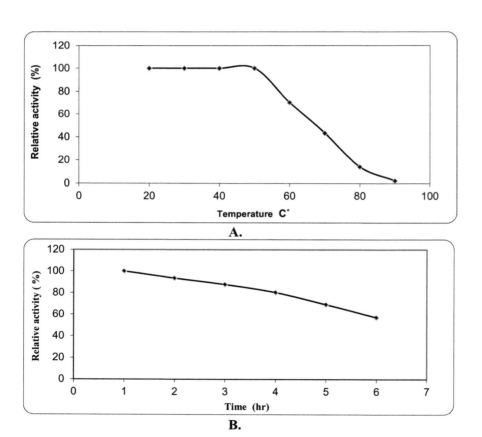

FIGURE 4.8 Thermal stability of β-galactosidase from *Aspergillus oryzae*.

Studies differed in determining the thermal stability of the β-galactosidase, according to the source of the enzyme. Shaikh et al. [31] indicated that the enzyme retained 67% of its activity at the optimal temperature for a period of 50 minutes. Todorova-Balvay et al. [37] observed that the enzyme produced from *Aspergillus oryzae* was able to retain its full activity when incubated at a temperature of 60°C for two hours.

It is observed that the optimal temperature of the enzyme activity is not a constant character of the enzyme being dependent on experimental conditions. The decline in the enzyme stability at prolonged period of time for the enzyme reaction at this degree was as a result of the effect of temperature in the synthesis of the enzyme. The greater is the period of time, the effect was increased. The true optimal temperature of the enzyme was at highest temperature where the enzyme could retain its activity within a period of time at least longer than the period in which the enzyme activity is assessed in a normal case [30, 40].

4.2.8 DETERMINATION OF MOLECULAR WEIGHT

Figure 4.9 describes the electrophoresis of standard proteins and β-galactosidase under study in the presence of SDS [2]. The relative movement Rm was measured for β-galactosidase and through this value it was possible to determine the molecular weight from the logarithm of molecular weight under the same conditions and was equal to 97.72 kDa. The molecular weight of β-galactosidase depends on the source of this enzyme and the estimation of the molecular weight with the electrophoresis method. Poly acryl amide gel in the presence of denaturation factors gives molecular weight per one unit of the enzyme molecule and it appears as a single protein bundle. This was also mentioned by Nagy et al. [21] that the enzyme from *Penicillium chrysogenum* appeared in one bundle and has four similar units of molecular weight of 66 kDa per unit. However, Todorova-Balvay et al. [37] have confirmed that the enzyme produced from *Aspergillus oryzae* using the electrophoresis method and the existence of denaturation factors showed one bundle and consists of a single unit only (Monomeric protein) with a molecular weight of 113 kDa. O'Connell et al. [23] also found that the enzyme produced from *Aspergillus niger* has a single bundle and was composed of one unit with molecular weight of 129 kDa.

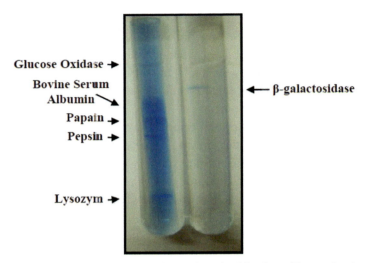

FIGURE 4.9 SDS-polyacrylamide gel electrophoresis of β-galactosidase molecular weight.

The molecular weight of the enzyme was also estimated by gel filtration dextran blue (Blue Dextran 2000) method. To determine the space volume between the particles, the void volume (Vo) and the standard protein recovery volumes were determined. The relationship is linear between the ratio of void volume and elution volume (Vo/Ve) and logarithm of molecular weight of standard proteins (Figure 4.10). After calculating the ratio between the recovery volume of each protein (Vo), the molecular weight of the enzyme was determined, which amounted to 103.51 kDa [2]. This value is roughly equivalent to the estimated value of 97.72 kDa by electrophoresis method, indicating that the enzyme produced in this study consists of only one unit, and these results are in agreement with those by Tanaka et al. [35], who observed that the enzyme β-galactosidase produced from *Aspergillus oryzae* had a molecular weight of 105 kDa using gel filtration method. On the other hand, Nagy et al. [21] stated that the molecular weight of β-galactosidase 270 kDa, produced from *Penicillium chrysogenum* and by gel filtration (Sephadex G-200) method. However, O'Connell et al. [23] have stated that the β-galactosidase from *Aspergillus niger* has a molecular weight of 123 kDa with a gel filtration method. It is worth mentioning that the molecular weight of β-galactosidase depends on the source and the method of estimation.

Technology of Solid State Fermentation in Dairy Products 85

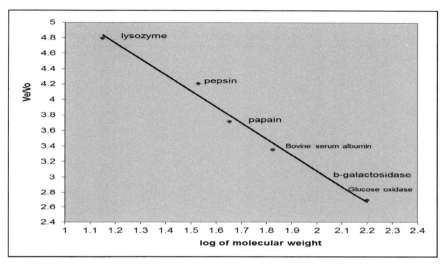

FIGURE 4.10 Logathimic standard curve of protein for β-galactosidase produced from *Aspergillus oryzae* and estimated by gel filtration method (G-200).

4.2.8.1 Effect of Metal Ions and Compounds on the Effectiveness of the Enzyme

The Table 4.2 shows the effects of metal ions (at two concentrations of 5 and 10 mM Molars) and compounds on the effectiveness of purified beta-galactose enzyme, which included all of $MnCl_2$, $CuCl_2$, $FeCl_2$, $MgCl_2$, $CaCl_2$, NaCl, and KCl. It can be observed that manganese and sodium ions had a stimulus role on the effectiveness of the enzyme at both concentrations comparing with the untreated enzyme. The remaining effectiveness reached 107.80 and 104.84%, respectively, at a concentration of 10 mM Molars, while remaining effectiveness was decreased at different rates when using magnesium, copper, iron and calcium with increased concentration; while adding potassium did not affect the effectiveness of the remaining enzyme.

These results agreed with other studies by Shaikh et al. [31], who found that the addition of divalent ions (Zn^{+2}, Ni^{+2}, Hg^{+2}, Fe^{+2}, Ca^{+2}, Mn^{+2}, Mg^{+2} and Cu^{+2}) to the purified enzyme from *Rhizomucor* spp. mold led to low activity at different rates except addition of cobalt caused 33% increase in the enzyme activity. Chen et al. [6] observed that the addition of magnesium and potassium ions at a concentration of 10 mM and

TABLE 4.2 Effects of Metal Ions on β-galactosidase Activity from *Aspergillus oryzae*

Metal ions	Concentration (mM)	Relative activity (%)
(control)	0	100
$MnCl_2$	5	100.92
	10	107.80
$CuCl_2$	5	77.88
	10	38.73
$FeCl_2$	5	95.16
	10	58.22
$MgCl_2$	5	90.14
	10	82.92
$CaCl_2$	5	96.65
	10	89.47
NaCl	5	101.02
	10	104.84
KCl	5	99.32
	10	100.41

manganese concentration at 2 mM were able to increase the activity of the purified enzyme from *Bacillus stearothermophilus* bacteria; whereas enzymatic activity was declined by adding Ca^{+2}, Cu^{+2}, Fe^{+2}, and Pb at different concentrations. White et al. [38] studied the effects of metal ions at 1–10 mM concentrations on the activity of raw β-galactosidase produced from *Streptococcus thermophilus* and *Lactobacillus delbrueckii* subsp. bulgaricus and a mixture of the two cultures. They found that the enzyme activity was inhibited by adding Zn^{+2}, Cu^{+2}, and Ca^{+2} whereas the activity was increased in the presence of Mg^{+2} and Mn^{+2}.

However, the effects of metallic bonding factors play a major role in the inhibition of the enzyme activity in which the metal ion forms an essential part of the composition of the active site or those enzymes that need in their activity the existence of this ion to complete its action to the fullest. The Table 4.3 shows a slight increase in the enzymatic activity when adding the mercaptoethanol. The enzymatic activity was increased by 0.93% at a concentration of 1 mM, but the increasing the concentration to 5 mM caused a decline in the activity, and that the slight increase in

Technology of Solid State Fermentation in Dairy Products

TABLE 4.3 Effects of Enzyme Inhibitors and Activators on β-galactosidase Activity [2]

Enzyme treatment	Concentration (mM)	Relative activity (%)
Control	0	100
EDTA	1	99.53
	5	98.23
2-mercaptoethanol	1	100.93
	5	94.54
Urea	1	78.32
	5	50.30

the enzyme activity is due to the possibility of participation of sulfhydryl groups at the active site of enzyme activation or at least its importance in maintaining the protein spatial structure. However, the compound EDTA did not affect the enzymatic activity at a concentration of 1 and 5 mM and this proves that β-galactosidase is not from the mineral enzymes (Metalo enzyme) in which a specific metal ion constitutes the basis and an important part in stimulating its activity. Since EDTA is one of the Chelating agents, it works to extract metal ions from the enzyme molecules through the formation of complexes with it, thus leading to the inhibition of activity, while a decline in the enzymatic activity occurred when using urea, and for the increase in concentration the remaining enzymatic activity reached 50.30%. This may be due to the fact that urea is one of denaturation factors that lead to the destruction of the natural form of the protein through the formation of hydrogen bonds in addition to the peptide bonds, which looses the secondary structural stability of the protein. This was confirmed by Haider et al. [10] when adding urea to a 1–5% concentration of the enzyme β-galactosidase purified from *Aspergillus oryzae* mold, since the enzyme had lost its full activity at a concentration of 5% for a period of two hours.

The galactose sugar and its effects are described in Figure 4.11. With an increase in sugar concentration, the enzymatic activity was decreased, and the enzyme had lost more than 41% of its activity at a concentration of 5% [2]. This indicates that the galactose is an inhibitor of the enzyme. Park et al. [26] observed the inhibiting effect of galactose sugar on the activity of the enzyme β-galactosidase purified of *Aspergillus oryzae*. With the

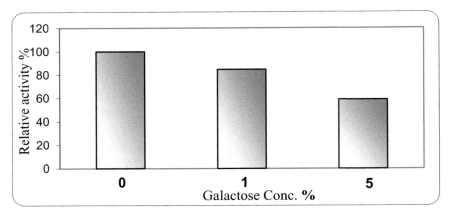

FIGURE 4.11 Effects of different concentrations of galactose on relative activity of β-galactosidase that was purified from *Aspergillus oryzae* A9.

increase in the substrate concentration (ONPG), the inhibiting effect of the galactose was reduced. The research conducted in this area has shown that β-galactosidase is inhibited competitively by the galactose sugar [12, 22].

These results are in agreement with those by Tanaka [35], who found that the EDTA and mercaptoethanol do not affect the enzyme activity of β-galactosidase of *Aspergillus oryzae*, when added at a concentration of 10 mM. A decrease in the enzymatic activity was recorded when adding galactose sugar at a concentration of 0.1 M. The results in this chapter also agreed with those reported by Chen et al. [6], who found an increase in enzymatic activity of β-galactosidase when mercaptoethanol was added at a concentration of 1 mM. EDTA also showed a slight decrease in the enzymatic activity as the enzyme retained a 97.4% of its activity.

4.2.9 KINETIC CONSTANTS

Figure 4.12 indicates the relationship between the reaction speed at different concentrations of ONPG acting as a synthetic substrate to determine the values of Michaelis constant (K_m) and the maximum speed V_{max}. It was noted that K_m was 1.435 mM and V_{max} was 1040.23 units/ml. A number of researchers has estimated the kinetic constants values for β-galactosidase, whereas Shaikh et al. [31] used ONPG and PNPG as reactive substances to produce exo-β-galactosidase from *Rhizomucor* sp. They found that the

Technology of Solid State Fermentation in Dairy Products 89

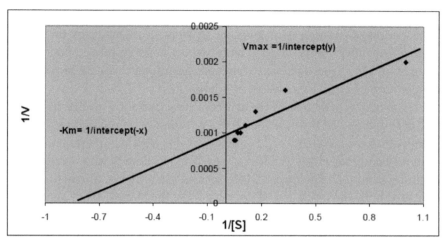

FIGURE 4.12 The Lineweaver–Burk plot.

K_m values were 0.785 and 0.39 mM, respectively. Also they indicated that the V_{max} values of the enzymes usually were increased with increasing of the length of reaction materials chain and it amounted to 232.1 mmol/min/mg. In addition, O'Connell et al. [23] found that K_m values were 1.74 and 1.14 mM, and V_{max} values were 137 and 5.033 micromoles/min/mg of the substrates ONPG and PNPG, respectively. The values of the kinetic constants variation is due to: the type and source of the enzyme; the conditions of activity assessment (when using the enzyme of the same source); the pH; the reaction temperature; the type of buffer solution; and ionic strength.

4.2.10 FOLLOW-UP OF LACTOSE HYDROLYSIS BY THIN LAYER CHROMATOGRAPHY (TLC)

The hydrolysis products of lactose were separated by β-galactosidase purified from *Aspergillus oryzae* A9. The reaction products of pure enzyme on lactose sugar at different periods ranged from ½ to 6 hours, by the manifestation of three types of stains that appeared at the beginning of the reaction and becoming more apparent with the increase in reaction time [2]. The decomposition products were inferred in comparison with lactose sugar, galactose and glucose with the help of the relative movement.

The Rm values and the stains were identical with Rm value for lactose, galactose and glucose, which show the hydrolysis of lactose by the action of enzyme under study. These results were similar to those obtained by Song et al. [34], who treated raw enzyme from the yeast with lactose at a temperature of 50°C for a period of 4 hours.

The products of lactose by the action of enzyme during the first half hour and until the end of the reaction were late stains with relative movement with a Rm value of 0.246–0.256, respectively, which suggests the possibility of being galacto-oligosaccharides. The increase in oligosaccharides resulting from the hydrolysis of lactose with the progress of reaction time indicated to the continuation of the enzyme in the hydrolysis of lactose by breaking the beta bonds (1–4), which gave products of a small molecular weight. This was in line with many studies that have confirmed the formation of oligosaccharides through the reactions of *Trans*-galactosylation as a result of the hydrolysis of lactose by β-galactosidase, which is affected by several factors: the type and concentration of the substrate; source of the enzyme; its concentration and temperature; pH; reaction time and the presence of inorganic ions [25].

4.3 METHODS TO IMMOBILIZE β-GALACTOSIDASE

Several methods were evaluated to select the best procedure to immobilize the enzyme by estimating the activity of the immobilization process. Table 4.4 shows the use superiority of Chitosan in immobilizing β-galactosidase compared with other substances by obtaining the highest activity in the immobilization process, reaching 76.60% with reference to the importance of addition of Gluteraldehyde, which is responsible for the formation of cross-linking and thus the increase in the activity of the process of immobilization due to prevention of enzyme leakage and increase stability. The method of the enzyme immobilization by the Silica Gel was less efficient as it only amounted to 31.95%. These findings are similar to these of Esawy [8] when studying the possibility of immobilizing Levansucrase enzyme that was purified partially from *Bacillus subtilis* bacteria on a set of immobilized substances. Among these materials, Chitosan with the presence of Gluteraldehyde at a concentration of 3% was found to cause 85.51% activity of enzyme

Technology of Solid State Fermentation in Dairy Products 91

TABLE 4.4 Efficiency of Different Methods to Immobilize β-galactosidase.

Method of immobilization	Enzyme activity U/ml	Efficiency of immobilization
Free enzyme	831.26	100
Adsorption on silica gel	265.58	31.95
Entrapment in Agar 2%	362.17	43.57
Entrapment in Agar 4%	400.16	48.14
Entrapment in Calcium alginate	477.55	57.45
Entrapment in Gelatin	493.02	59.31
Covalent Binding with Chitosan	636.74	76.60

immobilization; and the immobilization activity was 29% with Agar (a concentration of 1%) and was less than 5% with calcium alginate, respectively. Also Smaali et al. [32] used Chitosan, Amberlite MB1, Alginate with gelatin, Eudragit S-100 and polyacrylamide gel to immobilize the enzyme β-xylosidase. They noted the superiority of Chitosan to immobilize and it was 91% compared to other methods. The active groups represent hydroxyl, amine, amide, carboxyl, sulfhydryl, phenol and indole groups that are necessary for the reaction and immobilization process occurring between proteins (enzymes) and supporting material. Also, the immobilization process occurs as a result of formation of peptide bonds (Amide), alkylation process, by Schiffs base reactions with Gluteraldehyde or other reactions [14, 36]. These results are encouraging in immobilizing enzyme under study for the method of immobilizing enzyme covalently with Chitosan and immobilization activity was 50–80% of the original activity [20]. For this, method of enzyme immobilization was based on Chitosan, which is a natural and cheap source that can be used instead of costly commercial carriers in the subsequent experiments of the study.

The effect of the amount of enzyme added to the Chitosan granules on the activity of the immobilized enzyme was studied. As shown in Figure 4.13, the activity of the immobilized enzyme was increased significantly with the amount of enzyme added until the activity reached a value of 652.19 units/g of granules using 0.75 ml of the enzyme in the process of immobilization [17], and then the enzymatic activity was stabilized with increasing of the amount of β-galactosidase added.

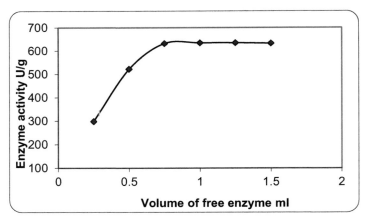

FIGURE 4.13 Effects of amount of enzyme added to the Chitosan granules on the activity of the immobilized enzyme.

These findings are in agreement with those by Makkar et al. [19] when studying the best concentration of the enzyme added to the process of immobilization on the silica material. They observed that with the increase in concentration of additive enzyme, the enzymatic activity was increased to reach a maximum after which activity was stabilized and remained constant. Also, Pan [24] studied the amount of β-galactosidase that is necessary to immobilize with magnetic Fe_3O_4-chitosan in concentration ranging from 0.3 to 0.7 mg.

For the activity of immobilized enzyme when incubated for 3 hours, it was found that the activity of the immobilized enzyme was increased with addition of 0.3 mg of enzyme reaching maximal activity at 0.5 mg, after that there was no increase in the enzymatic activity with the increase of concentration. This increase in enzymatic activity was due to the increase in the amount of enzyme with the stability of the substrate, quantity on which the enzyme works and its hydrolysis is greater in one period of incubation. However, the stability of activity is due to an increase in the enzymatic groupings and lack of active sites available to bind to as a result of the occurrence of saturation process. This was confirmed by Puri et al. [28], when 25, 50, 100, 150 and 175 units of raw β-galactosidase was added to the immobilization process. They had noted that the best activity of the immobilized enzyme was with addition of 150 units of the enzyme. Zhang et al. [41] also noted that the best enzymatic activity of immobilized β-galactosidase was with addition of 2.5 ml of the enzyme to 2 grams of chitosan granules.

4.3.1 DETERMINATION OF OPTIMAL pH FOR ACTIVITY AND STABILITY OF IMMOBILIZED ENZYME

The effects of optimal pH on the immobilized β-galactosidase activity were studied under pH range of 2 to 8. Figure 4.14A shows that the optimal pH activity was 5, which is similar to the free enzyme, which showed the highest activity of the enzyme. These results agree with similar studies of the free and immobilized enzyme, which proved that the free and immobilized β-galactosidase possess the same optimal pH [9, 10]. The effects of pH on

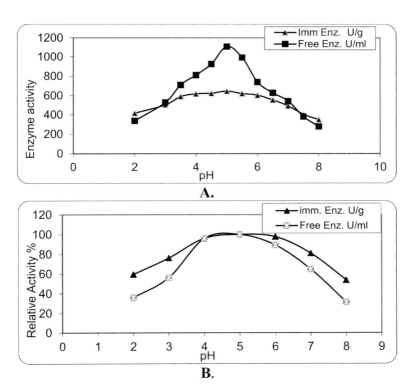

FIGURE 4.14 Optimal pH for the activity and stability of the immobilized enzyme: (Top A.) Enzyme activity; (Bottom B.) Relative activity.

the immobilized enzyme stability are shown in Figure 4.14B, which indicates that the immobilized enzyme expressed good stability in the range of 3.5–6 with more widening of curve compared to the free enzyme. It was noted that it has led to increased stability, which was more evident on the acidic side, due to the protection provided by the Chitosan gel network to the enzyme and redistribution of charges thereon by the effect of buffer solution.

Batra et al. [5] stated that β-galactosidase purified from *Bacillus coagulans* bacteria and immobilized by the substance calcium alginates was able to has retain 80% of its activity over a pH range of 6.5–11. Also Pan et al. [24] studied the stability of the immobilized enzyme by the substance magnetic Fe_3O_4-chitosan, and the stability at pH range of 5–7. They found that 55% of its activity at pH 8 was retained, while the free enzyme retained 26% of its activity at the same pH for a period of 6 hours.

Zhang et al. [41] indicated that the use of polycationic carriers in the immobilization process led to the displacement of the optimal pH of the immobilized enzyme toward the acidic side, due to the increased positive charges on the immobilized enzyme molecules and vice versa.

4.3.2 DETERMINATION OF OPTIMAL TEMPERATURE FOR IMMOBILIZED ENZYME ACTIVITY AND STABILITY

Figure 4.15A indicates that the immobilized enzyme reached the highest enzyme activity at a temperature of 50°C, which is the same optimal temperature of the free enzyme. It was confirmed by other studies, and the extent of the thermal curve after the immobilization process at higher temperatures was offset by an increase in the enzymatic activity compared with the free enzyme. Grosova et al. [9] studied the effect of heat on the free enzyme, immobilized and purified enzyme from *Aspergillus oryzae* mold and *Kluyveromyces lactis* yeast. They observed that the optimal temperature of the immobilized enzyme was at 55°C for mold and 50°C for yeast, respectively, and it was the same optimal temperature for the free enzyme. Neri et al. [22] did not notice any change in the optimal temperature for β-galactosidase purified from *Kluyveromyces lactis* yeast when immobilized on polysiloxane-polyvinyl alcohol magnetic, which was 50°C, although there was more enlargement in the curve of activity towards the temperature compared to the free enzyme. While Haider et al. [11] found

Technology of Solid State Fermentation in Dairy Products 95

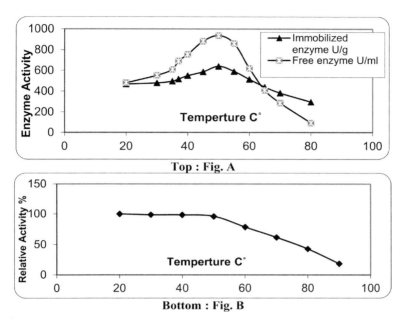

FIGURE 4.15 Optimal temperature for the immobilized enzyme activity and stability.

differences in the optimal temperatures between the free and immobilized enzyme (IE), and it was 50 for free and 60°C for IE, respectively.

Also Zhang et al. [41] observed that the enzyme immobilized with granules from chitosan and gum in the presence of Gluteraldehyde has an optimal temperature of 47°C after it was 37°C before immobilization. This difference was attributed to the thermal stability shown by the immobilized enzyme compared to the free one. The enzyme immobilization process has led to an increase in the thermal stability of the immobilized enzyme for all thermal temperatures, where it has retained approximately the full enzymatic activity at a temperature of 20–50°C for 60 minutes (Figure 4.15B). Also it has retained 78.65% and 61.73% of activity when incubation at temperatures of 60 and 70°C for the same period, respectively. The effect of the immobilization was evident in relatively high temperatures compared to the free enzyme. Figure 4.15B also indicates the extent of the thermal stability at a temperature of 50°C for a period of 6 hours, as the enzyme retained 70% of its activity after 6 hours of incubation compared to free enzyme.

This may be attributed to what Smith [33] stated that the immobilization substance works to protect the enzyme from high-temperature impact with increased thermal stability of the enzyme, as it is granted thereby the tolerance and protection from the denaturation process by the influence of heat. The results confirm with the findings of Batra et al. [5], who indicated that β-galactosidase immobilized on DEAE-cellulose and calcium alginates had kept up 82.8% of its activity at a temperature of 60°C for 15 hours and 50% of activity at temperatures of 65°C for a period of 9 hours for both immobilization methods, respectively. Neri et al. [22] also noted that the thermal stability of the enzyme β-galactosidase from the *Kluyveromyces lactis* yeast and immobilized in polysiloxane-polyvinyl alcohol magnetic was higher compared to the free enzyme at a temperature of 35°C for a period of 24 hours and the increase was attributed to the decline in the movement ability of protein molecules as a result of covalent link, which gives it rigidity at each of linking points to immobilizing substance and thus provide protection from effects of the surrounding environment. Pan et al. [24] stated that the immobilizing process of the enzyme in the magnetic Fe_3O_4-chitosan has improved its thermal stability as the enzyme retained 65% of the enzymatic activity at an incubation temperature of 50–60°C for a period of 6 hours while the free enzyme retained 20% of the original activity through the incubation under the same conditions.

4.3.3 EFFECTS OF INHIBITION ON ENZYME IMMOBILIZED BY CHITOSAN

The effect of inhibition was studied by the addition of galactose at different concentrations on the activity of immobilized enzyme as shown in Figure 4.16. The immobilized enzyme retained 93.15% and 77.01% of its activity at a concentration of 1% and 5%, respectively. This indicates that the immobilized enzyme was better than the free enzyme thus retaining the enzymatic activity due to the accumulation of the reaction resultants or inhibitors, and this was confirmed by Haider et al. [10], who observed that adding galactose with a concentration of 5% for one hour at a temperature of 37°C led to the loss of 72% of the free enzyme activity while immobilized enzyme retained 65% of its activity.

As the effect of inhibitors on the immobilized enzymes shall be much lower compared to the free enzymes, this phenomenon can be explained

Technology of Solid State Fermentation in Dairy Products 97

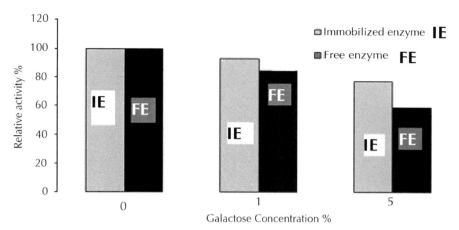

FIGURE 4.16 Effects of inhibition with galactose at different concentrations on the activity of immobilized enzyme

by the slow pace of penetration and arrival of these substances through the immobilizing substance network to the active sites. The quick penetration of the substrate and the arrival of these substances to the enzyme may be late and after the beginning of the enzymatic reaction. Also the new conditions operate after the immobilization process to hide the active site and block the influence of these compounds on the enzymatic reaction [7, 11].

4.3.4 EFFECT OF NUMBER OF TIMES OF USE FOR ENZYME IMMOBILIZATION IN CHITOSAN ON ENZYMATIC ACTIVITY

The effect of the number of times of use for the enzyme immobilized by Chitosan granules on the enzymatic activity was studied. Figure 4.17 shows that the immobilized enzyme lost 28.65% of its activity after the tenth use, and this decline in the enzymatic activity is attributable to the occurrence of enzyme infiltration or as a result of exposure to heat constantly, when enzyme immobilizing granules are used repeatedly.

The results were consistent with those indicated by Haider et al. [10], who observed that the enzyme immobilized by starch granules substance and calcium alginates retained 65% of its activity after the sixth time. Also Neri et al. [22] studied the effects of the number of times of use for β-galactosidase immobilized by polysiloxane-polyvinyl alcohol on the enzymatic activity; and they observed a decrease in enzymatic activity after using immobilized

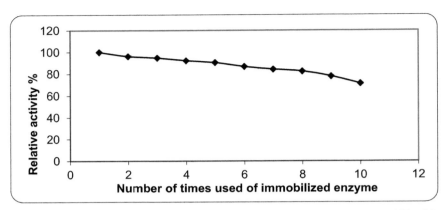

FIGURE 4.17 Effect of the number of times of use for the enzyme immobilized in Chitosan on enzymatic activity at 50°C for 15 min.

enzyme 10 times where it lost 14% of the activity at the tenth time. On the other hand, Pan et al. [24] found that enzyme immobilized by magnetic Fe_3O_4-chitosan was able to retain 92% of its activity after use for 15 times,

Also Zhang et al. [41] stated that the enzyme immobilized with granules prepared from chitosan and gum in the presence of Gluteraldehyde had retained 53% of its activity after 9 times of use. Puri et al. [28] also recorded that the raw enzyme produced from *Kluyveromyces marxianus* YW-1 yeast and immobilized by gelatin with stability in the hydrolysis of lactose after each use until the eighth use when the hydrolysis rate of 49% was observed; and after which the hydrolysis activity started to decline until its percentage reached 25% with the ninth time.

4.3.5 STABILITY OF FREE AND IMMOBILIZED ENZYME DURING STORAGE PERIODS

The Figure 4.18 shows that the immobilized enzyme retained its full activity at a temperature of 5°C, but it has lost less than 9% only of its activity at 25°C; while the free enzyme has lost more than 12% and 29%, respectively of its activity at a storage temperature of 5 and 25°C after 30 days of storage.

These results are in agreement with many of the research studies that have confirmed the superiority of the immobilized enzyme on the free enzyme in storage. Makkar et al. [18] noted that the enzyme immobilized in the substancepoly acryl amide retained its full activity after 20 and

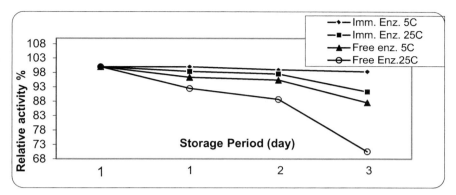

FIGURE 4.18 Effects of storage on the free and immobilized enzyme for a storage period up to 30 days.

30 days of storage at a temperature of 25 and 4°C, respectively, but it had lost 30% of its activity after 60 days at 4°C; while the free enzyme has lost 60% of its activity at temperature of 4°C for the same period and 27% at a temperature of 25°C for a storage period of 20 days.

Zhou et al. [42] also studied the effects of storage on the immobilized enzyme activity at a temperature of 4°C for 30 days, and they noticed that the enzyme retained 86% of its activity after the expiration of the storage period. However, Haider et al. [10] observed that the relative activity of the immobilized β-galactosidase at a temperature of 4°C for a period of 60 days was 85% for the immobilized β-Galactosidase, whereas the free enzyme retained 35% of its activity during the same period. This confirms that the immobilization of enzymes leads to an increase in its stability at the storage.

4.3.6 USE OF FREE AND IMMOBILIZED ENZYMES IN LACTOSE HYDROLYSIS

The Figure 4.19 (A, B, and C) indicated that the free enzyme was higher than the immobilized enzyme at the beginning of the reaction, however the increase of duration of reaction has led to an increase in the rate of hydrolysis of lactose for the immobilized enzyme compared to the free enzyme. In addition, the highest percentage of hydrolysis of the lactose content in the buffer acetate reached 66.2% and 81.54%; and the lowest was in skimmed milk at 40.72% and 61.84%, respectively. While in whey it was 49.35% and 66.47% for the free and immobilized enzymes, respectively.

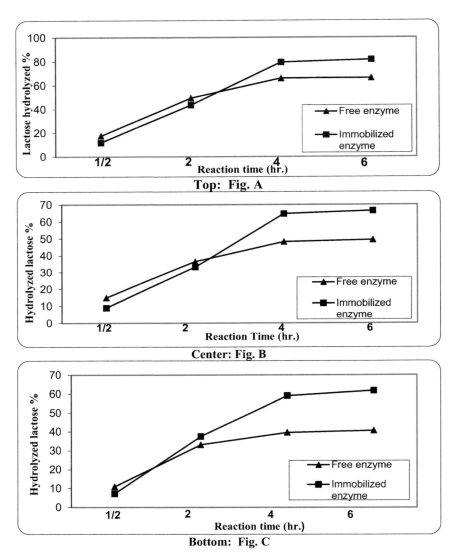

FIGURE 4.19 Lactose hydrolysis by free and immobilized enzyme at 50 °C for 6 hours in: Top – A) Acetate buffer, Center – B) Whey, and Bottom – C) Skim milk.

The proportion of lactose hydrolysis in whey and milk depends on the enzymatic activity of β-galactosidase and its concentration, pH, temperature and the time needed for the hydrolysis process. The increase in the rate of hydrolysis of the whey compared to the milk is due to the

Technology of Solid State Fermentation in Dairy Products 101

fact that the pH of the whey ranges between 4.5–5, which is the optimal for the operation of β-galactosidase purified from the *Aspergillus oryzae* compared to the milk with pH range between 6–6.8 [11]. However, the increase in the rate of hydrolysis of lactose when the immobilized enzyme is used compared to the free one is due to the stability shown by the immobilized enzyme towards the heat as well as towards the inhibition process occurring as a result of the release of galactose during hydrolysis.

These results are in consistent with the various studies on the decomposition of lactose. Batra et al. [5] have treated β-galactosidase immobilized on the substance DEAE-cellulose and calcium alginates in the hydrolysis of lactose prepared in buffer phosphate with a concentration of 5% at a temperature of 55°C for 48 hours. They obtained 92.5% rate of hydrolysis and 93.7%, but the rate of decomposition dropped to 57.1% and 69.7%, respectively, after 20 times of use. In addition, Neri et al. [22] used the enzyme produced from the *Kluyveromyces lactis* yeast and immobilized covalently in polyvinyl alcohol magnetic polysiloxane activated with Gluteraldehyde in the hydrolysis of lactose in low fat milk (skimmed milk); they observed 90% hydrolysis rate at a temperature of 25°C for a period of 120 minutes. Also Haider et al. [11] have used β-galactosidase produced from the *Aspergillus oryzae* mold immobilized with calcium alginates and starch in the hydrolysis of the lactose of milk and whey; and they observed the hydrolysis rate of 79% for 3 hours and 89% for 3 hours, respectively, while the hydrolysis rate was of 70% and 61%, respectively when the free enzyme was used under the same conditions. Puri et al. [28] showed that the rate of hydrolysis of the lactose in milk was 49% for the immobilized enzyme and produced from the *Kluyveromyces marxianus* yeast at a temperature of 40 °C for a period of 4 hours.

4.4 FUTURE PERSPECTIVES AND RESEARCH OPPORTUNITIES

This study can benefit those people, who suffer from lactose intolerance and lactase deficiency. They can intake dairy products containing it without any complications. The future studies can focus on the use of the immobilized enzyme to hydrolyze lactose by continuous method.

4.5 SUMMARY

Fifty local isolations of *Aspergillus oryzae* fungi were isolated from soil, cheese and whey; and underwent a purification operation and screening (primary and secondary). *Aspergillus oryzae* fungi was the most efficient in the production of beta-galactosidase enzyme. The crude enzyme was purified by using concentration with ammonium sulfate with a saturation of 65–80% and ion exchange using ion exchanger DEAE-sephadex A-50 and gel filtration using Sephadex G-100 with a 9.70 times purification and the enzymatic yield of 25.88%, respectively. The enzyme purity was revealed completely through one band. The characteristics of the purified enzyme under study were: optimum effective pH of the enzyme of 5; the optimal term stability of the enzyme range of 6.5–4. The results showed a thermal stability between 50–20°C for 15 minutes, and the optimum temperature of 50°C, at which the enzyme retained its full effectiveness during incubation for an hour. The activation energy value was found to be 6.19 kilocalories/mol while the Energy metamorphosis enzyme value was 52.31 kcalories/mol. The molecular weight of the enzyme reached 97.72 kDa using electric deportation technique while 103.51 kDa was reached with gel filtration technique. It was observed that the manganese and sodium ions have a tonic role in the effectiveness of the enzyme, while activity was decreased at different rates when adding magnesium and copper and iron. The efficiency of enzyme was not affected by adding potassium. Low enzymatic efficiency was observed with the addition of urea and sugar galactose. A slight increased presence of mercaptoethanol the EDTA had no impact on the effectiveness.

The Michals constant Km values, and top speed (Vmax) of the enzyme were 1.435 million Molars and 1040.23 units/ml, respectively. When using ONPG, degradation products of lactose was detected by TLC thin-layer technology, both glucose and galactose were detected. Use of Chitosan also showed superior restrictive efficiently of 76.60% and the best quality of the enzyme was 0.75 ml/0.5 g of Chitosan. The optimum pH of 5 and optimum temperature of 50°C for the immobilized enzyme similar to the free enzyme were observed. The enzyme curve appeared to be widening over the stability of both qualities. Low impact damping with galactose was noted on the effectiveness of the enzyme restricted compared to the free enzyme;

Technology of Solid State Fermentation in Dairy Products

and the immobilized enzyme remained effective by 71.35% after using for 10 times. The stability of the free and restrained enzyme was studied during extended storage. It has less than 9% of its effectiveness at 25°C, while the free enzyme lost more than 12% and 29% of its effectiveness when stored at 5°C and 25°C, respectively within 30 days of storage. When using free and immobilized enzyme in the analysis of lactose, it was found that the highest percentage of hydrolysis was present in buffer lactose acetate which reached 66.2% and 81.54% in the milk sorting 40.72% and 61.84%, while the whey gave a hydrolysis ratio of 49.35% and 66.47% while using free and immobilized enzyme in a row after 6 hours of incubation at 50°C.

The possibility of producing an enzyme with a high purity and with qualified characteristics for food applications, especially in dairy products as well as the possibility of restricting the enzyme covalently with Chitosan with a good specifications and quality storage with low impact damping on the effectiveness of enzyme, and the restricted enzyme have efficient analytical properties when applied in the dairy industry.

KEYWORDS

- ***Aspergillus oryzae***
- **characterization**
- **chitosan**
- **purification**
- **screening**
- **β-galactosidas**

REFERENCES

1. Aguilar, C., Sanchez, G., Barragan, P., Herrera, R., Hernandez, P., & Esquivel, J. (2008). Perspectives of solid state fermentation for production of food enzymes. *American Journal of Biochemistry and Biotechnology, 4*(4), 354–366.
2. Alrikabi, A. K., Majeed, G. H., & Al-Manhal, A. J. (2011). Study of Immobilization β-galactosidase purified from mold *Aspergillus oryzae* by solid state fermentations

and it's applications in some dairy products. *Basra Journal of Agriculture Science, 24*(1), 229–250.

3. Al-khafaji, Z. M. (2008). *Microbial Biotechnology.* University of Basrah, p. 763.

4. An, Z. (2005). *Handbook of industrial mycology.* New York, p. 642.

5. Batra, N., Singh, J., Joshi, A., & Sobti, R. C. (2005). Improved properties of *Bacillius coagaulans* β-galactosidase through immobilization. *Engineering in Life Sciences, 5*(4), 581.

6. Chen, W., Chen, H., Xia, Y., Zhao, J., Tian, F., & Zhang, (2008). Production, purification, and characterization of a potential thermostable galactosidase for milk lactose hydrolysis from *Bacillus stearothermophilus. Journal of Dairy Science, 91*, 1751–1758.

7. El-Shora, H. M. (2001). Properties and immobilization of urease from leaves of chenopodium album (C3). *Bot. Bull. Acad. Sci., 42*, 251.

8. Esawy, M. A., Mahmoud, D. A. R., & Fattah, A. F. A. (2008). Immobilization of Bacillus subtilis NRC33a levansucrase and some studies on its properties. *Brazilian Journal of Chemical Engineering, 25*(2), 237.

9. Grosova, Z., Rosenberg, M., Gdovin, M., Slavikova, L., & Rebroš, M. (2009). Production of D-galactose using β-galactosidase and Saccharomyces cerevisiae entrapped in poly(vinylalcohol) hydrogel. *Food Chemistry, 116*, 96.

10. Haider, T., & Husain, Q. (2008). Concanavalin a layered calcium alginate–starch beads immobilized β-galactosidase as a therapeutic agent for lactose intolerant patients. *International Journal of Pharmaceutics, 359*, 1–6.

11. Haider, T., & Husain, Q. (2009). Hydrolysis of milk/whey lactose by β-galactosidase: A comparative study of stirred batch process and packed bed reactor prepared with calcium alginate entrapped enzyme. *Chemical Engineering and Processing, 48*, 576–580.

12. Hatzinikolaou, D. G., Efstathios, K., Diomi, M., Amalia, D. K., Paul, C., & Dimitris, K. (2005). Modeling of the simultaneous hydrolysis–ultrafiltration of whey permeate by a thermostable β-galactosidase from *Aspergillus niger. Biochemical Engineering Journal, 24*, 161–172.

13. Klich, M. A. (2002). *Identification of Common Aspergillus Species.* 1st Edition. Wageningen, Netherlands. pp. 116.

14. Kosseva, M. R., Panesar, P. S., Kaur, G., & Kennedy, J. F. (2009). Use of immobilized biocatalysts in the processing of cheese whey. *International Journal of Biological Macromolecules, 45*, 437.

15. Laemmli, U. K. (1970). Cleavage of structural proteins during the assembly of the head of bacteriophage T4. *Nature, 227*, 680–285.

16. Lehmacher, A., & Bisswanger, H. (1990). Isolation and characterization of an extremely thermostable D-glucose/xylose isomerase from *Thermus aquaticus* HB8. *Journal of General Microbiology, 136*, 679–686.

17. Majeed, G. H. M., Ali, K. J., Alaa, J., & Al-Manhal, A. (2011). Purification, Characterization of β-galactosidase produced from a local isolate of *Aspergillus oryzae* by solid state fermentations. *Basra Sci. J., 37*(2), 69–80.

18. Makkar, H. P. S., Sharma, O. P., & Negi, S. S. (1981).Immobilization and properties of β-D-galactosidase from *Lactobacillus bulgaricus. Journal Bioscience, 3*(1), 7.

19. Mariotti, M. P., Yamanaka, H., Araujo, A. R., & Trevisan, H. C. (2008). Hydrolysis of whey lactose by immobilized β-galactosidase. *Brazilian Archives of Biology and Technology, 51*(6), 1233.

Technology of Solid State Fermentation in Dairy Products

20. Monsan, P., & Combes, D. (1988). Enzyme stabilization by immobilization. In: Klaus, M. (ed.). *Methods in Enzymology*. Academic Press, Inc.

21. Nagy, Z., Keresztessy, Z., Szentirmai, A., & Biro, S. (2001). Carbon source regulation of β-galactosidase biosynthesis in *Penicillium chrysogenum*. *Journal of Basic Microbiology*, *41*(6), 351–362.

22. Neri, D. F. M., Balco, V. M., Carneiro-da-Cunh, M. G., Carvalho Jr., L. B., & Teixeira, J. A. (2008). Immobilization of β-galactosidase from *Kluyveromyces lactis* onto a polysiloxane–polyvinyl alcohol magnetic (mPOS–PVA) composite for lactose hydrolysis. *Catalysis Communications*, *9*, 2334–2339.

23. O'Connell, S., & Walsh, G. (2010). A novel acid-stable, acid-active β-galactosidase potentially suited to the alleviation of lactose intolerance. *Applied Microbiology and Biotechnology*, *86*(2), 517–524.

24. Pan, C., Hu, B., Li, W., Sun, Y., Ye, H., & Zeng, X. (2009). Novel and efficient method for immobilization and stabilization of β-galactosidase by covalent attachment onto magnetic Fe_3O_4-chitosan nanoparticles. *Journal of Molecular Catalysis B: Enzymatic*, *61*, 208.

25. Park, A. R., & Oh, D. K. (2010). Galacto-oligosaccharide production using microbial β-galactosidase: Current state and perspectives. *Applied Microbiology and Biotechnology*, *85*, 1279–1286.

26. Park, Y. K., Santi, M. S. S., & Pastore, G. M. (1979). Production and characterization of β-galactosidase from *Aspergillus oryzae*. *Journal of Food Science*, *44*(1), 100–103.

27. Parker, J. N., & Parker, P. M. (2002). *The Official Patients Sourcebook on Lactose Intolerance*. USA. p. 200.

28. Puri, M., Gupta, S., Pahuja, P., Kaur, A., Kanwar, R., & Kennedy, J. F. (2010). Cell disruption optimization and covalent immobilization of β-D-galactosidase from Kluyveromyces marxianus YW-1 for lactose hydrolysis in milk. *Applied Microbiology and Biotechnology*, *160*, 98.

29. Saad, R. R. (2004). Purification and some properties of β-galactosidase from *Aspergillus japonicus*. *Annals of Microbiology*, *54*(3), 299–306.

30. Segel, I. H. (1976). *Biochemical Calculations*. 2nd Edition, John & Sons Inc., New York.

31. Shaikh, S. A., Khire, J. M., & Khan, M. I. (1997). Production of β-galactosidase from thermophilic fungus *Rhizomucor* sp. *Journal of Industrial Microbiology and Biotechnology, 19*, 239–245.

32. Smaali, I., Rémond, C., Skhiri, Y., & O'Donohue, M. J. (2009). Biocatalytic conversion of wheat bran hydrolysate using an immobilized GH43 β-xylosidase. *Bioresource Technology*, *100*, 338.

33. Smith, E. J. (2009). *Biotechnology*. 5th Edition, Cambridge University Press, pages 266.

34. Song, C., Chi, Z., Li, J., & Wang, X. (2010). β-galactosidase production by the psychrotolerant yeast *Guehomycespullulans* 17-1 isolated from sea sediment in Antarctica and lactose hydrolysis. *Bioprocess and Biosystems Engineering*, 1–7.

35. Tanaka, Y., Kagamishi, A., & Kiughi, A. (1975). Purification and properties of β-galactosidase from *Aspergillus oryzae*. *Journal of Biochemistry*, *77*, 241–247.

36. Tanriseven, A., & Doǧan, S. (2002). A novel method for the immobilization of β-galactosidase. *Process Biochemistry*, *38*, 27–30.

37. Todorova-Balvay, D., Stoilova, I., Gargova, S., & Vijayalakshmi, M. A. (2006). An efficient two step purification and molecular characterization of β-galactosidase from *Aspergillusoryzae*. *Journal of Molecular Recognition, 19*, 299–304.
38. Ustok, F. I., Tari, C., & Harsa, S. (2010). Biochemical and thermal properties of β-galactosidase enzymes produced by artisanal yogurt cultures. *Food Chemistry, 119*, 1114–1120.
39. White, A., Handler, P., & Smith, E. (1973). *Principles of Biochemistry*. McGraw-Hill Book Company, New York.
40. Whitaker, J. R. (1972). *Principles of Enzymology for the Food Science*. Marcel Dekker. Inc. New York, USA.
41. Zhang, S., Gao, S., & Gao, G. (2010). Immobilization of β-galactosidase onto magnetic beads. *Applied Biochemistry and Biotechnology, 160*, 1386.
42. Zhou, Z. K. Q., & Chen, D. X. (2001). Immobilization of β-galactosidase on graphite surface by glutaraldehyde. *Journal of Food Engineering, 48*, 69.

CHAPTER 5

NANO-PARTICLE BASED DELIVERY SYSTEMS: APPLICATIONS IN AGRICULTURE

DEEPAK KUMAR VERMA, SHIKHA SRIVASTAVA,
VIPUL KUMAR, BAVITA ASTHIR, MUKESH MOHAN, and
PREM PRAKASH SRIVASTAV

CONTENTS

5.1   Introduction 107
5.2   Classification of Nano-Delivery Systems 110
5.3   Applications of Nano-Particle Based Delivery
      Systems in Agriculture 118
5.4   Conclusions and Future Opportunities 123
Keywords 125
References 125

5.1   INTRODUCTION

In the recent past, various players in the field of technology have been exploring the possibility of using new science to sustain mankind, as

---

Reprinted with permission from Deepak Kumar Verma; Shikha Srivastava; Prem Prakash Srivastav; and Bavita Asthir. Nano particle based delivery system and proposed application in agriculture. In: Megh R. Goyal and R. K. Sivanappan, Eds., Engineering Practices for Agricultural Production and Water Conservation: An Interdisciplinary Approach; Apple Academic Press Inc., 2017, Chapter 16, pp. 325–348.

resources become scarce and demand increases. The emergence of new nano-devices and nano-materials opens up potential novel applications in agriculture and biotechnology. Use of such technology employs the fundamental knowledge of biology, chemistry, biochemistry, molecular biology, chemical engineering, biological engineering, agronomy, etc. to decipher the science and technological knowhow to rely on nano-based technology. The term "nano-technology" indicates a multi-disciplinary approach concerning materials, devices and systems in which at least one of three characteristic dimensions of their components is measured at the nano-metric scale (nm). The nano-metric scale is characterized by: (i) the atomic dimensions, (ii) the molecular dimensions, and (iii) the distance among the atoms in ordinary condensed matter.

The nano-meter sub-multiplies in the atomic world are more commonly expressed in Angstroms (A^0), where $1\ A^0 = 0.1\ nm = 10^{-10}\ m$. Lower resistance to electricity, lower melting point and faster reactions are among the basic properties to be shown by nano-devices [24]. Reduction in size also leads to appearance of surface effects related to the high number of surface atoms as well as to a high specific area, making these nano-materials important from the practical point of view.

Nano-based "smart delivery systems" are most promising and recent technologies that have been widely used in agriculture and medical field. Paul Ehrlich termed it as "magic bullets" loaded with specific analyte and drug to the target site, thus taking appropriate remedial action [18]. In these days, bio-degradable nano-particles are gaining much more attention due to the site specific delivery of various biological active compounds viz. micro-macronutrients, vitamins, plant growth regulators, plant staining agents and genes [46]. Proteins, polysaccharides and synthetic bio-degradable polymers (like chitosan based nano-material) are bio-degradable in nature and pose an eco-friendly to the environment. The selection of base material depends upon many factors like size, surface properties, encapsulating materials and its biocompatibility, etc. (Figure 5.1).

There are lots of emerging applications in agriculture describing its role as an essential and demanding role for smart use of nano-based materials (Table 5.1). Torney et al. [48] reported the use of gold-based nano-particle as suitable carrier materials for recombinant DNA to target into the plant cells for plant transformation and its genetic modification, which is found to be the recent use of the nano-based composite material in the field of agriculture.

Nano-Particle Based Delivery Systems

TABLE 5.1 Nano-Technology Based Advancements and Applications in Agriculture

Year	Nano-technology based advancement	Application in Agriculture	Ref.
2014	• Chitosan nano-formulation loaded with herbicide • Nano-based atrazine	• Control release of herbicide and weed management • Weed control and management of competitive weeds	[19]
2013	• Nano-gel formulation impregnated with pheromone • Silica conjugated nano-materials with naphthalinic acid	• Management of fruit fly and other bugs • As a plant growth enhancer	[7]
2012	• Nano-emulsion of neem oil • Antifungal nano-disks with amphotericin-B	• Larvicidal agent • Treatment of fungal infection in wheat and chick pea	[34]
2011	• Amino nano-composite • Zn-Cd optical nano-sensor	• Detection of organochlorine and organophosphorus in vegetables • Detection of pesticide	[4]
2010	• Nano-clay material • Gold nano-composite • Magnetic carbon nanotubes • Gold electrochemical sensors	• Soil enhancer and detection water retention capacity • Pesticide detection in vegetables • Smart delivery of agrochemicals • *Sclerotinia* pathogen detection	[43]
2009	• Polyethylene glycol based nano-material with essential oil • Cd-Te nano-structure • Polyaniline nano-based material • Nano bio-sensor • Gold nano-particle	• Control of pest infestation in vegetables • Detection of 2,4-D herbicide • Detection of parathion and chloropyrifos • Detection of storage infection • Detection of pesticide	[25, 53]
2008	• Nano-emulsions • Nano-starch particles	• Plant growth regulator and stress alleviator • Transporting DNA to transform plant cells	[2, 27]
2007	• Aliposome nano-sensor • Mesoporous silica nano-particle	• Pesticide detection • Targeted delivery of DNA to plant cell for transformation	[51]

TABLE 5.1 (Continued)

Year	Nano-technology based advancement	Application in Agriculture	Ref.
2006	• Methyl acrylic acid polymer nano-structure • TiO_2 Nano-structure on electrode • Porous silica nano-particles • TiO_2 coated with filters	• Detection of pesticides and its analysis in vegetables • Detection of parathion residues in vegetables • Controlled delivery system for water-soluble pesticide • Photo-catalytic degradation of agrochemicals in contaminated waters	[26, 32]
2005	• Zn-Al layered nano-composite	• Control release of herbicides and pesticides	[22]
2003	• Nano-silica based soil binder	• Prevention of soil run-off • Seed blending for germination	[22]

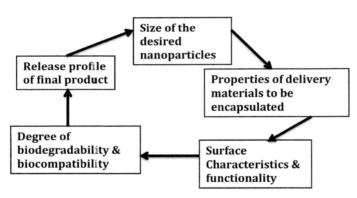

FIGURE 5.1 Factors influencing bio-degradable nano-materials preparation.

5.2 CLASSIFICATION OF NANO-DELIVERY SYSTEMS

Nano-delivery systems (Figure 5.2) are classified into three main groups: Emulsions, vesicular delivery system, and lipid based nano-particles. The two types of emulsions of nano-delivery systems are micro-emulsions (MEs) and nano-emulsions (NEs). The three types of vesicular nano-delivery systems are liposomes, niosomes and transferosomes. The group

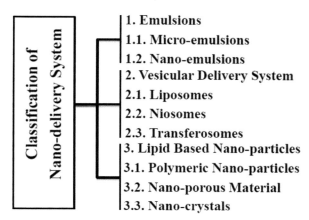

FIGURE 5.2 Flow diagram for classification of nano-delivery system.

of lipid-based nano-particles is composed of polymeric nano-particles and nano-porous material.

5.2.1 EMULSIONS

In a very simple form, emulsion is the mixture of two immiscible liquids, generally oil and water. Talking in nano-technology sense, these are classified as follows:

5.2.1.1 Micro-Emulsions

MEs are emulsion of water, oil, and amphiphile (Figure 5.3) with size range in the nano-meter (nm) and with diasteriomeric and thermodynamically stable liquid containing particle with diameters of 100 nm and less [3]. These are transparent emulsions, in which oil is dispersed in an aqueous medium containing surfactant with or without suitable co-surfactant. These favor formation of thermodynamically stable system with droplets of internal phase. Specific analyte/molecule carried in the micro-emulsion is in the solubilized form either in the oil or the aqueous phases. MEs are being employed for the delivery of drugs as it has the ability to improve the solubility and stability of drugs/chemicals [8].

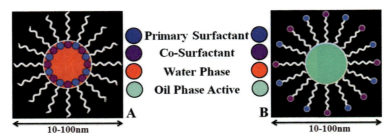

FIGURE 5.3 Structure of micro-emulsion: (A) water-oil micro-emulsion droplet, and (b) oil-water micro-emulsion droplet, (Source: http://www.enviroquestgpt.co.uk/content/micro-emulsion.html)

5.2.1.2 Nano-Emulsions

NEs droplets measure between 10 and 100 nm in diameter and are typically transparent due to their sizes being on a scale smaller than the ultraviolet-visible light range. Two types of methods are commonly used in NEs synthesis: mechanical or chemical processes. Mechanical processes employ sonicators and micro-fluidizers to break larger emulsion droplets in to smaller ones, whereas chemical method result in spontaneous formation of emulsion droplets due to hydrophobic effect of lipophilic molecules in the presence of emulsifiers [20, 31].

5.2.2 VESICULAR DELIVERY SYSTEM

5.2.2.1 Liposomes

Liposomes (Figure 5.4) are self-assembled nano-particles formed from phospholipids in the form of a bilayer [6]. In the liposomes, each sub unit is composed of one polar head group and other side with long chain of hydrophobic residues with microscopic size. Two monolayers fold back to back to form a liposome layer in a two dimensional manner. Bilayer formation occurs most readily when the cross-sectional areas of the head group and acyl side chain(s) are similar, as in glycerophospholipids and sphingolipids.

The hydrophobic portions in each monolayer, excluded from water, interact with each other. The hydrophilic head groups interact with water

Nano-Particle Based Delivery Systems

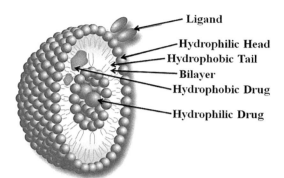

FIGURE 5.4 Structure of liposomes [6]. (Reprinted from Bei, D., Meng, J., & Youan, B. C. (2010). Engineering nanomedicines for improved melanoma therapy: progress and promises. *Nanomedicine (Lond.)*, *5*(9), 1385–1399. With permission from Future Medicine Ltd.)

at each surface of the bilayer. Because the hydrophobic regions at its edges are transiently in contact with water, the bilayer sheet is relatively unstable and spontaneously forms a third type of aggregate: it folds back on itself to form a hollow sphere, a vesicle or liposome (Figure 5.5). These vesicles consist primarily of phospholipids (synthetic or natural), sterols and an antioxidant [10, 11, 29, 45].

Size, number of lamellae and surface charge differentiate the liposome structures and are classified as oligo-, uni- or multi-lamellar and small, large or giant. Uni-lamellar liposomes contain a single layer and are further classified into various size ranges (Table 5.2). Multi-lamellar liposomes consist of many concentric lamellae, exhibiting an onion-like structure with diameter range between 500 nm and 5 μm [8].

Method for synthesizing liposomes includes the gentle hydration method and layer by layer electrostatic deposition. Liposomes have been developed for food application as a method for creating iron-enriched milk, antioxidant delivery, co-delivery of vitamin E and C with orange juice [30].

5.2.2.2 Niosomes

Niosomes are the vesicles composed of nonionic surfactant (Figure 5.6). They have been mainly studied because of their advantage compared to liposomes, i.e., higher chemically stability of surfactant than phospholipid,

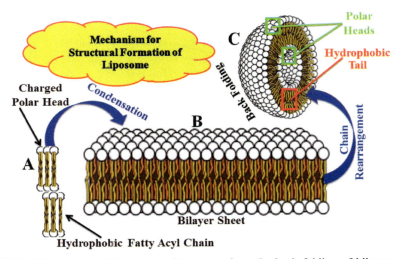

FIGURE 5.5 Structural formation of liposome from the back folding of bilayers. (A) Basic unit of liposome showing polar head and hydrophobic tail; (B) condensation of all basic unit to form bilayer; and (C) back folding and chain rearrangement of bilayer to form the liposome.

TABLE 5.2 Classification of Uni-Lamellar Liposomes

Uni-lamellar liposomes	Size (diameter)
Small	25–100 nm
Large	100–1 μm
Gaint	Greater than 1 μm

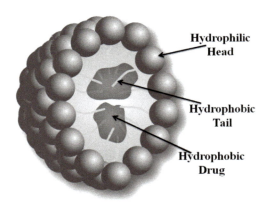

FIGURE 5.6 Structural of Niosome.

require no condition for preparation and storage. They have no purity related problems and manufacturing costs are also low [6, 23]. Niosomes mostly show an ability to increase the stable entrapment of input drugs, improved bioavailability and enhanced penetration [49].

5.2.2.3 Transferosomes

Transferosomes (Figure 5.7) are lipid vesicles containing large fractions of fatty acids. These are vesicles composed of phospholipid as their main ingredients with 10–25% surfactant and 3–10% ethanol [14, 23]. The inventers claim that transferosomes are ultra-deformable and squeeze through pores less than 1/10th of their diameter. Higher membrane hydrophilicity and flexibility both help transfersomes to avoid aggregation and fusion, which are observed with liposomes [35].

5.2.3 LIPID-BASED NANO-PARTICLES

Lipid based nano-particles, mainly solid lipid nano-particles (SLNs) and nano-structure lipid carriers (NLCs), are similar to emulsions, with the exception that the lipids are in a solid phase [32]. SLNs are colloidal particles containing highly purified triglyceride, composed mainly of lipids that are solid at room temperature [38]. SLNs have the potential

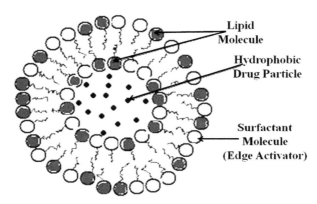

FIGURE 5.7 Structure of transferosomes.

to provide controlled release of various lipophilic components due to decreased mobility of bioactive in the solid matrix [33]. NLCs are lipid nano-particles with both crystallized and liquid phases of lipid used for nano-delivery. The inner liquid phase dissolves and entraps the bioactive to provide the advantage of chemical stability, controlled release and a higher loading efficiency [46]. There are several methods used to produce SLNs and NLCs as shown in Table 5.3. SLNs are developed for food applications with goal of improving shelf life, the stability of quercitin, β-carotene and α-tocopherol [8, 33, 46, 54].

5.2.3.1 Polymeric Nano-Particles

Polymeric nano-particles (PNs) are made up form polymers. These are matrices of polymers and entrapped molecules surrounded by an emulsifier or surfactant [13, 21, 37]. The material to be delivered, i.e., drugs/bioactive is manifested by various methods like in the form of dissolved substances, entrapped/encapsulated relying upon the method of preparation, nano-capsules, nano-particles or nano-spheres can be obtained. These particles, nano-spheres and nano-capsules are designed to protect the entrapped bioactive from degradation. In the recent years, biodegradable PNs found to the most potential drug delivery agent and as carriers of DNA in gene therapy and their ability to deliver proteins and peptides, etc. Besides these, natural polymers still enjoy popularity in delivery, and some of them are: gums, chitosan, gelatin, sodium alginate and albumic. It is reported that protein polymer (*zein*) nano-particles successfully entrapped essential oils with antioxidant properties [52]. Gelatin

TABLE 5.3 Methods of Production of SLNs and NLCs

Methods
Emulsification-sonication
Homogenization (hot and cold)
Micro-emulsion solvent emulsification-evaporation technique
Micro-fluidization
Ultra-sonication

Nano-Particle Based Delivery Systems 117

nano-particles encapsulated polyphenol antioxidants to protect their anti-oxidant activity and provide controlled release [42]. PNs are engineered by emulsification followed by solvent displacement or nano-precipitation or desolvation, salting out, and emulsion evaporation methods [39]. Other methods such as pH cycling, thermal treatment, atomization spraying, and the use of supercritical fluids are mostly applied to nano-particle synthesis for proteins and hydrocolloids.

5.2.3.2 Nano-Porous Materials

A typical and relatively simple porous system is one type of dispersion that is classically described as a material with gas-solid interface as the most dominant characteristic [1]. Accordingly, porous material might be classified by the size of pores or may be distinguished by different network materials. As per the classification based on the pore size, nano-particles range from 1 nm to 1000 nm. According to IUPAC, three distinctions can be made: 1) Micro-porous material (0–2 nm pores), 2) Meso-porous material (2–50 nm pores), and 3) Macro-porous material (>50 nm pores). Among these, the unique structural features of organically functionalized meso-porous silica nano-particles (MSNs) – such as their chemically and thermally stable structures, large surface area (>800 m^2g^{-1}), tunable pore sizes (2–10 nm) and well defined properties – have made them ideal guest for hosting guest molecules of various sizes, shapes and functionalities [1].

5.2.3.3 Nano-Crystals

Nano-crystals are crystals with less than 1 μm size but within the range of 10–400 nm. They are aggregates comprising several hundred to tens of thousands of atoms that combine into a "cluster". They consist of bio-active surrounded by a surfactant [16]. Nano-crystals improve the solubility of poorly water soluble drugs/chemicals by increasing the surface area to volume ratio, which increases the dissolution rate of bioactive *in vivo*. They can be made using mechanical or chemical methods [44]. Food grade starch and protein nano-particles are being developed for various applications in food and biomedical research [12, 15, 50].

5.3 APPLICATIONS OF NANO-PARTICLE BASED DELIVERY SYSTEMS IN AGRICULTURE

5.3.1 FOR EFFICIENT DELIVERY OF FERTILIZERS

Fertilizers play a pivotal role in enhancing agriculture production. Inorganic fertilizers are formulated in appropriate concentrations and combinations to supply three main nutrients: nitrogen (N), phosphorus (P) and potassium (K) for various crops. However, major part of these fertilizers are lost to environment though physical, chemical processes like leaching, volatilization, emission and cannot be absorbed by plants causing economic and resource losses [41]. To enhance the nutrient use efficiency, nano-fertilizers are emerging as alternative approach. For controlled and sustained release of fertilizer and pesticides, nano-fertilizers are expected to be far more effective with the high surface area to volume ratio. The use of nano-fertilizers helps plant roots and microorganisms to take nutrient ions from solid phase of minerals easily and in a sustainable way. Thus it enhances soil quality by decreasing toxic effects associated with the overdose application of fertilizer [24].

Nano-fertilizers are the encapsulated fertilizers in which NPK fertilizers are entrapped in nano-particles. Slow release of nutrients in the environment could be achieved by coating and cementing of nano and sub nano-composites [28]. For example: Nano-zeolite can act as a potential carrier for developing smart delivery fertilizers. Zeolites are group of naturally occurring minerals and has honey comb like porous layered crystal structure with the ability to exchange ions and catalyze reactions. Being made up of networks of interconnected tunnels and cages, it can be uploaded with nutrients (N, P and K) combined with other slowly dissolving ingredients containing phosphorus, calcium and a complete suite of minor and trace nutrients. Thus, nano-zeolite acts as a reservoir for nutrients that are released "on demand" [55]. It is reported that nano-composite based fertilizers consisting of N, P, K, micronutrients, mannose and amino acids have been developed that has increased the nitrogen uptake and utilization by grain crops.

Bio-degradable chitosan nano-particles have also gained great attention as controlled release for NPK fertilizers due to their bio-absorbable and bactericidal nature [36]. A number of studies indicate that these slowly released nano-fertilizers improve grain yield and has proved to be safe for germination of cereals. Besides the delivery of primary nutrients, sulfur coated nano-fertilizers have gained much attention due to their potential role in releasing sulfur in sulfur deficient soils. Materials such as kaolin and polymeric biocompatible nano-particles also have potential application in nano-fertilizers. When compared to bulk fertilizers, nano-fertilizers gain much more attention due to their small size, high specific surface area, reactivity, increased solubility and easy availability to plants.

Recent research has reported the potential of using polymeric nano-particles for coating of bio-fertilizer preparations for producing formulations resistant to desiccation [17]. In this way, nano-technology has opened a new opportunity in improving the nutrient uptake efficiency and minimizing costs of environmental protection.

5.3.2 DELIVERY OF AGROCHEMICALS

Indiscriminate use of agrochemicals (viz. pesticides and insecticides) since 20th century has caused environmental pollution, emergence of resistant pests and pathogens and loss of biodiversity. Encapsulation, controlled release methods and entrapment of agrochemicals has revolutionized the use of nano-composite material for agricultural use. Nano-encapsulated chemicals possess characteristics, such as: proper concentration with high solubility, stability and effectiveness, timely release in response to environmental stimuli, enhanced targeted activity and less eco-toxicity making production of crops higher and less injury to agricultural workers. Nano-particles within the 100–250 nm size range can be dissolved in water more effectively than existing ones due to their effective surface tension quality. Use of nano-emulsions can be a better choice in terms of application of uniform suspension of pesticides or herbicidal agents (either water or oil-based) and contain nano-particles in the range of 200–400 nm. Formulations in the form of gels, creams, liquids, etc. have been performed to assess the effectiveness of the sol-gel based nano-materials.

5.3.3 NANO-HERBICIDES

Elimination of weeds can be achieved by destroying their seed banks in the soil and prevent them from germinating when weather and soil conditions become favorable for their growth, and is one of the most promising technology. Molecular characterization of underground parts of a plant can be used for a new target domain and developing a receptor based herbicide molecule having specific binding property with nano-herbicide molecules like carbon nano-tubes capable of killing the viable and dormant underground propagates of weed seeds.

Target specific nano-encapsulated herbicide molecules is aimed for specific receptor in the roots of target weeds [9]. When these molecules enter into system by forming association with receptor, get translocated into its parts and inhibit glycolysis of food reserves in root system thus making specific weed to starve for food and get killed.

5.3.4 NANO-PESTICIDES

Nano-pesticide is any formulation that intentionally includes elements in the nm size range and/or claims novel properties associated with the small size. The most sophisticated approach to formulate nano-scale pesticides is encapsulating the nano-scale active ingredients within "envelope" or "shell" of nano-size. The aim of nano-pesticide formulation is to protect against premature degradation and its uniform suspensions of pesticides or herbicidal agents. Polymer based nano-pesticide formulation have received the greatest attention over the last two years, followed by formulation containing inorganic nano-particles (silica, titanium dioxide) and nano-emulsions [26].

Hydrophobic nano-silica embedded with surface materials has been successfully used for controlling incidence of insects/pests and diseases. Properly functionalized lipophilic nano-silica gets absorbed into the cuticular lipids of insects by physio-sorption and damages the protective wax layer and induces death by dessication [5].

Pesticides containing nano-scale active ingredients have been introduced in market with more research on the development of new nano-pesticides formulation that is coming to the world's market with innovative use of nano-materials in the field of agriculture. Syngenta, (world's most notable biotechnology company) is using nano-emulsions in its pesticide products. It also claims its Banner MAXX fungicide, which has extremely small particle size of about 100 nm. Due to small particle size, the chemical mixes completely in water. Additionally, the fungicide gets absorbed in such a way that it does not even gets washed off by rain or irrigation. Word's 4th ranking agrochemical corporation, BASF of Germany is conducting new research on pesticide delivery to plant root and rhizosphere with new nano-based composite materials, which involves an active ingredient whose ideal particle size is between 10–150 nm. Marketed under the name Karate ZEON is a quick release microencapsulation nano-pesticide formulation containing active compound, i.e., λ-cyhalothrin, which breaks open on contact with leaves. In contest, the encapsulated product "gutbuster" only bark opens to release its content when it comes into contact with alternative environment such as stomach of certain insect.

5.3.5 DELIVERY OF PLANT GROWTH REGULATORS (PGRs)

Plant growth regulators are natural as well as synthetic compounds capable of modifying the biochemical pathways and physiological processes in plant. Normally PGRs are synthesized throughout the plant and transported mainly by vascular system. But with any change in external environment, cell response for balancing PGRs by up or down regulation and under such circumstances, a wanted response cannot be achieved as cell naturally tries to balance the change that has occurred due to external PGR application. Therefore, a target delivery system with controlled release of PGRs is essential for the expected developmental response.

In a tissue culture experiment, addition of nano-formulation into culture medium shows a dose dependent effect on shoot length of *Asparagus officinalis*. Saponin (3.0 mg/l) and GA_3 (0.2 mg/l) formulation with variable duration of sonication (1, 3, 5, 7, 9, 11, 13, and 15 minutes) were evaluated for shoot elongation. It was observed that highest shoot lengths were found after 5 and 7 minutes of sonication [40].

Saponins form self-assembled nano-sized vesicles and can act as delivery agent for different compounds. Saponin based smart delivery system consists of PGRs, i.e., auxin, GA immobilized or encapsulated with nano-structures. The surface of delivery system is functionalized with effector molecules (signal domain) in such a way that it would enable the nano-formulation to recognize the target sites in plant tissues. This approach can lead to the safe passage of PGRs under diverse environmental conditions with site-targeted delivery [40].

5.3.6 DELIVERY OF GENETIC MATERIALS INTO PLANT TISSUES

Gene transformation in plants is normally carried out by Agrobacterium mediated and micro-projectile bombardment methods. These conventional methods suffer from many drawbacks such as low transformation, high cost, low resolution and specificity [47].

Nano-biotechnology offers a new set of tools to manipulate the genes using nano-particles, nano-fibers and nano-capsules. It is reported that when carbon nano-fibers surface modified with plasmid DNA were integrated with viable cells, successful delivery and integration of plasmid DNA took place. This integration was further confirmed from the gene expression. The process is nearly similar with microinjection method of gene delivery and hence it is possible with plant cells in which the treated cells could be regenerated into whole plant that would express the introduced trait.

DNA and RNA fragments can easily be cross-linked with metallic and non-metallic nano-material for delivery into plant cells. Due to their nano-size, materials can easily pass through nano-scale pores of biological membrane. Ultrasonication and mild vortexing can assist the transfer of nano-fibers to carry gene into cultured plant tissues. The application of fluorescent labeled starch nano-particles as transgenic vehicle has been reported in which the nano-particle biomaterial was designed in such a way that it binds and transports the gene across the cell wall by inducing instantaneous pore channels in cell wall, cell membrane and nuclear membrane. It is also possible to integrate different genes on the nano-particles at the same time. Being labeled with fluorescent marker, the imaging can be carried out with fluorescence microscope for further understanding of the movement of transferred genes.

The unique feature of organically functionalized MSNs such as chemical and thermal stability, large surface area, tunnel pore size and well defined surface properties made them ideal for hosting molecules of various size, shape and functionalities. The ability of these MSNs to penetrate cell wall opens up new ways to manipulate gene expression at isolated plant cell system.

It is reported that a honeycomb like MSNs system with 3 nm pores can transport DNA and chemicals into isolated plant cells and intact leaves [48]. In MSNs system, gene and chemical inducers are loaded in the meso-pores and encapsulated with covalently bound caps. These bound caps physically block the drug from leaching out and molecules entrapped in the pores are released by introduction of uncapping triggers such as dithio-threitol (DTT). Capping by surface-functionalized gold nano-particles (10–15 nm size) serves as a biocompatible capping agent and prevents the gene from leaching out. The advancement in MSN system can simultane-ously deliver both the gene and the chemical that trigger gene expression. It has been reported that MSN system can deliver DNA molecules carrying a marker gene and a chemical that is needed for transgene expression into plant simultaneously and release the encapsulated chemical in a controlled manner to trigger the expression of co-delivered transgene into the cell.

Further developments such as pore enlargement and multi-function-alization of MSN systems may offer new possibilities in target-specific delivery of proteins, nucleotides and chemicals in plant biotechnology.

5.4 CONCLUSIONS AND FUTURE OPPORTUNITIES

Despite contribution of highest gross domestic product to the world, agriculture is still in susceptible condition of meeting the demand of food supply of ever increasing population. There is also an urge need of environmental sustainability with food security without harming the soil health, quality and ecosystem. Anthropogenic as well as human and indus-trial activities have threatened the natural environment with huge input of toxic chemicals, xenobiotic and other non-biodegradable products. Also agricultural production is continuously being challenged by both biotic and abiotic losses which counts to be 40% of the total loss, still the food supply to the demand is at easy pace. Hence, use of excess agrochemicals for increasing the productivity leads to deterioration of soil quality, flora and fauna as well as incidence of higher herbicide and pesticide resistance in weeds and insects-pathogens. Therefore, there is an alarming need of ecosystem homeostasis and natural restoration through *"environmental 3 rule* (which includes *Reduce, Reuse and Recycle*), where deployment of eco-friendly technology and systems is most required. In this context, use of nano-technology has emerged as a basic tool for technological advancement mitigating the environmental claims, management of pest-pathogen stresses, detection of diseases, nutrient status in agricultural crops, enhancing crop yields, nano-based delivery systems for increased fertilization, water soluble pesticide, plant growth regulators, etc. Some of the most promising achievements incurred in the field of nano-technology: Use of nano-silica component for preventing soil run-off, Zn-Al nano-materials for controlled release of herbicides, Ti-based nano-tube for detection of diseases in vegetables and photo-activation based biodegradation of xeno-biotics, nano-emulsion for plant growth promotion and stress alleviation, poly-ethoxy glycol (PEG) based nano-material embedded with garlic oil for control of pest incidence, gold nano-particle for pesticide detection, carbon nano-tube for smart delivery of agro-chemicals.

Recently the essence of nano-technology has been tested in biotechnology involving use of nano-gold/titanium and most interestingly chitosan based nano-delivery system for plant transformation and genetic engineering for crop biotechnology has been a subject of immense interest among agricultural researcher that have completely replaced the traditional delivery system of genetic material to plant tissues. Use of nano-technology based electronic devices like sensors are also being used for the purpose of environmental monitoring for detection of contaminants and with due advancements, nano-technology has become an integrated part of agriculture. With the list of nano-based products as a tool for agricultural advancement, the mere on site (*in-situ*) application is still in juvenile stage. Exploration of nano-materials for delivery of metabolite of interest throughout the plant vascular system with target delivery approach is still a question and a possible solution can bring a revolution in the alleviation of stress in crops and their physiological disorders, ultimately helping to make agriculture sustainable. There is an ongoing research effort in formulating low cost, efficient, low dose, selective nano-materials and it's scaling up process for its development in agricultural production. Apart from its use, there is need of a better understanding of its post use effect

Nano-Particle Based Delivery Systems

and known to be a major hurdle in the research of nano-technology. With eyeing towards sustainable development, uses of nano-based materials have a promising future as well as of great value for making sustainable agriculture and eco-friendly environment.

KEYWORDS

- **bio-degradable**
- **encapsulation**
- **liposomes**
- **nano-capsule**
- **nano-delivery**
- **nano-fertilizer**
- **nano-fiber**
- **nano-materials**
- **nano-particles**
- **nano-pesticide**
- **nano-scale**
- **nano-size**
- **nano-technology**

REFERENCES

1. Afzali, A., & Maghsoodlou, S. (2015). Engineered Nanoporous Materials: A Comprehensive Review. In: E. M., Howell, B. A. Pethrick, R. A., & Zaikov, G. E. (Eds.). *Physical Chemistry Research for Engineering and Applied Sciences by Pearce, Volume-1, Principles and Technological Implications*. Apple Academic Press Inc., pp. 318–350.
2. Alemdar, A., & Sain, M. (2008). Isolation and characterization of nanofibers from agricultural residues – Wheat straw and soy hulls. *Bioresour. Technol., 99*, 1664–1671.
3. Amnon, C., & Haim, V. (2007). A microemultion based delivery system for the dermal delivery of therapeutics. *Inno. Pharmace. Tech., 23*, 68–72.
4. Bakar, N. A., Salleh, M. M., Umar, A. A., & Yahaya, M. (2011). The detection of pesticides in water using ZnCdSe quantum dot films. *Adv. Nat. Sci. Nanosci. Nanotech., 2*, 233–238.

5. Barik, T. K., Sahu, B., & Swain, V. (2008). Nanosilica-from medicine to pest control. *Parasitol. Res., 103,* 253–258.
6. Bei, D., Meng, J., & Youan, B. C. (2010). Engineering nanomedicines for improved melanoma therapy: progress and promises. *Nanomedicine (Lond.), 5*(9), 1385–1399.
7. Bhagat, D., Samanta, S. K., & Bhattacharya, S. (2013). Efficient management of fruit pests by pheromone nanogels. *Sci. Rep., 3,* 1294.
8. Bonifácio, B. V, Silva, P. V., Ramos, M. A. S., Negri, K. M. S., Bauab, T. M., & Chorilli, M. (2014). Nanotechnology-based drug delivery systems and herbal medicines: a review. *Int. J. Nanomed., 9,* 1–15.
9. Chinnamuthu, C. R , & Kokiladevi, E. (2007). Weed management through nano-herbicides. In: Chinnamuthu, C. R., Chandrasekaran, B., & Ramasamy, C. (Eds.). *Application of Nanotechnology in Agriculture,* Tamil Nadu Agricultural University, Coimbatore, India.
10. Chorilli, M., Leonardi, G. R., Oliveira, A. G., & Scarpa, M. V. (2004). Lipossomas em formulações dermocosméticas [Dermocosmetic liposome formulations]. *Infarma., 16*(7–8), 75–79.
11. Chorilli, M., Rimério, T. C., Oliveira, A. G., & Scarpa, M. V. (2007). *Estudo da estabilidade de lipossomas unilamelares pequenos contendo cafeína por turbidimetria* [Study of the stability of small unilamellar liposomes containing caffeine turbidimetric]. *Rev. Bra.s Farm., 88*(4), 194–199.
12. De-Mesquita, J. P., Donnici, C. L., Teixeira, I. F., & Pereira, F. V. (2012). Bio-based nanocomposites obtained through covalent linkage between chitosan and cellulose nanocrystals. *Carbohydr. Polym., 90*(1), 210–217.
13. Des Rieux, A., Fievez, V., Garinot, M., Schneider, Y. J., & Preat, V. (2006). Nanoparticles as potential oral delivery systems of proteins and vaccines: a mechanistic approach. *J. Control. Release., 116*(1), 1–27.
14. Dubey, V., Mishra, D., & Jain, A. A. (2006). Transdermal Delivery of a Pineal Hormone: Melatonin Via Elastic Liposomes. *Biomater., 27,* 3491–3496.
15. Flauzino Neto, W. P., Silverio, H. A., Dantas, N. O., & Pasquino, D. (2013). Extraction and characterization of cellulose nanocrystals from agro-industrial residue soy hulls. *Ind. Crops Prod., 42,* 480–488.
16. Florence, A. T. (2005). Nanoparticle uptake by the oral route: Fulfilling its potential? *Drug Discov. Today: Technol., 2*(1), 75–81.
17. Ghormade, V., Deshpande, M. V., & Paknikar, K. M. (2011). Perspectives for nanobiotechnology enabled protection and nutrition of plants. *Biotechnol. Adv., 29,* 792–803.
18. Gonzalez-Melendi, P., Fernandez Pacheko, R., Coronado, M. J., Corredor, E., Testilano, P. S., Risueno, M. C., Marquina, C., Ibarra, M. R., Rubiales, D., & Perez-De-Luque, A. (2008). Nanoparticles as smart treatment-delivery systems in plants: assessment of different techniques of microscopy for their visualization in plant tissues. *Ann. Bot., 101,* 187–195.
19. Grilloa, R., Pereira, A. E. S., Nishisaka, C. S., de, L. R., Oehlke, K., Greiner, R., & Fraceto, L. F. J. (2014). Chitosan nanoparticle based delivery systems for sustainable agriculture. *Hazard. Mater., 278,* 163–171.
20. Guzey, D., & McClements, D. J. (2006). Formation, stability and properties of multilayer emulsions for application in the food industry. *Adv. Colloid Interface Sci., 128–130,* 227–248.

21. Hunter, A. C, Elsom, J., Wibroe, P. P., & Moghimi, S. M. (2012). Polymeric particulate technologies for oral drug delivery and targeting: a pathophysiological perspective. *Nanomed., 8*(Suppl. 1), 5–20.
22. Hussein, M. Z., Yahaya, A. H., Zainal, Z., & Kian, L. H. (2005). Nanocomposite-based controlled release formulation of an herbicide, 2,4-dichlorophenoxyacetate in capsulated in zinc-aluminum-layered double hydroxide. *Sci. Technol. Adv. Mater, 6,* 956–962.
23. Kombath, R. V., Minumula, S. K., Sockalingam, A., Subadhra, S., Parre, S., Reddy, T. R., & David, B. (2012). Critical issues related to transfersomes – novel Vesicular system. *Acta Sci. Pol., Technol. Aliment., 11*(1), 67–82.
24. Kundu, S., Huitink, D., & Liang, H. (2010). Formation and catalytic application of electrically-conductive Pt nanowires. *J Phys Chem C., 114*(17), 7700–7709.
25. Lisa, M., Chouhan, R. S., Vinayaka, A. C., Manonmani, H. K., & Thakur, M. S. (2009). Gold nanoparticles based dipstick immunoassay for the rapid detection of dichlorodiphenyltrichloroethane: an organochlorine pesticide. *Biosens. Bioelectron., 25,* 224–227.
26. Liu, F., Wen, L. X., Li, Z. Z., Yu, W., Sun, H. Y., & Chen, J. F. (2006b). Porous hollow silica nanoparticles as controlled delivery system for water soluble pesticide. *Mat. Res. Bull., 41,* 2268–75.
27. Liu, J., Wang, F., Wang, L., Xiao, S., Tong, C., Tang, D., Liu, X., & Cent, J. (2008). Preparation of fluorescence starch-nanoparticle and its application as plant transgenic vehicle. *South Univ. Technol., 15,* 768–773.
28. Liu, X., Feng, Z., Zhang, S., Zhang, J., Xia, Q., & Wang, Y. (2006a). Preparation and testing of cementing nano-subnano composites of slow or controlled release of fertilizers. *Sci. Agr. Sin., 39,* 1598–1604.
29. Madrigal-Carballo, S., Lim, S., Rodriguez, G., Vila, A. O., & Krueger, C. G. (2010). Biopolymer coating of soybean lecithin liposomes via layer-by-layer self-assembly as novel delivery system for ellagic acid. *J. Funct. Foods, 2*(2), 99–106.
30. Marsanasco, M., Marquez, A. L., Wagner, J. L., Alonso, S., & Chiaromoni, N. S. (2011). Liposomes as vehicles for vitamins E and C: an alternative to fortify orange juice and offer vitamin C protection after heat treatment. *Food Res. Int., 44*(9), 3039–3046.
31. McClements, D. J., & Li, Y. (2010). Structured emulsion-based delivery systems: controlling the digestion and release of lipophilic food components. *Adv. Colloid Interface Sci., 159*(2), 213–228.
32. McMurray, T. A., Dunlop, P. S. M., & Byrne, J. A. (2006). The photocatalytic degradation of atrazine on nanoparticulate TiO_2 films. *J. Photochem. Photobiol A: Chem., 182,* 43–51.
33. Mehnert, W., & Mader, K. (2012). Solid lipid nanoparticles. *Adv. Drug Deliv. Rev., 64,* 83–101.
34. Milani, N., McLaughlin, M. J., Stacey, S. P., Kirby, J. K., Hettiarachchi, G. M., Beak, D. G., & Cornelis, G. (2012). Fate of Zinc Oxide Nanoparticles Coated onto Macronutrient Fertilizers in an Alkaline Calcareous Soil. *J. Agric. Food Chem., 25,* 3991–3998.
35. Mozafari, M. R., & Khosravi-Darani K. (2007). An Overview of Liposome Derived Nanocarrier Technologies. In: Mozafari, M. R. (Ed.). *Nanomaterials and Nanosystems for Biomedical Applications,* Springer. pp. 113–123.

36. No, H. K. (2007). Applications of chitosan for improvement of quality and shelf life of foods: a review. *J. Food Sci., 72*(5), 87–100.
37. Plapied, L., Duhem, N., des Rieux, A., & Preat, V. (2011). Fate of polymeric nanocarriers for oral drug delivery. *Curr. Opin. Colloid Interface Sci., 16*(3), 228–237.
38. Puri, A., Loomis, K., Smith, B., Lee, J. H., Yavlovich, A., Heldman, E., & Blumenthal, R. (2009). Lipid-based nanoparticles as pharmaceutical drug carriers: from concepts to clinic. *Crit. Rev. Ther. Drug Carrier Syst., 26*(6), 523–580.
39. Sabliov, C. M., & Astete, C. E. (2008). Encapsulation and controlled release of antioxidants and vitamins. In: Garti, N (Ed.). *Delivery and Controlled Release of Bioactives in Foods and Nutraceuticals,* Cambridge: Woodhead Publ., pp. 297–330.
40. Saharan, V. (2010). Effect of gibberellic acid combined with saponin on shoot elongation of *Asparagus officinalis. Biologia. Plant, 54*(4), 740–742.
41. Saigusa, M. (2000). Broadcast application versus band application of polyolefin-coated fertilizer on green peppers grown on andisol. *J. Plant Nutr., 23*, 1485–1493.
42. Shutava, T. G., Balkundi, S. S., & Lvov, Y. M. (2009). Epigallocatechin gallate/gelatin layer-by-layer assembled films and microcapsules. *J. Colloid Interface Sci., 330*(2), 276–283.
43. Singh, S., Singh, M., Agrawal, V. V., & Kumar, A. (2010). An attempt to develop surface plasmon resonance based immunosensor for Karnal bunt (*Tilletia indica*) diagnosis based on the experience of nano-gold based lateral flow immuno-dipstick test. *Thin Solid Films, 519*, 1156–1159.
44. Sun, B., & Yeo, Y. (2012). Nanocrystals for the parenteral delivery of poorly water-soluble drugs. *Curr. Opin. Solid State Mater. Sci., 16*(6), 295–301.
45. Tamjidi, F., Shahedi, M., Varshosaz, J., & Nasirpour, A. (2013). Nanostructured lipid carriers (NLC): A potential delivery system for bioactive food molecules. *Inno. Food Sci. Emerg. Technol., 19*, 29–43.
46. Tarafdar, J. C., Sharma, S., & Raliya, R. (2013). Nanotechnology: Interdisciplinary science of applications. *Afr. J. Biotech., 12*(3), 219–226.
47. Taylor, N. J., & Fauquet, C. M. (2002). Microparticle bombardment as a tool in plant science and agricultural biotechnology. *DNA Cell Biol., 21*, 963–977.
48. Torney, F., Trewyn, B. G., Lin, V. S., & Wang, K. (2007). Mesoporous silica nanoparticles deliver DNA and chemicals into plants. *Nat. Nanotechnol., 2*, 295–300.
49. Tripathi, P. K., Choudary, S. K., Srivastva, A., Singh, D. P., & Chandra, V. (2012). Niosomes: an study on novel drug delivery system-A review. *Inter. J. Pharmac. Res. Develop., 3*, 100–106
50. Tzoumaki, M. V., Moschakis, T., & Biliaderas, C. G. (2011). Mixed aqueous chitin nanocrystal–whey protein dispersions: microstructure and rheological behavior. *Food Hydrocoll., 25*(5), 935–942.
51. Vamvakaki, V., & Chaniotakis, N. A. (2007). Pesticide detection with a liposome-based nano-biosensor. *Biosens. Bioelectron., 22*, 2848–2853.
52. Wu, Y., Luo, Y., & Wang, Q. (2012). Antioxidant and antimicrobial properties of essential oils encapsulated in zein nanoparticles prepared by liquid–liquid dispersion method. *Food Sci. Technol., 48*(2), 283–290.
53. Yao, K. S., Li, S. J., Tzeng, K. C., Cheng, T. C., Chang, C. Y., Chiu, C. Y., Liao, C. Y., Hsu, J. J., & Lin, Z. P. (2009). Fluorescence Silica Nanoprobe as a Biomarker for Rapid Detection of Plant pathogens. *Adv. Mater. Res., 79*, 513–516.

54. Zambrano-Zaragoza, M. L., Mercado-Silva, E., Ramirez-Zamorano, P., Cornejo-Villegas, M. A., Gutierrez-Cortez, E., & Quintanar-Guerrero, D. (2013). Use of solid lipid nanoparticles (SLNs) in edible coatings to increase guava (*Psidium guajava L.*) shelf-life. *Food Res. Int., 51*(2), 946–953.
55. Zhang, F., Wang, R., Xiao, Q., Wang, Y., & Zhang, J. (2006). Effects of slow/controlled-release fertilizer cemented and coated by nano-materials on biology. II: Effects of slow/controlled-release fertilizer cemented and coated by nano-materials on plants. *Nanosci., 11*, 18–26.

CHAPTER 6

GREEN SYNTHESIS OF SILVER NANOPARTICLES FROM ENDOPHYTIC FUNGUS ASPERGILLUS NIGER

POONAM RANI and VEDPRIYA ARYA

CONTENTS

6.1 Introduction 131
6.2 Synthesis of Nanoparticles 132
6.3 Material and Methods 135
6.4 Results and Discussion 139
6.5 Summary 141
Keywords 142
References 142

6.1 INTRODUCTION

Nanotechnology is emerging as a rapidly growing field with its application in Science and Technology to manufacture new materials at the nanoscale level [10, 31]. Biotechnology is an integration between biotechnology and nanotechnology for developing biosynthetic and environmental-friendly technology for synthesis of nanomaterials. Nanoparticles are clusters of atoms in the size range of 1–100 nm. "Nano" is a Greek word synonymous to dwarf meaning extremely small. The use of nanoparticles is gaining impetus in the present century as they possess defined chemical, optical, and mechanical properties. The metallic nanoparticles as they show good antibacterial properties due to their large surface area to volume ratio. Due to the growing microbial resistance against metal ions, antibiotics, and the development of resistance strains, other sources of antimicrobials are demanded [12]. Different types of nanomaterials like copper, zinc, titanium, magnesium, gold [13] alginate and silver have come up but silver nanoparticles have proved to be most effective because of antimicrobial efficacy against bacteria, viruses and other eukaryotic micro-organisms [12].

Silver nanoparticles are used as drug disinfectant and have some risks as the exposure to silver can cause agyrosis and argyria also; and are toxic to mammalian cells [12, 13]. Use of silver ion or metallic silver as well as silver nanoparticles can be exploited in medicine for burn treatment, dental materials, textile fabric, water treatment, sunscreen lotions, etc. and possess low toxicity to human cells, high thermal stability and low volatility [8].

In this chapter, endophytic fungus *Aspergillus niger* was isolated from leaves of *Ficus panda* for the synthesis of silver nanoparticles and to study the antibacterial activity against clinical human pathogenic microbial strains.

6.2 SYNTHESIS OF NANOPARTICLES

There are physical, chemical, and biological methods to synthesize silver nanoparticles. The physical and chemical methods are numerous, and many of these methods are expensive or use toxic substances which are major factors that make them 'not so favored' methods of synthesis. There are various physical and chemical methods, of which the simplest method involves the chemical method of reduction of the metal salt $AgBF_4$ by $NaBH_4$ in water. The obtained nanoparticles with the size range of 3 to 40 nm are characterized by transmission electron microscopy (TEM) and UV-visible (UV-vis) absorption spectroscopy to evaluate their quality [33]. An alternate method to synthesize silver nanoparticles is to employ biological methods of using microbes and plants. There are also many more techniques of synthesizing silver nanoparticles:

Green Synthesis of Silver Nanoparticles

- electrochemical method;
- microwave synthesis of silver nanoparticles;
- thermal decomposition in organic solvents [9];
- chemical and photo reduction in reverse micelles [22];
- spark discharge [34];
- cryochemical synthesis;
- Sonodecomposition to yield silver nanoparticles.

Physical method of silver nanoparticles synthesis includes different techniques:

- lithography;
- ultraviolet irradiation;
- aerosol technology, laser ablation; and
- photochemical reduction of metals.

The problem with most of the chemical and physical methods of nano-silver production is that they are extremely expensive and also involve the use of toxic, hazardous chemicals, which may pose potential environmental and biological risks. Therefore among these, biological method is cheap and eco-friendly. There are three major sources of synthesizing silver nanoparticles: bacteria, fungi, and plant extracts.

The first evidence of bacteria synthesizing silver nanoparticles was established using the *Pseudomonas stutzeri* strain that was isolated from silver mine [15]. Silver nanoparticles are also synthesized from *Bacillus licheniformis* [17] and *Proteus mirabilis* [24]; fungi *Fusarium oxysporum* [7], *Aspergillus flavus* [35] *and Penicillium* sp. [28]; and from plants *Medicago sativa* [11] *Azadirachta indica* [27], *Aloe vera* [5], *Desmodium trifolium, Coriandrum sativum* [25], and *Aerva lanata*. The major advantage of using plant extract for silver nanoparticles synthesis is that they are easily available, safe and non toxic in most cases; and have a broad variety of metabolite that can aid in the reduction of silver ions and are quicker than microbes in the synthesis.

6.2.1 ENDOPHYTIC FUNGUS

From thousands of years, mankind has used natural product, chemicals produced by plants, fungi, bacteria and other living organism in a variety of application: drugs, food and hallucinogens. Now a day's large number of medicines is also prepared from fungi, plants and bacteria. But from all of these, fungi play an important role for formation of useful drugs which are used for curing number of diseases. Fungi are eukaryotic organisms that differ from bacteria and other prokaryotes in many ways. The simplest type of fungi is unicellular yeast. Depending upon the morphology, the fungi are divided into many categories. But from all these, endophytic fungi for preparation of useful product for mankind has shown scope.

The term 'endophyte' includes a family of microorganisms that grows intra and intercellular in the tissues of higher plants without causing any symptoms on the plants in which they live. These microorganisms may produce a large number of novel natural products for medical, agriculture and industrial uses such as antibiotics, anticancer reagents, biological control agents and other useful bioactive compounds [20, 30]. The endophytes may provide protection and survival condition to their host plant by producing a plethora of substances which once isolated and characterized, may also have potential for use in industry, agriculture and medicine [29]. Studies have shown that the endophytes are not host specific and a single endophyte can survive a wide range of hosts. A large number of fungi have been isolated from the different parts of the same plant which differ in their ability to utilize different substances [4]. Therefore, a number of fungi can be isolated from different plant belonging to different genera and grow under different climatic conditions. The host and endophytes relationship varies from host to host and endophytes. Studies indicate that host plant and endophyte relationship is able to maintain the pathogen host antagonism [26]. Plant endophytic fungi have been found in each plant species, and it is estimated that there are over one million fungal endophytes in the nature [21]. Drefyuss and Chapala [6] have estimated that there are 1.3 millions species of endophytic fungi alone, the majority of which are likely to be found in tropical ecosystem. The greatest fungal diversity probably occurs in tropical forest, where highly diverse populations of angiosperms exist [2]. Endophytic relationship may have begun from the time, when higher plants first appeared (millions of years ago).

Evidences for plant-associated fungi have been discovered in fossilized tissues of stem and leaves [32]. As a consequence of these long term association, some of these microorganisms may have developed genetic stem that allow the exchange of information between themselves and the higher

Green Synthesis of Silver Nanoparticles

plant. This exchange would allow the fungi to cope with the environmental condition more efficiently and perhaps increase compatibility with the plant host. Moreover, the dependant evolution of endophytic fungi may have allowed them a better adaptation with the plant so that the fungi could contribute in the relationship by performing protective function against pathogen and insect [14, 29]. To make these contributions to their plant host, endophytic fungi produce bioactive secondary metabolites [29].

According to the plantendophyte co-evolution hypothesis [16], It might be possible for endophytes to assist the plant in chemical defense by producing bioactive secondary metabolites [4]. Most endophytic fungi were first studied in plants from temperate climates. These fungi have also recently been isolated from many tropical hosts. A better understanding of fungal diversity may prove crucial in fungi utility and cultivation. The development of a high-throughput, dilution-to-extinction approach has been developed for fungi that significantly increased the assessment of the diversity of cultured fungi.

6.3 MATERIAL AND METHODS

6.3.1 PLANT MATERIALS

The *Ficus panda* leaf samples were collected from the botanical garden of Guru Nanak Girls College, Ludhiana, India. The *Ficus panda* is from the family Moraceae and needs a lot of light to have a vigorous growth.

6.3.2 ISOLATION AND IDENTIFICATION OF ENDOPHYTIC FUNGUS

The collected samples were thoroughly washed under running tap water for 5–8 min (4–5 times). The materials were surface sterilized by immersing sequentially in 70% ethanol for 3 min, 0.5% sodium hypochlorite for 1 min and again in 70% ethanol for 30 sec. Finally, samples were rinsed three times in sterile distilled water. The excess water was dried in laminar air flow chamber. Then with a sterile scalpel, the leaves of 0.5 cm size were carefully dissected and placed on Petri plates containing Potato Dextrose

Agar. The media was supplemented with streptomycin sulfate (100 mg/l) to suppress the bacterial growth. The plates were then incubated at 28°C until endophytic fungi emerged. Based on literature and other morpho-taxonomic features, selected isolate was identified as *Aspergillus niger*.

Aspergillus niger, isolated as endophyte from *Ficus panda plant*, was used in the study as a reducing agent for the aqueous silver ion solution. The fungal biomass used for biosynthetic experiment was grown aerobically in liquid medium containing (g/l): KH_2PO_4 7.0 g, K_2HPO_4 2.0 g, $MgSO_4.7H_2O$ 0.1 g, $(NH_4)_2SO_4$ 1.0 g and yeast extract 0.6 g and glucose 10.0 g. The flasks were inoculated with spores and incubated at 28°C on a rotary shaker (150 rpm) for 72 hours. After incubation, the biomass was filtered using Whatman filter paper no. 1 and then extensively washed with sterile distilled water to remove any medium component. The fungal biomass was used for preparation of extract [19].

6.3.3 PREPARATION OF FUNGAL EXTRACT OF ENDOPHYTIC FUNGUS

The 21 g (wet weight) of fungal biomass was diluted 10 times in double distilled water and were given hot percolation treatment. In hot percolation treatment, fungal biomass is diluted in double distilled water and heated at 40°C for 2–3 hours till resultant mixture boils completely and then was kept undisturbed for 10 minutes. The resultant mixture was then filtered out using Whatman filter paper no. 1 in conical flask and then filtrate was kept in water bath at 60°C till reduced volume of filtrate was obtained. This filtrate so obtained from hot percolation treatment was used as raw extract for synthesis of silver nanoparticles.

6.3.4 SYNTHESIS OF SILVER NANOPARTICLES (AGNPS) OF BY AQUEOUS EXTRACT OF ENDOPHYTIC FUNGUS

6.3.4.1 Preparation of 1 M $AgNO_3$ Stock Solution

For each experimental set, fresh stock of $AgNO_3$ solution was prepared. 1 mM $AgNO_3$ (MW 169.88) 0.169 gm was dissolved in 1000 ml double distilled water resulting in 1000 ml $AgNO_3$ stock solution.

Green Synthesis of Silver Nanoparticles

6.3.4.2 AgNO$_3$ Treatment

Experimental set up was prepared by adding 5 ml of raw extract with 25 ml AgNO$_3$ in each flask and were analyzed under different reaction conditions for silver nanoparticles synthesis. Different reaction factor analysis silver nanoparticles were:

- **pH:** The pH of the AgNO$_3$ and extract (5 ml) solution was varied [9–11] to analyze the exact pH at which maximum nanoparticles are observed. The incubation at 37° was done for one hour and results were recorded by UV-Vis spectrometer by taking absorbance at 420 nm.
- **Temperature:** The temperature of incubation of AgNO$_3$ and extract was varied to know the temperature at which maximum nanoparticles observed. The temperature range (0°, 37°, and 100°C) and extract concentration was 5 ml constant for reaction and the result were recorded after one hour on UV-Vis spectrophotometer at 420 nm.

6.3.5 TEM ANALYSIS

The samples were subjected to TEM analysis in the SAIF, Punjab University, Chandigarh. This procedure was performed to confirm the size range of silver nanoparticles.

6.3.5.1 Antibacterial Testing

Bacterial strains: The pathogenic bacterial strains were *Escherichia coli, Enterococcus* sp., *Micrococcus* and *Vibrio cholera*. These stains were brought into laboratory in peptone water and maintained continuously by sub-culturing in nutrient agar.

Inoculum preparation: Isolated colonies were inoculated into 4 ml peptone water and incubated at 37°C for 2 hours. The turbidity of actively growing bacterial suspension was adjusted to match the turbidity standard of 0.5 McFarland units prepared by mixing 0.5 ml of 1.75% (w/v) barium chloride dehydrate with 99.5 ml 1% (v/v) sulfuric acid. This turbidity was equivalent to approximately $1\text{-}2 \times 10^8$ colony forming units per mm (cfu/ml). The 2 hours grown suspension was used for further testing.

Antibacterial bioassay: The antimicrobial activities of the extracts were determined by Agar well diffusion method according to NCCLS standards. Muller Hinton Agar (MHA) (HiMedia, Mumbai) was used for the antibacterial activity test. Under aseptic conditions in the Bio-safety chamber, 15 ml of MHA medium was dispensed into pre-sterilized petri dishes to yield a uniform depth of 4 mm and inoculated by the bacterial culture, respectively. The wells were prepared in MHA plates with help of gel punching machine. Streptomycin sulfate was used as positive control and water as negative control. The wells were marked as 1, 2, 3, 4 and blank:

1. 40 µl of conc. of extract
2. 60 µl of conc. of extract
3. 80 µl of conc. of extract
4. 100 µl of conc. of extract

6.3.5.2   Positive Control (Streptomycin 10 µg/ml)

The wells were spaced far enough to avoid reflections wave from the edges of the petri dishes and overlapping rings of inhibition. Finally, the petri dishes were incubated for 18–24 hours at 37°C for bacteria. The diameter of zone of inhibition as indicated by clear area which was devoid of growth of microbes was measured.

6.3.6   *UV-VIS SPECTROPHOTOMETRIC ANALYSIS*

UV-VIS spectrophotometric analysis is one of the simple and mostly used method to characterize the presence of nanoparticles in extracts based on change in color. Extracts were prepared from endophytic fungus by hot percolation. After giving hot percolation treatment, the prepared extracts were augmented with $AgNO_3$ treatment and then were kept on various conditions such as temperature (0°, 37°, 100°C) and pH [9, 10, and 11]. After giving various variation treatments, change in color from orange to dark brownish red were observed. This color showed the presence of silver nanoparticles or reduction of Ag^+ of $AgNO_3$ to Ag^0 (silver nanoparticles). After observing changes in color of the extracts, they were scanned from 420–680 nm spectrophotometric analysis keeping raw extract (without silver nitrate treatment) as

Green Synthesis of Silver Nanoparticles

FIGURE 6.1 Silver nanoparticles resulted change in color in endophytic fungal extract augmented with $AgNO_3$ after 1 hour incubation at 0°, 37°, 100°C.

blank and maximum absorbance at 420 nm due to Surface Plasmon Resonance (SPR) of silver nanoparticles (Figure 6.1).

6.3.7 STATISTICAL ANALYSIS

Statistical evaluation for UV-Vis analysis: statistical evaluation will be done by Graph Pad Prism 5.0 software.

6.4 RESULTS AND DISCUSSION

6.4.1 EFFECTS OF TEMPERATURE

It was interpreted that fungal extract incubated at 100°C temperature have more number of silver nanoparticles. It was observed that increase in temperature from 0° to 100°C resulted in increase in number of silver nanoparticles due to their property of Surface Plasmon Resonance (SPR) (Table 6.1). In accordance with the above reported results, Kasture et al. [18] used temperature variation from 30° to 90°C and observed that the temperature increase surface Plasmon peak began to evolve and at 90°C a well-defined

TABLE 6.1 Effects of Temperature Variations on Absorbance of AgNO$_3$ Treated Endophytic Fungal Extract (at 420 nm)

Temperature	Absorbance at 420 nm
0° C	0.11
37° C	0.10
100° C	0.12

and sharp peak was observed. In contrast to these observations, Safekordi et al. [23] reported that at temperature 5°C, more number of nanoparticles were produced than at 90°C, because with increase in temperature, size of particles became larger and their dispersion became wide.

6.4.2 EFFECTS OF pH VARIATIONS

It was interpreted that fungal extract incubated at pH 11 produced more number of silver nanoparticles. It was observed that with increase in pH, there was change in color and absorbance was increased due to SPR (Figure 6.2). In accordance with these results, Amaladhas et al. [1] and

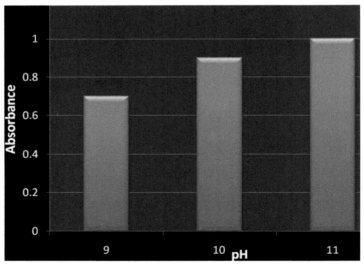

FIGURE 6.2 Effects of pH variations on absorbance (at 420 nm) observed in extract from endophytic fungal.

Green Synthesis of Silver Nanoparticles

many others have shown that alkaline pH solutions provide stability in synthesis of nanoparticles as compared with acidic pH solutions. Silver nanoparticles show high absorbance and narrow peak at high pH due to the formation of monodispersed and smaller sized nanoparticles.

6.4.3 TEM ANALYSIS OF SILVER NANOPARTICLES

The results showed 10–15 nm of silver nanoparticles under transmission electron microscope (TEM). The maximum number of nanoparticles was produced at pH of 11 and temperature of 100°C.

6.4.4 ANTIBACTERIAL ACTIVITY OF SYNTHESIZED SILVER NANOPARTICLES

Silver nanoparticles produced from endophytic fungal extract were assayed for their potential antibacterial activity against clinically isolated pathogens. These nanoparticles showed significant antibacterial activity against *Enterococcus sp., E. coli, Staphylococcus albus* and *Klebsiella pneumonia* (Table 6.2).

6.5 SUMMARY

This study paves the new way of synthesis of silver nanoparticles via biological route. It is an environment friendly and cheaper method as compared to traditional chemical synthesis approaches. *Ficus panda* is considered as an important medicinal plant. The endophytic fungus isolated from

TABLE 6.2 The Antibacterial Activity of Silver Nanoparticles Synthesized at pH 11 and Temperature 100°C

Bacterial Strain	Zone of inhibition
Escherichia coli	17.5±0.33
Staphylococcus albus	21.09±0.89
Enterococcus sp.	11.04±0.54
Klebsiella pneumoniae	12.11±1.21

this plant was used in this study for the synthesis of silver nanoparticles. The silver nanoparticles are reported to possess great antibacterial action. The other pharmacological activities from these nanoparticles are subject to under investigation by the authors.

KEYWORDS

- **nanoparticles**
- **antimicrobial**
- **silver nitrate**
- **green synthesis**
- ***aspergillus niger***
- ***ficus panda***

REFERENCES

1. Amaladhas, T. P., Sivagami, S., Devi, T. A., Ananthi, N., & Velammal, S. P. (2012). Nanoscience and Nanotechnology. *Adv. Nat. Sci., 3*, 1–13.
2. Arnold, A., Maynard, Z., Gilbert, G., & Kursar, T. (2000). *Are tropical Endophytes hyperdrivesde? Ecology letter, 3*(4), 267–274.
3. Carroll, G. C. (1988). Fungal endophytes in stems and leaves: from latent pathogen to mutualistic Symbiont. *Ecology, 69*, 2–9.
4. Carroll, G. C. (1991). Beyond pest deterrence alternative strategies and hidden costs of endophytic mutualisms in vascular plants. In: Andrews, J. A., & Hirano, S. S. (eds.), *Microbial Ecology of Leaves,* New York: Springer-Verlag, pp. 358–375.
5. Chandran, S. P., Chaudhary, M., Pasricha, R., Ahmad, A., & Sastry, M. (2006). Nano-triangles and silver nanoparticles using Aloe vera plant extract. *Biotechnology Program, 22*, 577–583.
6. Dreyfuss, M., & Chapela, I. (1994). Potential of fungi in the discovery of low molecular weight pharmaceutical. In: Gullo V. P. (ed.), *The Discovery of Natural Products with Therapeutic Potential, 26*, 49–80.
7. Duran, N., Marcato, P. D., Alves, O., & Souza, G. (2005). Mechanistic aspects of biosynthesis of silver nanoparticles by several *Fusarium oxysporum* strains. *J Nanotechnol, 3*, 8.
8. Duran, N., Marcato, P. D., De'Souza, G. I. H., Alves, O. L., & Esposito, E. (2007). Antibacterial effect of silver nanoparticles produced by fungal process on textile fabrics and their effluent treatment. *J Biomed Nanotechnol., 3*, 203–208.

Green Synthesis of Silver Nanoparticles

9. Esumi, K., Tano, T., Torigue, K., & Meguro, K. (1990). Preparation and characterization of bimetallic Pd-Cu colloids by thermal decomposition of their acetate compounds in organic solvents. *Chem. Mater.*, *2*, 564–566.
10. Feynman, R. (1959). Lecture at the California Institute of Technology.
11. Gardea-Torresdey, J. L., Gomez, E., Peralta-Videa, J. R., Parsons, J. G., Troiani, H., & Jose-Yacaman, M. (2003). Alfalfa sprouts: a natural source for the synthesis of silver nanoparticles. *Langmuir*, *19*, 1357–1361.
12. Gong, P., Li, H., He, X., Wang, K., Hu, J., & Tan, W. (2007). Preparation and antibacterial activity of Fe304 Ag nanoparticles. *Nanotechnology*, *18*, 604–611.
13. Gu, H., Ho, P. L., Tong, E., Wang, L., & Xu, B. (2003). Presenting vancomycin on nanoparticles to enhance antimicrobial activities. *Nano Letter*, *3*(9), 1261–1263.
14. Gunatilaka, A. A. L. (2005). Natural products from plant-associated microorganisms: distribution, structural diversity, bioactivity, and implications of their occurrence. *J Nat Prod*, *69*, 509–526.
15. Haefeli, C., Franklin, C., & Hardy, K. (1984). Plasmid-determined silver resistance in Pseudomonas stutzeri isolated from silver mine. *J Bacteriol*, *158*, 389–392.
16. Ji, H. F., Li, X. J., & Zhang, H. Y. (2009). Natural products and drug discovery: Can thousands of years of ancient medical knowledge lead us to new and powerful drug combinations in the fight against cancer and dementia? *EMBO Rep*, *10*, 194–200.
17. Kalimuthu, K, Babu, R. S., Venkataraman, D., Mohd, B., & Gurunathan, S. (2008). Biosynthesis of silver nanocrystals by Bacillus licheniformis. *Colloids Surf B*, *65*, 150–153.
18. Kasture, M. M., Patel, P., Prabhune, A. A., Ramana, C. V., Kulkarni, A. A., & Prasad, B. L. V. (2008). Synthesis of silver nanoparticles by sophorolipids: Effects of temperature and sophorolipid structure and the size of particles, *Chem. Sci.*, *120*(6), 515–520.
19. Kharwar, R. N., Verma, V. C., & Gange, A. C. (2010). Biosynthesis of antimicrobial silver nanoparticle by endophytic fungus *Aspergillus clavatus*. *Nanomedicine*, *5*(1), 33–40.
20. Li, E., Jiang, L., Guo, L., Zhang, H., & Che, Y. (2008). Pestalachlorides A-C, antifungal metabolites from the plant endophytic fungus *Pestalotiopsis adusta*. *Bioorg Med Chem*, *16*, 7894–7899.
21. Petrini, O. (1991). Fungal endophytes of tree leaves In: Andrews J. H., & Hirano S. S., (eds.). *Microbial Ecology of Leaves*. New York; Spring-Verlag, pp. 179–197.
22. Pileni, M. P. (2000). Fabrication and physical properties of self-organized silver nanocrystals. *Pure Appl. Chem*, *72*, 53–65.
23. Safekordi, A. A., Attar, H., & Ghorbani, H. R. (2011). Optimization of Silver Nanoparticles Production by E. coli and the study of reaction kinetics. *ICCEES*, 346–350.
24. Samadi, N., Golkaran, D., Eslamifar, A., Jamalifar, H., Fazeli, M. R., & Mohseni, F. A. (2009). Intra/extracellular biosynthesis of silver nanoparticles by an autochthonous strain of Proteus mirabilis isolated from photographic waste. *J Biomed Nanotechnol*, *5*(3), 247–253.
25. Sathyavati, R., Krishna, M. B., Rao, S. V., Saritha, R., & Rao, D. N. (2010). Biosynthesis of silver nanoparticles using Coriandrum sativum leaf extract and their application in nonlinear optics. *Adv Sci Lett*, *3*(2), 138–143.
26. Schulz, B., Mmert, R., Dammann, A. K., Aust, U., & Strack, D. (1999). The endophyte-host interaction: a balanced antagonism. *Mycol. Res*, *103*, 1275–1283.

27. Shankar, S. S., Rai, A., Ahmad, A., & Sastry, M. (2004). Rapid synthesis of Au, Ag and bimetallic Au core–Ag shell nanoparticles using Neem (*Azadirachta indica*) leaf broth. *J Colloid Interface Sci, 275*, 496–502.

28. Singh, D., Rathod, V., Ninganagouda, S., Herimath, J., & Kulkarni, P. (2013). Biosynthesis of silver nanoparticle by endophytic fungi *Penicillium* sp. isolated from *Curcuma longa* (turmeric) and its antibacterial activity against pathogenic gram negative bacteria. *Journal of Pharmacy Research, 7*(5), 448–453.

29. Strobel, G., & Daisy, B. (2003). Bioprospecting for microbial endophytes and their natural products. *Microbiol Mole Biol Rev, 67*, 491–502.

30. Tan, R. X., & Zou, W. X. (2001). Endophytes: a rich source of functional metabolites. *Nat Prod Rep, 18*, 448–459.

31. Taniguchi, N. (1974). On the Basic Concept of 'Nano-Technology'. In: *Proceedings of the International Conference on Production Engineering. Tokyo, Part II*, Japan Society of Precision Engineering (JSPE), pp. 18–23.

32. Taylor, J., Hyde, K., & Jones, E. (1999). Endophytic fungi associated with the temperate palm, *Trachycarpus fortune*, within and outside its natural geographic range. *New Physiologist, 142*, 335–346.

33. Thirumalai, A. V., Prabhu, D., & Soniya, M. (2010). Stable silver nanoparticle synthesizing methods and its applications. *J. Bio. Sci. Res., 1*, 259–270.

34. Tien, D. C., Tseng, K. H., Liao, C. Y., & Tsung, T. T. (2007). Colloidal silver fabrication using the spark discharge system and its antimicrobial effect on *Staphylococcus aureus*. *Med. Eng. Phys, 30*, 948–952.

35. Vigneshwaran, N., Ashtaputre, N. M., Varadarajan, P. V., Nachane, R. P., Paralikar, K. M., & Balasubramanya, R. H. (2007). Biological synthesis of silver nanoparticles using the fungus Aspergillus flavus. Mater Lett, 66, 1413–1418.

PART II

NOVEL PRACTICES IN AGRICULTURAL PROCESSING

CHAPTER 7

PRACTICES IN BIOLEACHING: A REVIEW ON CLEAN AND ECONOMIC ALTERNATIVE FOR SAFE AND GREEN ENVIRONMENT

SUNITA DEVI, BINDU DEVI, and SEEMA VERMA

CONTENTS

7.1	Introduction	148
7.2	Terminology	149
7.3	Historical Background of Bio-Leaching	150
7.4	Advantages of Bio-Leaching Process	151
7.5	Microbial Diversity in Bio-Leaching Environments	152
7.6	Leaching Techniques	153
7.7	Mechanisms of Bio-Leaching and Bio-Film Development	161
7.8	Factors Affecting the Bio-Leaching Process	164
7.9	Environmental Aspects of Bio-Leaching Process	168
7.10	Future Prospects	171
7.11	Summary	172
Keywords		172
References		173

7.1 INTRODUCTION

Metal ore deposits throughout the world are diminishing at a steady rate. Non-renewable raw materials need to be used in judicial and economic manner for sustainable development and the demand for primary resources need to be minimized [39]. Besides improving the available techniques of mining, it is important to employ new and innovative technologies to develop new resources for metals [2, 5, 7].

There has been an increasing accumulation of heavy metal contents in the soils causing worldwide concern. This increase has been caused by industrial, agricultural, mining and domestic activities. Being non-degradable by chemical or biological means, these heavy metals are difficult to eradicate from the environment and are therefore indestructible [2, 34, 48]. The interaction of these metals with the enzymes inhibits the metabolic processes resulting in toxicity. The soil-accumulation of heavy metals (such as: cadmium (Cd), chromium (Cr), copper (Cu), iron (Fe), lead (Pb), manganese (Mn), mercury (Hg), nickel (Ni), zinc (Zn), etc.) have been known to cause toxicity in Humans, plants, animals and aquatic life. Metals do not undergo break down by a natural microbial process. Their oxidation or reduction or transformation to different redox stages by organic metabolites can be effective in the cleaning up process [62].

Microbial leaching/bio-leaching is a clean technology that offers an economic substitute for the mining industry at low cost and less capital inputs requirements as compared to conventional methods [14]. Although the process is slow, yet it is more environment – friendly and consumes less energy as compared to physico-chemical processes which consume more energy, prove expensive for the recovery of metal ions from low and lean grade ores and also release harmful gases (like Sulphur dioxide) that cause environmental hazards [7, 28, 52, 57]. Moreover, while physico-chemical processes result in generating unwanted acid and metal pollution, bio-mining process has the advantage of leaving a residue which is less chemically active on its exposure to rain and air [61, 76].

It allows the cycling of metals by a process similar to natural biochemical cycles and helps in reducing the demand for resources such as ores, energy or landfill space [28, 83]. Moreover, this process has the potential to offer a much needed step-change in the technology for processing low grade [80, 81]. Microbes such as bacteria and fungi convert metal compounds into water-soluble forms and acts as biocatalysts for the process called microbial leaching or bio-leaching [12, 70]. Additionally, microbiological solubilization processes can be used to recover metal values from industrial wastes which can then serve as secondary raw materials.

This chapter reviews the technology of bio-leaching.

7.2 TERMINOLOGY

Bio-leaching is caused by the dissolution of metals from their mineral sources by the microorganisms that exist in nature. Microorganisms may also be used to transform elements so that the elements can be removed from a material by filtering water through it [4]. The term bio-oxidation is also used for bio-leaching [13, 37]. Bio-leaching generally implies using microorganisms to convert solid metal values into their water soluble forms. Thus, copper sulfide (Cu_2S) is oxidized to copper sulfate ($CuSO_4$) by microbial means and metal values are obtained in the aqueous phase. The remaining solids are dispensed away.

Microbial oxidation of host minerals containing metal compounds of interest is described as "bio-oxidation". In this process, metal values remain in the solid residues in a more concentrated form. Bio-oxidation in gold mining operations is a kind of pretreatment process to remove pyrite or arsenopyrite, at least partly. Such a process of refining of solid materials and removal of unwanted impurities from them is called bio-beneficiation [36]. The process of mobilization of elements from solid materials with bacteria and fungi is described as "bio-mining", "bio-extraction", or "bio-recovery" [50, 60]. The process of bio-mining involves an economical metal recovery in large scale operations of mining industries using applications of microbial metal mobilization processes.

"Bio-hydrometallurgy" is an interdisciplinary field and covers bio-leaching or bio-mining processes [64]. It combines different kinds of aspects of biotechnology, chemical engineering, geochemistry, geology, hydrometallurgy, microbiology (especially geo-microbiology) and mineralogy, etc. By this process, metals and metal-containing materials are treated by wet processes. In this, aqueous solutions play a major role in the extraction and recovery of metals from their ores [58]. Rarely, the term "bio-geotechnology" is also used instead of bio-hydrometallurgy [31].

7.3 HISTORICAL BACKGROUND OF BIO-LEACHING

Going into the history of bio-leaching, the technology made rapid strides in the 1980s'. The first commercial tank bio-leaching plant was established near Barberton in South Africa at the Fairview Goldmine [1].

It was as far back as 100–200 BC and perhaps even earlier that the Chinese used the technology of leaching of metal, particularly copper from its ore (bio-leaching) and bio-accumulation, i.e., the precipitation of copper from solution [56]. But it was not until the 1940s' that metal solubilization with the help of microorganisms was practiced. Further research later on has helped to understand better the mechanisms involved in the process [52].

Countries like Spain, Sweden, Germany and China and some others have for centuries put to use the technology of bio-oxidation of sulfide for copper recovery [30]. However, it is the Rio Tinto mines of southwestern Spain which have in fact been the cradle of bio-leaching technology and bio-hydrometallurgy [52]. Though these mines have been exploited since the pre-Roman times for copper, gold and silver, it was in the 1980s' that bio-leaching was used in these mines where large volumes of low grade copper had been left for natural decomposition for 1 to 3 years [65].

The copper mine of Rio Tinto was probably the first large-scale operation where the major role had been played by microorganisms. Colmer and Hinkel in 1947 were able to isolate bacteria of *Thiobacillus* genus from acid mine water and demonstrate the role played by the microorganisms in the bio-mining process [45]. Later, *Thiobacillus ferrooxidans* and *T. thiooxidans* were isolated and characterized [74, 75]. *T. ferrooxidans* and *T. thiooxidans* have later been renamed as *Acidithiobacillus ferrooxidans* and *Acidithiobacillus thiooxidans,* respectively [42].

Way back in 1980, biohydrometallurgy found its initiation as commercial application which was designed to facilitate microbial activity for copper leaching from heaps. Since then after 1980, a number of

Practices in Bioleaching 151

copper bioleach operations have been put in place [16]. For instance, using bio-leaching, Lo Aguirre mine in Chile produced about 16,000 tons of ore between 1980 and 1996 and Quebrada in Northern Chile has the capacity to process 17,300 tons of sulfide ore in a day [19].

Now-a-days, South Africa, Ghana, Australia and Peru collectively have nine operating mines [32]. There are a number of companies (Table 7.1), which are using the bio-oxidation processes for metal extraction [53]. While a few of them have grown as commercial units, some are still in the early stages of experimentation and existing as pilot plants [3, 16].

7.4 ADVANTAGES OF BIO-LEACHING PROCESS

When conventional methods prove uneconomical for metal recovery from mining and other industrial waste products, then microbial leaching processes find greater use and applicability. There are many advantages of bio-leaching operations as compared to the conventional mining processes [35], such as:

TABLE 7.1 Major Companies That Have Developed Bio-Oxidation Processes

Company	Process	Application
BacTech Enviromet	BacTech/Mintek Process	Agitated tank oxidation and leaching of copper sulfides
BHP Billiton, Ltd.	BIOCOP™	Agitated tank oxidation and leaching of copper sulfides
	BIONIC™	Agitated tank oxidation and leaching of nickel sulfides
	BIOZINC™	Agitated tank oxidation and leaching of zinc sulfides
Newmont Mining	BIOPROTM Process	Heap bio-leaching of sulfidic refractory gold ores
GeoBiotics, Inc.	GEOCOAT™ Process	Heap leaching sulfide mineral concentrates
Gold Fields, Ltd	BIOX™ Process	Agitated tank oxidation of refractory gold ores

152   Engineering Interventions in Agricultural Processing

- Applicable for bulk complex concentrate.
- Dust and SO_2 free process.
- Economically viable for extraction of metal from low grade ores and major mine waste.
- Environment friendly process.
- Flexible for the treatment for mineral resources with a variety of metals and in variable concentrates.
- Low energy input and capital cost.
- Low to moderate capital investment.
- Process is simple to operate and maintain.
- Provide essential pre-treatment for refractory ores for selective leaching.
- Simple stepwise expandability by a single reactor or in modules of reactors.
- Use of naturally occurring key components, i.e., microorganisms, water and air.
- Work at ambient pressure and temperature.

7.5 MICROBIAL DIVERSITY IN BIO-LEACHING ENVIRONMENTS

There is now an increasing awareness in the mining industries about exploiting microbial activity. Therefore, it has now come to prove as a viable economic option for the treatment of specific mineral ores rather than just being a technology with promise [1]. For bio-leaching, Iron and sulfur oxidizing bacteria *Acidithiobacillus ferrooxidans,* sulfur oxidizing *A. thiooxidans* and *A. caldus* and iron oxidizing *Leptospirillium* spp. (*L. ferriphilum* and *L. ferrooxidans* are the most important ones at moderate temperatures [21, 33].

Other important bacteria used in the bio-leaching process are *Acidiphilum, Sulfobacillus, Ferroplasma, Metallospaera, Thermothrix* and *Acidianus* [27]. These bacteria are characterized by some shared common features: Gram negative, iron oxidizing and sulfur oxidizing chemolithoautotrophs growing autotrophically by fixing atmospheric CO_2 and for also being heterotrophs by using peptone and yeast extracts [82]. Quite a number of potential bacterial species that have the ability to accumulate metals from the aquatic environment have now been identified through

Practices in Bioleaching 153

many studies and research findings. The *Bacillus* spp. has found much wider use in commercial bio-sorption processes because of its high potential of metal sequestration [17]. Similarly, the use of *Pseudomonas* spp. *Zooglearamigera* and *Streptomyces* spp. for bio-sorption of metals has also been reported. *Rhodobacter sphaeroides*, *Alcaligenes eutrophus* and *Staphylococcus saprophyticus* are some of the other species that have been used in various other research projects. Bio-sorption helps to detoxify the aquatic environment from the heavy metal where they accumulate in the bacterial cells [52].

Microorganisms are the backbone for the bio-leaching activity. Very rich microbial diversity is found in the bio-leaching environment which includes bacteria, fungi, algae and some of the yeasts [47]. However, recently some protozoa species were reported to be the key players in bio-leaching environments (Table 7.2).

Detailed investigations have been carried out by employing molecular methods such as DNA-DNA hybridization, 16S-rDNA sequencing, polymerase chain reaction (PCR) based methods with primers derived from rRNA sequencing and immunological techniques. It has been found from such studies that the microorganisms that comprise these bio-leaching communities constitute very wide variety. This in turn results in complex microbial interaction (synergism, mutualism, competition and predation) and nutrient flow [41].

7.6 LEACHING TECHNIQUES

Leaching techniques can be divided into three main areas depending upon the working volume.

7.6.1 LABORATORY SCALE (0–10 DM³)

There are two main groups into which the laboratory-level leaching techniques can be placed: (i) A qualitative or semi quantitative analysis of an ore using microbes. Manometric and stationary flask techniques belong to this category, and (ii) Assessment of quantitative measures using an analytical approach including air-lift percolator and shake flask techniques [9, 28, 64].

154 Engineering Interventions in Agricultural Processing

TABLE 7.2 Diversity of Microorganisms in Bio-Leaching Environments

Source	Organism	Metals leached
Bacteria microbial group		
Acid mine drainage (AMD)-impacted sites	*A. ferroxidans*	Zn and Al
Battery wastes	*Leptospirillum* spp.and *Sulfobacillus* spp.	Ni, Cd and Li
Catalysts	*T. ferrooxidans*	Cu, Pb, V, and Zn
Contaminated sediments like anaerobic sludge, river sediments, etc.	*A. thiooxidans, S. thermosulfido oxidans*	Cu, Co, Ni, Zn, Cr, Mn and Fe
Copper-containing ores, electronic scrap and spent auto mobile catalytic converters	*Chromobacterium violaceum* and *Pseudomonas fluorescens*	Ni, Au, Pt and Cu
Electronic waste	*A. ferrooxidans, A. thiooxidans, Sulfobacillus thermosulfidooxidans*	Cu, Al, Fe and Ni
Industrial wastewater	*Bacillus licheniformis-*ATCC12759, *Brevibacillus laterosporus* ATCC64 and *P. putida* ATCC31483	Ti, Co, Pb, V, Zn, Cd, Ni, Cu and Mn
Lead sulfide bearing minerals	*T. ferrooxidans, T. thiooxidans* and *L. ferrooxidans*	Pb
Metal sulfides and Pyrite	*T. ferrooxidans, T. thiooxidans,* and *L. ferrooxidans*	Cu and U
Pyrite-arsenopyrite ore concentrates	*A. ferrooxidans, A. thiooxidans, A. caldus, Leptospirillum ferrooxidans* and *L. Ferriphilum*	Fe, S and As
Treated wood	*Bacillus licheniformis* CC01 and *P. putida*	Cu, As, Cr
Fungi Microbial group		
Aluminosilicate (95% spodumene)	*Rhodotorula rubra*	Al, Si
Bioreactor	*Aspergillus niger* and *Penicillium simplicissimum*	Cu, Sn, Al, Ni, Pb and Zn
Municipal solid waste (MSW) Incineration fly ash	*A. niger*	Al, Zn, Cu, Fe, Mn, Pb, Cu, Zn, Cr and Cd
Nickeliferous lateritic ore	*A. wentii*	Co and Ni

Practices in Bioleaching   155

TABLE 7.2   (Continued)

Source	Organism	Metals leached
Algae Microbial group		
Acid mine drainage (AMD)-impacted sites	*Ulothrix, Microspora, Klebsormidium* and *Tribonema*	Zn, Cu, Cd, Pb, Fe, Al and As
Filamentous green algae	*Cladophora glomerata* and *Oedogonium rivulare*	Zn, Pb, Cd, Co
Freshwater diatom	*Planothidium lanceolatum*	Zn, Cd, Co
Inorganics	*Phormidium* spp.	Cd, Zn, Pb, Ni and Cu
Pretreated algal biomass	*Spirogyra* sp., *Spirullina* sp.	Ni, Cr, Fe, Mn, Cu, Zn
Streamer communities	*Tribonema viride, T. minus, T. ulotrichoide*	Fe, Mn, Ni, Zn and U
Protozoa Microbial group		
Industrial wastewater	*Aspidisca* spp., *Trachelophyllum* spp. and *Peranema* spp.	Ti, Co, Pb, V, Zn, Ni, Cd, Cu and Mn
Urban sewage	*Drepanomonas revoluta* BQ1, *Uronema nigricans* BQ2 and *Euplotes* sp. BQ3	Cd, Zn, Cu

7.6.1.1   Manometric Technique

Manometric techniques for estimating exchange of gases have been used in the study of both chemical and biological reactions for generations. A wide variety of techniques have been employed and many types of apparatuses have been developed. The most widely used respirometer is commonly called the "Warburg instrument". The respirometer is based on the principle that at constant temperature and constant gas volume any changes in the amount of a gas can be measured by changes in its pressure [64].

7.6.1.2   Stationary Flask Technique

This technique, besides being simple in experimental sense, is cost effective also in the use of equipments. Filtered air is not allowed to enter by plugging the opening with absorbent cotton. Sample sterilization is done by ultraviolent light at ambient temperature. After this, the culture medium, substrate and inoculums are introduced into the flask. In order to ensure entrance of air into the liquid medium area of air-liquid contact should be increased to the maximum. This technique is useful for determining interaction between metals and microbes as well as for understanding the physiology of the microorganisms. However, it is limited to use [64].

7.6.1.3 Percolator Leaching

Initially, air-lift percolators were used for bacterial leaching for experiments. In simple terms, a percolator is a glass tube fitted with a sieve plate at the bottom and filled with ore particles. The ore packing is irrigated or flooded with a nutrient inoculated with bacteria. The leach liquor that percolates through the column is pumped up by compressed sterile air to the column top for recirculation. At the same time, the stream of air aerates the system. Liquid samples taken at intervals help to monitor the progress of the leaching process. The pH measurements, microbiological investigations and chemical analysis of metals that have passed into solution are used to determine the state of the leaching process [10].

7.6.1.4 Shake Flask Technique

This technique shows more advantages than the stationary flask technique. In order to fully mix and homogenize the medium, Erlenmeyer flasks containing growth medium are fixed and incubated in rotary shaker. Afterwards, dissolution of atmospheric gases such as oxygen and carbon dioxide is increased as these are essential for the growth of microorganisms. It is necessary to keep the experimental conditions constant for accurate measurement of kinetic parameters [31, 44, 64].

7.6.2 PILOT PLANT (<10 M³)

Experiments at the laboratory level are useful because they provide useful information in a relatively short time. But they are not reliable models for commercial scale plants. For commercial scale tests, different conditions come into play. Particle size is a critical factor in metals dissolution. Large particle size and reduced exposed surface area results in the accessibility of solution and microorganisms to valuable metals, getting decreased. This causes decreased metal stabilization [64]. Even otherwise, it is difficult to apply the controlled environmental conditions to commercial-scale plants. For a pilot plant leaching in columns and in agitated tanks or reactors are the main concepts employed [46, 73].

7.6.2.1 Column Leaching

Column leaching which is used as a model for heap or dump leaching processes operates on the principle of percolator leaching. For making columns materials like glass, plastic, lined concrete or steel are used, depending on their size. The capacities of these columns can be several kilograms to a few tons. For taking samples or in order to install special instruments for the measurement of pH, oxygen or carbon dioxide, most column systems have devices at various distances [10]. This is useful in obtaining information on how to optimize the leaching conditions and what to expect in heap or dump leaching.

7.6.2.2 Agitation Tank Leaching Process

This process is called as stirred tank process. In this process, stirred tank bioreactors are used. For leaching of this type, intermediate to high grade ore types of resources or raw materials are used. Mechanical agitation is employed for leaching the ore concentrates taken in a tank. In this kind of process minerals are treated highly aerated and continuous-flow reactors placed in series. Viewed from the perspective of process-engineering, since bio-leaching involves complex biochemical reactions, it would be performed in the best way in reactor. In this way good control of the pertinent variables would be possible which would in turn lead to better performance [59].

7.6.3 COMMERCIAL SCALE (>10 M³)

The mining industry has major applications of commercial scale bioleaching. It was in the early 1950s that these techniques were first patented for the copper industry. Later it was used on a commercial scale for the mining operations of gold, uranium and zinc from low grade ores:

7.6.3.1 In-Situ Leaching

In-situ leaching (ISL) is employed on underground ores to directly recover useful minerals and metals.

The method is categorized under solution mining into two parts: leaching (drawing out the mineral from the earth) and fluid recovery [6, 20, 38]. It also comprises of drawing out minerals from the untouched ore. It is combined with mineral recovery operation time and again to pull out the minerals from recovered fluid solution or leachate. With the aid of gravity flow and pumping, a large volume of solution is circulated. The extraction operation generates leachates known as "pregnant solutions" whereas the returning fluids to the extraction operation are known as "barren solution". There is permeability of the ore-body which can be increased by fragmenting of ores in place, called "rubblizing [76]" that serves as the main determining factor for this method (Figure 7.1). ISL was generally used for the extraction of uranium, copper and gold [38].

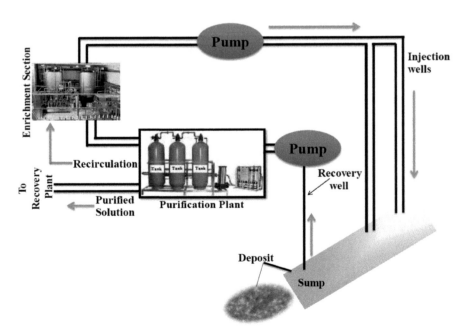

FIGURE 7.1 Process flow for *in-situ* leaching.

7.6.3.2 Dump Leaching

The piling up of uncrushed waste rock in dumps forms the basis of dump leaching [70]. It is mainly used for the extraction of minerals from low grade ores. The bigger rocks are splintered by blasting in the pit and are carried to dumps as large fragments. These dumps contain run of mine ore in million tones. The optimum conditions required for the growth of micro-organisms, which will oxidize the mineral for extraction of the metal, acidified water is spread on the top surface that ultimately percolates in the dump. Dump leaching was popularly used for the extraction of minerals from copper sulfide ores [14].

7.6.3.3 Heap Leaching

Heap leaching is slightly different than dump leaching in a way that the bigger rocks are crushed into smaller particles and agglomerated in revolving drums containing acidified water. It results in the conditioning of the ore, i.e., provides required conditions for the growth of the microorganisms. The conditioned ore is spread on specially engineered pads which consist of perforated plastic drain lines to improve drainage of the mineral containing solution from the bottom of the ore (Figure 7.2). It is also aerated to ensure the optimum growth of micro-organisms [15, 76]. Heap leaching is used for the leaching of copper sulfide and to pre-treat gold ores to extract the occluded gold from sulfide minerals [14].

The dump and heap leaching processes are analogous in two ways: Firstly, both make use of lixiviant at the top of dump and heap. Secondly, both include the recovery of mineral under the effect of gravity [60, 70].

7.6.3.4 Vat Leaching

Presently, vat leaching is mainly operated to extract minerals from oxide ores. Large sized tanks which are equipped with agitators retain ore slurry and solvent for several hours in vat leaching (Figure 7.3). It is used to carry out cyanidation for ores with high gold content and to extract precious metals from ores [70, 76].

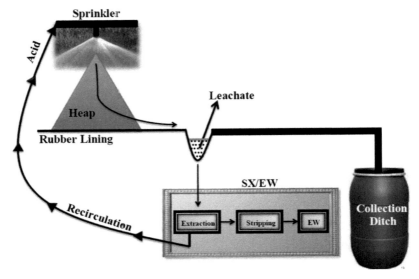

FIGURE 7.2 Process flow for heap leaching.

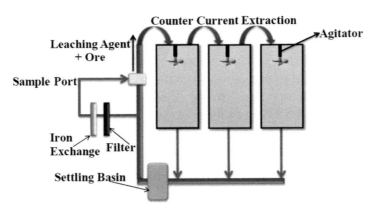

FIGURE 7.3 Process flow for vat leaching.

7.6.3.5 Reactor Leaching

This method involves elevated installation and higher costs of operation. Therefore, it is more expensive and its use is restricted to the products of conventional concentration processes. In this process the ore to be leached stays for some time in reactors consisting of well-stirred tanks

[64]. Microbial leaching was responsible for 10 per cent of world copper obtained by microbial leaching in 1980 [40]. According to other estimates the figure given is 15–30% [18]. As regards gold, of the 2.1 kt, about 20% was obtained by microbial activities [13].

7.7 MECHANISMS OF BIO-LEACHING AND BIO-FILM DEVELOPMENT

Breaking down of minerals into constituent minerals is the basis of biomining. This provides energy to the microorganisms involved (Figure 7.4). Extraction of metal from the sulphidic ores, concentrate and tailing was initially described by two mechanisms: one as 'direct leaching or enzymatic oxidation' and the other one as indirect leaching or non-enzymatic oxidation of metal'. Later on, the third mechanism came into existence that is galvanic conversion. As regards direct enzymatic oxidation of the Sulphur moiety of heavy metal sulfides, originally known as 'direct mechanism' of biological metal sulfide oxidation, it is widely accepted that this mechanism does not exist.

The non-enzymatic metal sulfide oxidation mechanism by ferric ions combined with enzymatic (re)oxidation of the regenerated ferrous ions, or the 'indirect mechanism' consists of two sub-mechanisms: 'contact' and 'non-contact mechanisms. Basically, the planktonic bacteria exercise the non-contact mechanism by oxidizing ferrous ions in solution. The ferric ions thus obtained come into contact with a mineral surface and are reduced to respective sulfates and re-enter the cycle.

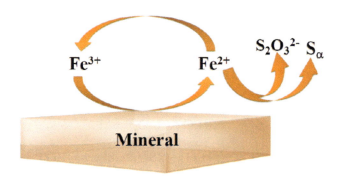

FIGURE 7.4 Mechanism of bio-leaching.

Strictly speaking, this is simply the indirect mechanism designated earlier. The contact mechanism involves the process of the cells attaching to the surface of sulfide minerals. Thus, the electrochemical processes that cause the dissolution of sulfide minerals take place at the interface between the bacterial cell-wall and the mineral sulfide surface. In both contact and non-contact mechanisms, bacteria contribute to mineral dissolution by generating the oxidizing agent, the ferric ion and oxidizing the sulfur compounds that result from the dissolution [63].

It is now well recognized that bio-leaching is essentially a biochemical process in which ferric iron and protons (H^+) cause solubilization of minerals. Microorganisms are responsible for triggering the leaching chemicals, i.e., lixivient. They also make available the space where leaching reactions take place. An exopolysaccharide or exopolymeric substance (EPS) layer is produced by the leaching organisms. Owing to this they adhere to the surface.

The bio-oxidation of iron happens in a very rapid and efficient way within this EPS layer rather than in the bulk solution. Thus, the EPS layer provides the reaction space. Not all metal sulfides have similar mineral dissolution reaction. It can vary depending on the nature of the sulfides. The oxidation of different metal sulfides proceeds by one of the two main pathways via different intermediates.

7.7.1 THIOSULFATE PATHWAY

The oxidation of acid insoluble metal sulfides such as pyrite (FeS_2) and molybdenite (MoS_2) occurs through this pathway. These minerals are solubilized when ferric iron attacks the acid-insoluble metal sulfides along with the formation of thiosulfate as the main or first intermediate and sulfate as the main end product [24, 68]. Therefore, this pathway is known as 'thiosulfate pathway'. This mechanism, using pyrite as a mineral substrate can be represented as:

$$FeS_2 + 6Fe^{3+} + 3H_2O \rightarrow S_2O_3^{2-} + 7Fe^{2+} + 6 H^+ \tag{1}$$

$$S_2O_3^{2-} + 8Fe^{3+} + 5H_2O \rightarrow 2SO_4^{2-} + 8Fe^{2+} + 10 H^+ \tag{2}$$

7.7.2 POLYSULFIDE MECHANISM

The solubilization of the acid soluble metal sulfides such as sphalerite (ZnS), chalcopyrite (CuFeS$_2$) or galena (PbS) takes place through this pathway. In this case, the acid-soluble metal sulfides are solubilized through a combined attack by ferric iron and protons, with the formation of elemental sulfur as the main intermediate [24, 66]. This elemental Sulphur is relatively stable but may be oxidized to sulfate by the Sulphur-oxidizing microbes present in the system such as *A. thiooxidans* or *A. caldus*.

$$MS + Fe^{3+} + H^+ \rightarrow M^{2+} + \tfrac{1}{2} H_2Sn + Fe^{2+} (n \geq 2) \qquad (3)$$

$$\tfrac{1}{2} H_2Sn + Fe^{3+} \rightarrow \tfrac{1}{4} S_8 + Fe^{2+} + H^+ \qquad (4)$$

$$\tfrac{1}{4} S_8 + \tfrac{1}{2} O_2 + H_2O \quad \underline{\text{Microorganism}} \quad SO_4^{2-} + 2 H^+ \qquad (5)$$

Thus, it can be inferred that the ferrous iron produced in both the pathways, gets re-oxidized to ferric iron by the iron-oxidizing microorganisms such as *A. ferrooxidans* or the species of *Leptospirillum* or *Sulfobacillus* if present in the system.

$$2Fe_2 + \tfrac{1}{2} O_2 + 2 H^+ \quad \underline{\text{Microorganisms}} \quad 2 Fe^{3+} + H_2O \qquad (6)$$

The fundamental role of the microorganisms in the solubilization of metal sulfides is to provide sulfuric acid for a proton attack and to keep the high concentration of ferric iron (strong oxidizing agent) for an oxidative attack on the mineral sulfides [26].

In the bio-leaching of metal sulfides, EPS produced by *A. ferrooxidans* is reported to play crucial role. The EPS allows the cells to adhere to the surface of metal sulfides, resulting in the augmentation of bio-leaching processes [22, 67, 72]. It is due to the deposition of ferric ions in the EPS layers that this augmentation or increase takes place. This provides a zone of high oxidation activity at the close vicinity of the mineral surface. No enzymatic reaction between the cell and metal sulfides is involved in this situation. Therefore, contact leaching was proposed to describe the process [63].

The third mechanism of sulphidic mineral oxidation is known as 'Galvanic interaction'. It is an intrinsic phenomenon of mixed sulphidic minerals that operates automatically in a heterogeneous system or wherever two or more different environments coexist. The role of the galvanic interaction in the bio-leaching of mixed sulphidic minerals has been reported extensively. In this phenomenon, the minerals with a comparative lower rest potential values behave as an anode and therefore, undergo dissolution, i.e., oxidation whereas, the minerals with higher rest potential values act as a cathode at which reduction of oxygen takes place [78]. During galvanic interaction, following electro chemical reactions takes place:

Anodic reaction on active sulfide sites:

$$MS \rightarrow M^+ + S^\circ + 2e^- \tag{7}$$

where, M = bivalent metal.

Cathode oxygen reduction on noble mineral:

$$O_2 + 4 H^+ + 4 e^- \rightarrow 2H_2O \tag{8}$$

The metals are arranged as per their dissolution behavior under the influence of galvanic interaction [25, 54]. These biological processes are used for extraction of metal if they are allowed to function in control manner at desired place and time. But if they are not under control, as can be seen from equations 1, 2, 4 and 5, they are responsible for acid generation at mining and dumping sites. Many of these biological processes in and around mines generate metals, sulfates and dissolved solids which are in turn responsible for environmental pollution.

7.8 FACTORS AFFECTING THE BIO-LEACHING PROCESS

Many physical, chemical, mineralogical and biological factors influence the process of bio-leaching as is a complex phenomenon [49, 79]. Primarily the efficiency of the organisms determines the effectiveness and efficacy of the leaching process. Secondly, the chemical and mineralogical composition of the ore to be leached is also determining factor to some extent. In order to maximize the yield of metal extraction leaching conditions should

Practices in Bioleaching 165

be in tune with the optimum growth conditions of the leaching bacteria [11]. Some details are illustrated below for vital factors that influence the bio-extraction of metals from minerals.

7.8.1 BIOLOGICAL FACTORS

7.8.1.1 Nutrients

Bio-mining bacteria are mainly chemolithoautotrophs hence, relatively less-demanding and require only inorganic compounds for growth. Inorganic compounds such as ammonium sulfate $[(NH_4)_2SO_4]$, dipotassium hydrogen phosphate (HK_2O_4P), potassium chloride (KCL), magnesium sulfate $(MgSO_4)$ and calcium nitrate $(Ca(NO_3)_2)$ are required by these organisms to synthesize their cell material. Generally, these mineral nutrients are obtained from the environment as well as from the material to be leached out. However, a desired quantity of iron and sulfur compounds, supplemented with some amounts of ammonium, phosphate and magnesium salts is necessary for their optimum growth. Water is mandatory as a carrier of O_2, CO_2 and nutrients. It also acts as the transporting agent of the bacterial species. Additionally, it is required to solubilize the metal sulfate produced during the process [11, 78].

7.8.1.2 Oxygen and Carbon Dioxide

Metal leaching bacteria are mostly aerobic and chemolithotrophic in nature, an adequate supply of oxygen is a pre-requisite for their optimum growth and high activity in leaching process. It also serves as an electron acceptor during the oxidation of reduced Sulphur compounds to sulfate and ferrous to ferric ions. The amount of carbon dioxide present in the air provides sufficient amount of carbon for biomass generation. An adequate supply of both O_2 and CO_2 to the leaching system is ensured through aeration [31, 59].

7.8.1.3 pH

Since most of the bacteria used in metal leaching are acidophiles, they function best at pH range of 1.8 to 2.2. To ensure optimum growth and activity of leaching bacteria, the adjustment of pH to an optimum value is a necessary condition. It is decisive for the solubilization of the metals too. The substrate is serves as a source of energy which is available only in the form of electrons for the metabolic activity of leaching bacteria in an inorganic system. Exchange of hydrogen ions as well as electrons takes place during the biological oxidation of ferrous iron and metal sulfides. It implies that, the pH has a definite effect on their metabolism [71].

7.8.1.4   Redox Potential

Usually, *A. ferrooxidans, A. thiooxidans* and *L. ferrooxidans* and one or two moderate thermophiles are used in the bio-leaching process. Indirect assessment of their biological activity can be done by estimating the redox potential. Generally, their activity is responsible for oxidation of pyrite and ferrous to ferric iron, which ultimately results in continuous increase in redox potential. Many times, the redox potential lies in the range as high as 750–850 mV depending on the ferric-ferrous ratio. Thus, it can also be correlated with the extraction of metals. Moreover, it is necessary to maintain the redox potential to optimize metal dissolution and control iron precipitation, as the behavior of metal sulfide and ferric precipitation depend on the redox potential of system [78].

7.8.1.5   Temperature

Being a mesophile, the optimum temperature for the oxidation of ferrous iron and sulfide by *A. ferrooxidans* is between 28–30°C [10]. A decrease in metal extraction occurs at a temperature lower than optimum. Recently, uses of extreme and moderate thermophiles such as *Sulfolobus, Acidiphilum, Thiobacilluscaldus*, etc. for metal extraction from various minerals have been increasing at a steady rate. Thermophilic bacteria play a significant role in the bio-leaching process. However, in certain instances, use of moderate thermophiles was proved to be quite successful. When thermo tolerant and/or thermophilic organisms are involved, a temperature range 40–80°C is reported to be the most favorable depending on the species of the organism [59].

7.8.1.6 Heavy Metals

There is an accompanying increase in metal concentration in the leachate as the leaching of metal sulfides progresses. Generally speaking, the leaching organisms, in particular Thiobacilli, exhibit a high degree of tolerance to heavy metals and various strains are resistant to 1000 mM nickel, 800 mM copper, 500 mM cadmium, 84 arsenic or 1071 mM zinc. Altogether different sensitivities may be exhibited to heavy metals by different strains of some species. If the concentrations of metals or substrates are increased in a gradual manner, it is possible to adapt individual strains to higher concentrations of metals and to specific substrates [29].

A. ferrooxidans, being an obligate chemolithotroph, derives its energy from ferrous iron and Sulphur compounds. An inhibition in the growth of this organism was observed in the presence of organic compounds. The presence of surfactant carried over with pulp particles decreases the surface tension and reduces the mass transfer of oxygen. At present, solvent extraction is the preferred way for the concentration and recovery of metals from the pregnant solution. A combination of bacterial leaching and solvent extraction, results in the enrichment of the solvent in the aqueous phase. They have to be removed recirculation of the barren solution to the leaching operation [71]. In order to ensure proper microbial activity, development of such organisms that may be resistant to such organic additives is being attempted. Attempts are in progress to develop such microorganisms that are resistant to such organic additives so that microbial activity is not hampered.

7.8.2 MINERALOGICAL FACTORS

7.8.2.1 Mineral substrate

The mineralogical composition of the leaching substrate is of primary importance. Often, inherently present various toxic substances in sulfide mineral which are leached out during bio-oxidation of minerals have been found to inhibit the oxidation ability or growth of bacteria. These inhibitors affect the bacteria adversely which is likely to cause lowering of mineral oxidation rate. The pH registers an increase with the high carbonate content of the ore. This in turn results in inhibition or suppression of bacterial activity thereby causing cost escalation of the process by acid consumption. It is the porosity of the mineral particles which causes the penetration of leach liquor as well as the organisms inside the particle [11].

7.8.2.2 Particle Size

Contact mechanism is considered to be a significant method of metal bio-extraction, thus surface area plays a vital role. During the leaching process, ore or concentrate particles serve as substrates, so the leaching activity is proportional to the available surface. With the reduced particle size there is a corresponding increase in the total particle surface area. It is possible therefore to procure higher yield of metals without any change in the total mass of the particles. An optimum particle size for bioreactor leaching is considered to be about 42 μM. For heap and dump leaching it is in inches and may vary in minerals [71, 79].

7.8.2.3 Pulp Density

An increment in the total mineral surface area can also be achieved by an increase in the pulp density. Metal extraction may get increased with high pulp density but so will there be an increase in the dissolution of certain compounds which produce inhibitory or toxic effects on the growth of leaching bacteria. Gas diffusion, mixing and therefore the activity of the organism is adversely impacted by uncontrolled increase in the pulp density [23, 77]. Some other biotic and abiotic factors that affect the bio-leaching process are depicted in Table 7.3 [78].

7.9 ENVIRONMENTAL ASPECTS OF BIO-LEACHING PROCESS

Besides the commercial value of bacterial leaching, bacterial interaction with sulfide minerals is a significant factor in the formation of acid mine drainage (AMD). To address the issue of remediation of sites, several technical attempts have been made. Adequate understanding of the mechanisms of bacterial oxidation of sulfide minerals can go a long way

Practices in Bioleaching

TABLE 7.3 Factors Affecting Bacterial-Mineral Oxidation

Factors	Parameters affecting bio-leaching
Abiotic	Temperature, pH, redox potential, CO_2 and O_2 contents, nutrient availability, oxygen availability, pressure, surface tension, homogenous mass transfer, Fe (III) concentration, presence of inhibitors, etc.
Biotic	Microbial diversity, population density, microbial activities, metal tolerance, spatial distribution of microorganisms, attachment to ore particles, metal tolerance, adaptation abilities of microorganisms and inoculum
Mineralogical	Mineral composition, mineral type, grain size, mineral dissemination, surface area, porosity, hydrophobic galvanic interactions, and formation of secondary minerals
Processing	Leaching mode (*in situ,* heap, dump or tank leaching), Pulp density, stirring rate (in case of tank leaching operation), heap geometry (in case of heap leaching)

in bringing about improvements in the design and operation of bacterial leaching plants and may also help to prevent AMD which is a major environmental problem [78].

Metals that gain entry into the waterways through natural sources as well as owing to man-made discharges are strongly linked with sediment. Since sulfide is produced from microbial reduction of sulfate under anoxic conditions and reacts with heavy metals to form insoluble metal sulfides. The metals are present in the form of these sulfides and they persist in the sediments till the existence of reducing conditions. However, as a result of oxidation of metal sulfides, the metals get solubilized by exposing the contaminated sediments to atmospheric oxygen. As a consequence, if the dredge sediments are not handled or disposed of properly, uncontrolled release of toxic heavy metals into the environment may occur.

Although there have been several studies for suitable chemical methods for the removal of heavy metals from contaminated sludges and sediments, yet the practical application of these methods, in spite of good metal extraction through some of these methods, has been difficult in view of the costs involved, difficulties of operation and quantitative requirements. In this facet, it has been shown that the microbial leaching process is capable of eluting heavy metals associated with solids. Further, microbial leaching has found successful application in respect of anaerobically digested municipal sludges and contaminated sediments [43].

Since, bio-hydrometallurgical processes have lower capital and energy inputs, therefore bio-hydrometallurgical techniques of metal remediation by bio-leaching holds considerable advantage in economic terms. As a rule of thumb, bioprocesses cost one-third to one-half the cost incurred from using conventional chemical and physical remediation technologies [8].

7.9.1 IN INDIAN SCENARIO

India has cradled a very ancient and glorious civilization. Its heritage of mining and metallurgy is no less so and goes back to the pre-Harappan period (4000–2000 BC). Mother Nature has endowed India with a rich mineral wealth and for over 3000 years has been the sole source of diamond for the whole world, not to speak of its oldest heritage of zinc technology and copper, lead and zinc, etc. However, very scanty information is available in literature on the exploration of these mineral resources until independence. In the present scenario, India ranks second in the resources of manganese ore, third in barytes and iron ore, fourth in chromite, bauxite and magnetite, and fifth in coal and lignite in the world. Besides a host of minor and atomic minerals, India produces 65 minerals including fuel, metallic and non-metallic minerals.

In India, the consumption of these mineral resources is increasing steadily with increasing population and it is expected to rise by 6–9% annually. There is a huge fissure between demand and supply of metals from domestic sources, which is currently met by imports. In spite of cheap labor, strong technological base, liberalized market and vast consumer support, there exists a wide gap between the existing potential and the potentials to be exploited for economic metal growth. This gap could be attributed to the fact that India has a variety of untapped mineral resources which are largely sub-surface and hidden due to lack of modern technology, adequate exploratory efforts and intensive mineral investigation. Despite of these limitations and conditions, bio-extraction technology holds great promise and practical significance in India. Although, many research efforts on various aspects of bio-leaching of different types of

Indian ores have been made, yet no fruitful commercial application of bio-technology has so far been made in the country. However, some reports are available in the literature on the biological extraction of zinc from sphalerite concentrate and ore, copper extraction from chalcopyrite and gold from refractory gold ores and concentrates [25, 26]. There are also reports on the bio-extraction of copper, zinc and lead from polymetallic minerals and concentrates [51, 69], copper and nickel from bulk copper-nickel sulfide flotation concentrate and manganese from ocean nodules [53]. The only demonstration bio-oxidation plant for the refractory gold ore was commissioned in 2001 in Karnataka [55]. A pilot scale bio-leaching process is reported to treat polymetallic bulk concentrate [77, 78]. Chemical heap leaching was erected at Malanjkhand site for copper extraction from a lean grade ore. Laboratory and pilot scale bioreactor leaching for multi metal ore and concentrate obtained from Orissa and Gujarat has been developed.

Bio-leaching technology can be a 'boon' to a country like India where the low-grade mineral reserves are easily available and the warm climatic conditions favor the growth of the leaching organisms.

7.10 FUTURE PROSPECTS

Bio-leaching can be a promising approach, which will help in reducing the environmental pollution due to mining tailings, acid mine drainage, and release of sulfur into atmosphere. It can prove to be an alternative to traditional mining techniques which contributes to environmental pollution. Not only low grade ores but metal waste can also be considered as a source for metals by bio-mining. The future of bio-mining is thought-provoking, which is set to increase its application commercially because it not only offers operational simplicity but also offers low capital investment, low operating cost and shorter construction time.

It is seen that most of the research based on the type of micro-organisms which help in bio-leaching, is limited to institutes, universities and academic departments. For bio-leaching technology to be applied commercially, there is a dire need of cooperation and synchronization between mining companies and these research institutes. Also the government agencies need to demonstrate the technology at a large scale which would help in convincing the mining companies to adopt the technology for a better environment and better tomorrow.

For quicker and better results, either by genetic mutation or selection or genetic engineering, genetic improvement of bio-leaching bacteria rather than the conventional methods of screening and adaptation is called for. In-fact as regards *A. ferrooxidans* and *A. thiooxidans*, progress in the development of a genetic system for them has been indeed quite appreciable.

7.11 SUMMARY

Bio-leaching technology has a vast potential to make the environment safer and greener and can serve as a promising alternative to traditional mining techniques which contribute to environmental pollution. The major players in the bio-leaching process include bacteria (thriving at moderate temperatures to high), algae, fungi and yeasts. However, recently some species of protozoa have also been reported to be the key players in bio-leaching. Mainly three types of techniques: Laboratory-scale ($0-10$ dm^3), Pilot-plant (<10 m^3) and Commercial-scale (>10 m^3) are employed in leaching operations. The main determinants in the effectiveness of leaching process are the efficiency of the microorganisms and to some extent the chemical and mineralogical composition of the ore to be leached. Initially, leaching process was described by three mechanisms: direct leaching, indirect leaching and galvanic conversion but the direct mechanism has become obsolete now. India has a variety of untapped mineral resources which need intensive investigation with the involvement of modern technology and adequate exploratory efforts.

KEYWORDS

- **bio-hydrometallurgical processes**
- **bio-hydrometallurgy**
- **bio-leaching**
- **bio-sorption**

- **biotechnology**
- **chemolithoautotrophs**
- **clean technology**
- **geomicrobiology**
- **hydrometallurgy**
- **mechanisms of bioleaching**
- **microbial leaching**
- **microbial oxidation**
- **microbiological solubilization processes**
- **rubblizing**
- **thermophilic bacteria**
- **thiosulfate pathway**
- **Warburg instrument**

REFERENCES

1. Acevedo, F. (2002). Present and future of bioleaching in developing countries. *Biotechnology Issues for Developing Countries, 5*, 196–199.
2. Anjum, F., Bhatti, H. N., Asgher, M., & Shahid, M. (2010). Leaching of metal ions from black shale by organic acids produced by *Aspergillus niger*. *Applied Clay Science, 47*, 356–361.
3. Anon. (2002). Bioleaching Moves forward. *The Mining Journal*, May 31, pp. 392.
4. Atlas, R. M., & Bartha, R. (1997). *Microbial ecology- Fundamentals and applications*. 4th Ed., Menlo Park: Addison Wesley Longman.
5. Baba, A. A., Adekola, F. A., & Lawal, A. J. (2007). Investigation of Chemical and Microbial Leaching of Iron ore in Sulphuric acid. *Journal of Applied Sciences and Environmental Management, 11*, 39–44.
6. Barlett, R. W. (1992). *Solution Mining: Leaching and Fluid Recovery of Materials*. 2nd Ed., Gordon and Breach Science Publishers, The Netherlands.
7. Bo-Fu., H., Z., Rubing, Z., & Guanzhou, Q. (2008). Bioleaching of chalcopyrite by pure and mixed cultures of *Acidithiobacillus* spp. and *Leptospirillum ferriphilum*. *International Biodeterioration and Biodegradation, 62*, 109–115.
8. Bosecker, K. (2001). Microbial leaching in environmental cleanup programs. *Hydrometallurgy, 59*, 245–248.
9. Bosecker, K. (1994). Mikrobielle Laugung (Leaching). In: P. Prëave, U. Faust, W. Sittig, & D. A. Sukatsch (Eds.), *Handbuch der Biotechnologie*, R. Oldenbourg, Munich, pp. 835–858.

10. Bosecker, K. (1997). Bioleaching: metal solubilization by microorganisms. *FEMS Microbiology Reviews, 20,* 591–604.
11. Bosecker, K. (2006). Bioleaching: metal solubilization by microorganisms. *FEMS Microbiology Reviews, 20,* 591–604.
12. Brierley, C. L. (1982). Microbiological mining. *Scientific American,* 247, 42–51.
13. Brierley, C. L. (1997). Mining biotechnology, Research to commercial development and beyond. In: Rawling, D. E. (Ed.). *Biomining: Theory, Microbes and Industrial Processes.* Springer-Verlag, Berlin, pp. 3–7.
14. Brierley, C. L. (2008). How will biomining be applied in future? Transaction of Nonferrous Metals. *Society of China, 18,* 1302–1310.
15. Brierley, C. L., & Brierley, J. A. (1999). Bioheap processes-operational requirements and techniques. In: G. W. Jergensen (Eds.), *Copper Leaching, Solvent Extraction and Electrowinning Technologies.* Littleton. Colorado: Society of Mining Engineers. pp. 17–27.
16. Brierley, C. L., & Brierley, J. A. (2001). Present and future applications of biohydrometallurgy. *Hydrometallurgy, 59,* 233–239.
17. Brierley, J. A., Brierley, C. L., & Goyak G. M. (1986). A new wastewater treatment and metal recovery technology. In: Lawrence, R. W., Branion. R. M. R., & Ebner, H. G. (Eds.), *Fundamental and Applied Biohydrometallurgy.* Amsterdam: Elsevier. pp. 291–300.
18. Brombacher, C., Bachofen, R., & Brand H. (1997). Biohydrometallurgical processing of solids: a patent review. *Applied Microbiology and Biotechnology, 48,* 577–587.
19. Bustos, S., Castro, S., & Montealegre, S. (1993). The Sociedad Minera Pudahuel bacterial thin-layer leaching process at Lo Aguirre. *FEMS (Federation of European Materials Societies) Microbiological Reviews, 11,* 231–235.
20. Canterford, J. H. (1982). *Solution mining: general principles and Australian practice.* Jobson's Min. Yearb. *215,* 226.
21. Coram, N. J., & Rawling D. E. (2002). Molecular relationship between two groups of the genus *Leptospirillum* and the finding that *Leptospirillum* sp. nov. dominates South African commercial biooxidation tanks that operate at 40°C. *Applied and Environmental Microbiology, 68,* 838–845.
22. Costerton, J. W. Cheng, K. J., Geesey, G. G., Ladd, T. I., Nickel, J. C., Dasgupta, M., & Marrie, T. J. (1987). Bacterial Biofilms in Nature and Disease. *Annual Reviews of Microbiology, 41,* 435–464.
23. Das, B. K., Roy, A., Koschorreck, M., Mandal, S. M., Wendt-Potthoff, K., & Bhattacharya, J. (2009). Occurrence and role of algae and fungi in acid mine drainage environment with special reference to metals and sulfate immobilization. *Water Research, 43,* 883–894.
24. Das, T., Ayyappan, S., & Chaudhury, G. R. (1999). Factors affecting bioleaching kinetics of sulfide ores using acidophilic micro-organisms. *BioMetals,* 1–10.
25. Dave, S. R. (1980). *Microbiological and Bioleaching Studies on Metallurgical Bacteria Cultured from Indian Sulphidic Mine Water.* PhD Thesis, University of Mysore, Mysore, India.
26. Dave, S. R. (2008). Selection of *Leptospirillum ferrooxidans* SRPCBL and development for enhanced ferric regeneration in stirred tank and air lift column reactor. *Bioresource Technology, 99,* 7803–7806.

Practices in Bioleaching

27. Demergasso, C., Gallegillos, P., Escudero, L., Zepeda, V., Castillo, D., & Casamayor, E. O. (1996). Molecular characterization of microbial populations in low grade copper ore bioleaching test heap. *Hydrometallurgy, 80*, 241–253.
28. Devasia, P., & Natarajan, K. A. (2004). Bacterial Leaching-Biotechnology in the Mining Industry. *Resonance, 9*, 27–34.
29. Dopson, M., Baker-Austin, C., Koppineei, P. R., & Bond, P. L. (2003). Growth in sulphidic mineral environments: metal resistance mechanisms in acidophilic microorganisms. Microbiology, *Environmental Science and Technology, 5*, 179–182.
30. Ehrlich, H. L. (2001). Past, present and future of hydrometallurgy. *Biohydrometallurgy, 59*, 127–134.
31. Farbiszewska, T., Cwalina, B., Farbiszewskabajer, J., & Dzierzewicz. (1994). The use of bacterial leaching in the utilization of wastes resulting from mining and burning lignite. *Acta Biologica Cracoviensia Series Botanica, 36*, 1–9.
32. Fields, G. (2010). *Bioleaching Technology for Gold.* URL http://www.bactech.com/green/Overview.asp.
33. Fouchera, S., Bruneta, F. B., D'Huguesa, P., Clarensb, M., Godonc, J. J., & Morin, D. (2003). Evolution of the bacterial population during the batch bioleaching of a cobaltiferous pyrite in a suspended solids bubble column and comparison with a mechanically agitated reactor. *Hydrometallurgy, 71*, 5–12.
34. François, F., Lombard, C., Guigner, J. M., Soreau, P., Jaisson, F. B., Martino, G., Vandervennet. M., Garcia, D., Molinier, A. L., Pignol, D., Peduzzi, J., Zirah, S., & Rebuffata, S. (2012). Isolation and Characterization of Environmental Bacteria Capable of Extracellular Biosorption of Mercury. *Applied and Environmental Microbiology, 78*, 1097–1106.
35. Gilbertson, B. P. (2000). Creating value through innovation: biotechnology in mining. *Mineral Processing and Extractive Metallurgy, 109*(2), 61–67.
36. Groudev, S. N. (1999). Biobeneficiation of mineral raw material. *Minerals and Metallurgical Processes, 16*, 19–28.
37. Hansford, G. S., & Miller, D. M. (1993). Biooxidation of a gold-bearing pyrite-arsenopyrite concentrate. *FEMS Microbiology Reviews, 11*, 175–182.
38. Haque, N., & Norgate, T. (2013). The greenhouse gas footprint of in-situ leaching of uranium, gold and copper in Australia. *Journal of Cleaner Production-Science Direct, 122*, 83–96.
39. Ilyas, S., Anwar, M. A., Niazi, S. B., Ghauri, M. A., & Khalid, A. M. (2007). Microbial leaching of iron from pyrite by moderate thermophile chemolithotropic bacteria. *Journal of Research (Science), 18*, 159–166.
40. Ingledew, W. J. (1990). Acidophiles. In: C. A. Edwards (Ed.), *Microbiology of extreme environments.* Open University Press: Milton Keynes. pp. 33–54.
41. Johnson, D. B. (1998). Biodiversity and ecology of acidophilic microorganisms. *FEMS Microbiology Reviews, 27*, 307–317.
42. Kelly, D. P., & Wood, A. P. (2000). Reclassification of some species of *Thiobacillus* to the newly designated genera *Acidithiobacillus* gen. nov., *Halothiobacillus* gen. nov and *Thermithiobacillus* gen. nov. *International Journal of Systematic and Evolutionary Microbiology, 50*, 511–516.

43. Kim, S. D., Bae, J. E., Park, H. S., & Cha, D. K. (2005). Bioleaching of cadmium and nickel from synthetic sediments by *Acidithiobacillus ferrooxidans. Environmental Geochemistry and Health, 27,* 229–235.

44. Lee, J. U., Sung-Min, K., Kyoung-Woong, K., & In, S. K. (2005). Microbial removal of uranium in uranium-bearing black shale. *Chemosphere, 59,* 147–154.

45. Le-Roux, N. W. (1970). Mineral Attack by Microbiological Processes. In: J. A. D. Miller (Ed.), *Microbial Aspects of Metallurgy.* New York: American Elsevier Publishing Co. pp. 173–182.

46. Lizama, H. M., Harlamovs, J. R., McKay, D. J., & Dai, Z. (2005). Heap leaching kinetics are proportional to the irrigation rate divided by heap height. *Minerals Engineering, 18,* 623–630.

47. Lopez-Archilla, A. I., Marin, I., & Amlls, R. (1993). Bioleaching and interrelated acidophilic microorganism from Rio Tinto, Spain. *Journal of Geomicrobiology, 11,* 223–233.

48. Lucious, S., Reddyy, E. S., Anuradha, V., Vijaya, P. P., Ali, M. S., Yogananth, N., Rajan, R., & Kalitha, P. (2013). Heavy Metal Tolerance and Antibiotic Sensitivity of Bacterial Strains Isolated From Tannery Effluent. *Asian Journal of Experimental Biology and Sciences, 4,* 597–606.

49. Lundgren, D. G., & Silver, M. (1980). Ore leaching by bacteria. *Annual Reviews of Microbiology, 34,* 263–283.

50. Mandl, M., Hrbac, D., & Docekalova, H. (1996). Inhibition of Iron (II) oxidation by arsenic (III, V) in *Thiobacillus ferrooxidans:* Effects on arsenopyrite bioleaching. *Biotechnology Letters, 18,* 333–338.

51. Menon, A. G. (1995). *Biotechnology of Complex Sulfide Ore Processing.* PhD Thesis, Gujarat.

52. Mishra, D., Dong-Jin, K., Jong-Gwan, A., & Young-Ha, R. (2005). Bioleaching: A microbial Process of Metal Recovery. *Metals and Materials International, 11,* 249–256.

53. Natarajan, K. A. (2006). *Biotechnology for Metal Extraction, Mineral Beneficiation and Environmental Control.* Proceedings of the International Seminar on Mineral Processing Technology, Chennai, India, pp. 68–81.

54. Natarajan, K. A. (1998). Microbes, Minerals and Environment. *Geological Survey of India, 149,* 1959–1970.

55. Natarajan, K. A., Modak, J. M., & Raichur, A. M. (2001). Bioreactor engineering for treating refractory gold-bearing concentrates: An Indian experience. In: Ciminelli, V. S. T., & Garcia, Jr. O. (Eds.), *Biohydrometallurgy: Fundamentals, Technology and Sustainable Development. Part A, IBS.* Amsterdam: Elsevier, pp. 183–190.

56. Needham, L., & Gwei-Djen. (1974). *Chemistry and Chemical technology: Part II.* University Press Cambridge, pp. 25.

57. Orell, A., Claudio, A. N., Rafaela, A., Juan, C. M., & Carlos, A. J. (2008). Life in blue: Copper resistance mechanisms of bacteria and Archaea used in industrial biomining of minerals. *Biotechnology Advances, 28,* 839–848.

58. Parker, S. P. (1992). *Concise Encyclopedia of Science and Technology.* Mc-Graw Hill, New York.

59. Pradhan, N., Nathsarma, K. C., Rao, K. S., Sukla, L. B., & Mishra, B. K. (2008). Heap bioleaching of chalcopyrite: A review. *Minerals Engineering, 21,* 355–365.

60. Rawlings, D. E. (1997). *Biomining: Theory, Microbes and Industrial Processes.* Springer-Verlag, Berlin, Heidelberg, New York.

61. Rawlings, D. E., David, D., & Plessis, C. D. (2003). Biomineralization of metal-containing ores and concentrates. *Trends in Biotechnology*, *21*, 38–44.
62. Ren, W. X., Li, P. J., Geng, Y., & Li, X. J. (2009). Biological leaching of heavy metals from a contaminated soil by *Aspergillus niger*. *Journal of Hazardous Materials*, *167*, 164–169.
63. Rohwerder, T., Gehrke, T., Kinzler, K., andSand, W. (2003). Bioleaching review Part A: Progress in bioleaching: fundamentals and mechanisms of bacterial metal sulfide oxidation. *Applied Microbiology and Biotechnology*, *63*, 239–248.
64. Rossi, G. (1990). *Biohydrometallurgy*. McGraw-Hill Book Co., Hamburg, Germany.
65. Salkield, L. U. (1987). *Geotechnical Engineering*. Kluwer Academic Publisher, USA. pp. 230.
66. Sand, W., Gerke T., Hallmann R., Rhode K., Sobokte B., & Wentzien S. (1993). *In-situ* bioleaching of metal sulfides: the importance of *Leptospirillum ferrooxidans*. *Biohydrometallurgical Technologies*, *1*, 15–27.
67. Sauer, K., Camper, A. K., Erhlich, G. D., Costerton, J. W., & Davies, D. G. (2002). *Pseudomonas aeruginosa* displays multiple phenotypes during development as a biofilm. *Journal of Bacteriology*, *184*, 1140–1154.
68. Schippers, A., & Sand, W. (1999). Bacterial leaching of metal sulfides proceeds by two indirect mechanisms via thiosulfate or via polysulfides and sulfur. *Applied Environmental Microbiology*, *65*, 319–321.
69. Shah, T. J. (2005). *Development of Extremophilic Bioleaching Consortia for Sulphidic Mineral Bioprocessing*. PhD Thesis, Gujarat University, Ahmedabad, India.
70. Siddiqui, M. D. H., Kumar, A., Kesari, K., & Arif, J. M. (2009). Biomining: A Useful Approach Toward Metal Extraction. *American-Eurasian Journal of Agronomy*, *2*, 84–88.
71. Simona, C., & Micle, V. (2011). Consideration Concerning Factors Influencing Bio-leaching Processes. *ProEnvironment*, *4*, 76–79.
72. Stoodley, P., Sauer, K., Davies, D. G., & Costerton, J. W. (2002). Biofilms as complex differentiated communities. *Annual Reviews of Microbiology*, *56*, 187–209.
73. Szubert, A., Sadowski, Z., Gros, C. P., Barbe, J. M., & Guilard, R. (2006). Identification of metalloporphyrins extracted from the copper bearing black shale of Fore Sudetic Monocline (Poland). Minerals Engineering, featuring selected papers presented at Process Systems for the Metallurgical Industries' 05 Symposium, Vol. 19, Cape Town, South Africa. pp. 1212–1215.
74. Temple, K. L., & Colmer, A. R. (1951). The autotrophic oxidation of iron by a new bacterium: *Thiobacillus ferrooxidans. Journal of bacteriology*, *62*, 605–611.
75. Temple, K. L., & Delchamps, E. W. (1953). Autotrophic bacteria and the formation of acid in bituminous coal mines. *Applied Microbiology*, *1*, 255–258.
76. Thosar, A., Satpathy, P., Nathiya, T., & Rajan, A. P. (2014). Biomining: A revolutionizing technology for a safer and greener environment. *International Journal of Recent Scientific Research*, *5*, 1624–1632.
77. Tipre, D. R. (1999). *Scale-Up of Bioextraction Process for the Polymetallic Concentrate*. PhD Thesis, Gujarat University, Ahmedabad, India.
78. Tipre, D. R., & Dave, S. R. (2008). Bioleaching process for Cu-Pb-Zn bulk concentrate at high pulp density. *Hydrometallurgy*, *75*, 37–43.

79. Torma, A. E. (1977). The role of *Thiobacillus ferrooxidans* in hydrometallurgical processes. *Advances in Biochemical Engineering, 6*, 1–37.
80. Tzeferis, P. G., & Agatzinin-Leonardou, S. (1994). Leaching of nickel and iron from Greek non-sulfide nickeliferous ores by organic acids. *Hydrometallurgy, 36*, 345–360.
81. Valix, M., Usai, F., & Malik, R. (2001). Fungal bio-leaching of low grade laterite ores. *Minerals Engineering, 14*, 197–203.
82. Wei-Min, Z., Chang-Bin, W., Ru-Bing, Z., Pei-Lei, H., Guan-Zhou, Q., Guo-Hua, G., & Hong-Bo, Z. (2009). Isolation and Identification of moderately thermophillic acidophilic iron-oxidizing bacterium and its bioleaching characterization. *Transactions of Nonferrous Metals Society of China, 19*, 222–227.
83. Xu, T. J., & Ting, Y. P. (2009). Fungal bioleaching of incineration fly ash: Metal extraction and modeling growth kinetics. *Enzyme and Microbial Technology, 44*, 323–328.

CHAPTER 8

PRACTICES IN SEED PRIMING: QUALITY IMPROVEMENT OF OIL SEED CROPS

HASNAIN NANGYAL and NIGHAT ZIAUDIN

CONTENTS

8.1 Introduction 179
8.2 Types of Seed Priming 180
8.3 Biochemical Changes Induced by Seed Priming 184
8.4 Conclusions 187
8.5 Summary 187
Keywords 188
References 188

8.1 INTRODUCTION

Oil seeds are extremely susceptible to the unforgiving ecological conditions. It is estimated that their oil content promptly oxidizes, which break down the seed wellbeing away. A standout amongst the most vital viewpoints for oil seed creation is quick development and great seedling foundation in field. On the other hand, germination and emergence are vital issues in plant development and they have critical impact on the subsequent phases of plant development in field. Quick and uniform field emergence of the seeds is fundamental to accomplish high returns with having high quality and quantity in yearly harvests [21].

Pre-sowing seed treatments like chemical and physical treatments are known to improve seed performance in the field. Priming treatments are successfully applied either to poor germinating seed lots or to seeds which are sown under different stress conditions. Priming treatments are widely used in horticultural crops. Consequently, primed seeds are equipped with advanced germination and exhibit improved germination rate and uniformity. Moreover, seed priming is often implicated in improving the stress-tolerance of germinating seeds. Priming initiates metabolic activities, such as: protein, RNA, and DNA synthesis, DNA replication, and β-tubulin accumulation.

The priming permits part of the metabolic procedures vital for germination to happen before the germination starts. In preparing, seeds absorb moisture for distinctive arrangements with high osmotic potential. This keeps the seeds from engrossing enough water for radicle distension, therefore suspending the seeds in the slack stage. Seed preparation has been ordinarily used to diminish the time between seed sowing and seedling development and to synchronize emergence. In seed preparation, the osmotic weight and the period – for which the seeds are kept up in contact with the film – are adequate to permit pre-germinative metabolic procedures to occur inside the seeds up to a level before radicle emerges. Systems for developing seed and affecting drying up resistance in seed are likewise carried out. Ideally the semipermeable film is used as a round or c polygonal cross-segment container, which is pivoted to the seeds on its inward surface and the arrangement is held between its external surface and a body to which the layer is fixed in a watertight passage. Seed preparation has a vital part in increasing the yield by 37, 40, 70, 22, 31, 56, 50, and 20.6% of wheat, grain, upland rice, maize, sorghum, pearl millet, and chick peas, respectively [17].

8.2 TYPES OF SEED PRIMING

There are several types of seed priming, such as hydropriming, osmopriming, halopriming, solid matrix priming, physical priming, and biochemical priming.

8.2.1 HYDROPRIMING

Hydropriming is a basic hydration strategy of pre-germination digestion systems without real germination. It is a standout amongst the most even minded, basic, financial and transient ways to deal with the impacts of dry season [22] and other abiotic weights on seedling development and crop establishment. Hydro prepared seeds more often than not have early, higher and synchronized germination inferable from diminishment in the slack time of imbibition generally required much time [12] and development of germination improving metabolites. Water deficiency amid introductory phase results in inconsistent seedling emergence and plant foundation and in extreme cases, complete restraint of seedling emergence might likewise arise [12]. The greatest challenge on this front is to enhance the effectiveness and efficiency of water being utilized as a part of existing frameworks. Researchers showed that essentially absorbing seeds in plain water before sowing could build the speed and homogeneity of germination and emergence, prompting better plant population, and seedlings to develop considerably more vivaciously [32].

8.2.2 OSMOPRIMING

One noteworthy requirement to seed germination is soil saltiness, a typical issue in inundated regions, with low precipitation [22]. Soil saltiness might influence the germination of seeds either by making an osmotic potential outside to the seed counteracting water uptake, or through the negative impacts of Na and Cl particles on the germination seeds [23]. Another significant limitation to seed germination is water deficiency, as water shortfall conditions may influence altering so as to seedling development antagonistically starch digestion system and the vehicle of sucrose in chick pea seedling [16]. In any case, development controllers like gibberellic corrosive and kinetin might somewhat invert impacts of water deficiency stress amid germination by instigating changes in the exercises of catalysts of sugar digestion system [20]. A few strategies have been accustomed to preconditioning of seeds trying to enhance germination and seedling foundation of numerous vegetables. These incorporates substitute wetting and drying, pre-germination and controlled hydration by method for anosmatic, for example, polyethylene glycol (PG). This strategy for controlled hydration is called preparing or osmo-molding [24]. The broadly useful of seed preparation is to a point where germination procedures are started, however not finished. Treated seeds are generally re-dried to essential dampness before use, however they would display fast germination when reimbibed under ordinary or stress conditions [1]. Preparation of seeds in osmoticums (e.g., mannitol, PG and sodium chloride (osmopriming) and in water (hydro preparing)) has been found to be conservative, straightforward and a protected procedure for expanding the limit of seeds to osmotic modification and upgrading seedling foundation and crop generation with focus on conditions [26].

8.2.3 HALOPRIMING

Halopriming alludes to absorbing of seeds arrangement of inorganic salts (NaCl, KNO$_3$, CaCl$_2$, CaSO$_4$ and so on). Various studies have demonstrated a huge change in seed germination, seedling development and foundation, and crop yield in saline soils in light of halopriming. Khan et al. [24] assessed the reaction of seeds prepared with NaCl arrangement (1 mM) at different saltiness levels 0, 3, 6 and 9 dSm^{-1} in connection with early development and they found that seed preparation with NaCl was observed to be a better treatment as compared to non-prepared seeds if there should arise an occurrence of hot pepper for enhancing the seedling force and stand foundation under saline conditions. Preparing with NaCl and KCl was useful in uprooting the injurious impacts of salts [18]. Rice seeds treated with a blended salt were to develop more quickly than unprimed seeds under saline conditions [8]. Seed germination is advanced by halopriming additionally invigorate resulting development, thus improving product yield [14, 33].

Sedghi et al. [34] demonstrated that with expanding saltiness, germination attributes, for example, germination percentage, rate and plumule length were diminished. However, seed preparation with GA$_3$ and NaCl indicated lower lessening. In the greater part of the saltiness levels, prepared seeds had more germination rate and plumule length than the control. The most astounding new radicle and dry weight in pot marigold was seen at 7.5 dSm^{-1} saline stress level. It appears that higher germination rate in pot marigold demonstrates higher resistance to saltiness than sweet fennel.

Practices in Seed Priming

8.2.4 SOLID MATRIX PRIMING

Solid matrix priming (SMP) includes the utilization of a wet natural or inorganic material [29], which reproduces the normal imbibition occurring in the dirt [25]. The substrate must have qualities, such as [24]: (i) low matric potential, (ii) high seed wellbeing, (iii) high particular surface (i.e., high surface to volume ratio), (iv) negligible water dissolvability, (v) high adhesiveness to seed surface, and (vi) high ability to hold water. The materials utilized were peat or vermiculite, or some business substrates, for example, celite. The seed is put on or blended with the hydrating substrate, which step by step saturates the seed [25]. To enhance the control of imbibition, immaculate water may be supplanted by an osmotic arrangement, as in osmotic preparation [24].

8.2.5 PHYSICAL PRIMING

There are a few signs that numerous physiological components are included in seed preparation, for example, the repair of the age related cell and sub-cellular harm that can gather amid seed improvement [5, 7] and a progression of metabolic occasions of imbibition that set up the radicle distension [13]. Gamma beams with ionizing radiation are the most vigorous type of electro-attractive radiation. It has a vitality level of around 10 kilo electron volts (keV) to a few hundred keV. Hence, they are more frequently used than alpha and beta beams [22]. In another exploration led by Silvia Neam and Marariu, attractive field treatment (120 mT) of tomato seeds under introduction times of 5 min and 10 min brought an important increment in radicle and plumule length, leaf region, and dry weight of harvest plants. With a specific end goal to get the most elevated harvest potential in yield and/or quality, superb seeds that deliver quick and uniform seedling development are required [2]. The attractive incitement of wheat seeds was able to increase the speed of germination. It appears that physical treatment systems may be utilized as bio-stimulators as a part of agrarian therapeutic plants generation, for example, pot marigold [3, 30]. Magneto-gathering in plants relies on attractive presentation dosage, time of introduction and the dampness content that builds germination of seeds [3].

8.2.6 BIOCHEMICAL PRIMING

Biopreparation treatment is possibly noticeable to affect significant changes in plant qualities and to energize more uniform seed germination and plant development connected with growth and microorganism coatings. Natural components, for example, growths and microscopic organisms are utilized as a part of bio-preparation, which incorporates: parasites and enemy microorganisms and the most vital of all are *Trichoderma* and *Pesodomonas*. Application of *Trichoderma* fungi in agriculture has three beneficial effects for plants: (i) it can colonize plant roots and its rhizosphere; (ii) *Trichoderma* fungi can control plant pathogens though parasitism, and antibiosis production, and promote systemic resistance; and (iii) it improves plant health through increasing plant growth. Ultimately, *Trichoderma* fungi stimulates root growth and improve plant growth [37]. Studies have shown that *Trichoderma* treatment of tobacco seeds increased fresh weight of root and leaf area index (300%), lateral roots (300%) and the leaves (140%).

To supply nutrient and water requirements, plants are dependent on soil. Soil functions as a means for plant root development, excludes carbon, hydrogen, oxygen and even nitrogen. In nutrient cycling, soil microbes (bacteria, fungi, algae, protozoa and actinomycetes) have a great influence [27]. Soil bacteria and rhizobacteria comprise important soil biocontrol agents and their interaction with plant root systems can culminate to a wide range of secondary metabolites with a positive effect on plant growth. Resulting secondary metabolites of soil rhizobacteria and plant root system can enhance the availability of mineral and nutrient to the plant, improve plant nitrogen fixation ability and increase plant health by bio-control of phyto-pathogens [35].

8.3 BIOCHEMICAL CHANGES INDUCED BY SEED PRIMING

8.3.1 EFFECTS ON DNA

Guaranteeing DNA uprightness is of key significance to maintain a strategic distance from mistakes of replication and union of DNA. A solid increment in DNA blend just happens toward the end of germination (Stage III) in both primed and unprimed seed [12]. On the other hand, studies have

Practices in Seed Priming

demonstrated few impacts of preparing the DNA combination likewise in Stage II. Leek (*Allium porrum*) seed displayed a little increment of DNA substance in the development of life amid this stage, on account of plastid-and mitochondrial-DNA replication for the repairing. From that point, an expansion in DNA was observed for 14 days after the treatment in both prepared and unprimed treatments, when the seed had entered the irreversible germination stage [5]. It additionally created the impression that the pre-replication repair of harmed DNA favors DNA union [37]. Notwithstanding its vital part, the measure of DNA, which is required in the repair procedures, is just 20–30% of the aggregate DNA orchestrated amid preparing. Preparing may have no immediate impact on cell division, yet progresses its starting (G_1 and G_2 period of mitosis) from Stage III to Stage II of seed imbibition [28]. This development is empowered by an amassing of β-tubulins in prepared seed, in which are proteins included to keep up the cell cytoskeleton and shaping the microtubules important to cell division [11]. The collection of tubulins is connected with the synchronization of cells on the G_2 stage; in the consequent Stage III, cell division takes place in all cells.

8.3.2 EFFECTS ON RNA

Osmopriming builds the RNA content in developing the life. Bray et al. [5] demonstrated that the amassing included rRNA (Ribosomal RNA: 85% of aggregate RNA), in a turnover between corruption of harmed rRNA and union of new rRNA, while the level of mRNA (messenger RNA: 0.5% of aggregate RNA) stayed consistent. The rRNA is as much important to repair cell harms as DNA. Preparing permits the recuperation of rRNA honesty [10], thus guaranteeing a right coding of amino acids for the blend of protein amid seed germination.

8.3.3 EFFECTS ON PROTEIN SYNTHESIS

Protein amalgamation, a vital essential for germination, begins a few minutes after hydration [8]. At this stage, osmopriming discourages protein combination in the incipient organism and hold tissues, contrasted with seed absorbed water for the same time. At that point, preparing impels a more elevated amount in the ensuing germination. Then again, preparing does not seem to affect the amalgamation of a particular protein, as it has been exhibited by the subjective examination of protein example in pea seeds [11].

8.3.4 EFFECTS ON ENZYMES

Osmopriming affects the amalgamation and initiation of chemicals catalyzing the breakdown and preparation of stored substances [37]. This event has been seen in a few animal species. In sugar beet, the debasement result of the stored globulin 11-S was noted to cumulate subsequent to osmopriming [19]. Different catalysts are enacted for the activation of stored starches (α- and β-amylases) and lipids (isocitrateliase) [36]. This impact is connected with the water shortage instigated by osmopriming, which should decide an activation of stored proteins [37]. Seed crumbling amid capacity is another area of compound movement identified with preparing. This weakening is connected with the collection of active oxygen species (AOS): Hydrogen peroxide (H_2O_2), superoxide anion (O_2^-) and hydroxyl radical (OH^-). AOS respond with most natural atoms, bringing about oxidation and carboxylation of amino corrosive deposits and DNA change [4]; and they likewise respond with polyunsaturated unsaturated fats found in cell films, prompting lipid peroxidation and resulting disturbance of layer coordinate [4]. In plants, the resistance framework incorporates cell reinforcement compounds, that is, foragers of AOS, for example, superoxide dismutase (SOD), catalase (CAT) and glutathione reductase (GR) [4]. Preparing seems to fortify this safeguard framework: actually, the treatment was connected with an expansion in CAT expression in Arabidopsis [15] and sunflower (*Helianthus annuus*), and CAT movement in soybean [21] and maize.

8.3.5 DRYING-BACK AND SEED LONGEVITY

Preparing is associated with pragmatic enthusiasm for ensuing seed taking care of and stockpiling. Direct seeding of prepared seed is regularly unfeasible (e.g., wet seeds spanning inside of seeders), hazardous (e.g., downpours postponing the seeding date), or unreasonably expensive at

Practices in Seed Priming

a little homestead scale. Hence, drying back is important to permit seed capacity subsequent to preparing and leads to a significant stage, as the advantages accomplished with preparing may be lost [29]. The impacts of preparing on seed life span show up to some degree conflicting: the treatment was appeared to upgrade seed life span in pepper [15] and onion (*Allium cepa*) [12]

8.4 CONCLUSIONS

Preparing permits the synchronization of the metabolic occasions in a seed part, enhancing the velocity and consistency of field population. Amidst the treatment, seed harm because of different metabolic and hereditary occasions may be repaired through the culmination of Stages I and II of seed imbibition. Ideal preparing guarantees an immaculate replication and interpretation of nucleic acids, advances the actuation of catalysts assembling hold proteins, and readies the phones for division. A noteworthy downside is that seed life span is jeopardized by preparing, in spite of the fact that this relies on upon the conditions amid drying-back; a moderate drying-back counters the falling apart procedures, empowering a recuperation of seed life span after the treatment. Taking everything into account, the essential instruments of preparing are for the most part recognized, though the supporting physiological procedures are not generally clear and depicted in an adequate number of animal groups. This reflects in the differentiating impacts here and was observed in the wake of preparing, which are in charge of the vulnerabilities as yet encompassing this procedure. It indicates that giving further insights on the subject is the best way to accomplish significant and solid advantages of priming.

8.5 SUMMARY

A review of the scientific literature indicates that seed priming plays major role in the development of seeds. Also in primed seeds, mean germination time was less dependent on temperature, which is consistent with the effects expected from the treatment. Priming effects are mainly influenced by osmotic potential, temperature and time. Major biochemical processes (repair of damaged DNA and RNA, preparation for cell division and increased antioxidant activity) are involved in treatment effects to an extent which is not fully ascertained in literature. A reduction in seed storage life is the major disadvantage of priming and the principal constraint to its diffusion, since dehydration to the initial moisture (drying-back) is needed to allow seed storage. Seed behavior during drying-back, the role of the raffinose family oligosaccharides in cell membrane integrity and the expression of antioxidant enzymes in germinating seeds need to be further elucidated in a sufficient number of species, to promote a more reliable use of this technique of priming.

KEYWORDS

- biological priming
- halopriming
- hydropriming
- osmopriming
- physical priming
- seed priming
- solid matrix priming

REFERENCES

1. Ashraf, S., & Bal, U. (2008). Effects of the commercial product based on *Trichoderma harzianum* on plant, bulb and yield characteristics of onion. *Scientia Horticulturae, 116,* 219–222.
2. Artola, A., Carrillo-Castaneda, G., & Santos, G. D. L. (2003). Hydro-priming: A Strategy to increase *Lotus corniculatus L.* Seed vigor. *Seed Science and Technology, 31,* 455–463.
3. Bahram, M., Reza, M., Ghorbanian, T., & Sahar, B. K. (2015). Magnetic field Induction Stimulates Marigold Growth Characteristics Responsible for its Productivity under Greenhouse induction. *Biological Forum – An International Journal, 7,* 1070–1074.
4. Bailly, C., Benamar, A., Corbineau, F., & Come, D. (2005). Antioxidant systems in sunflower (*Helianthus annuus L.*) seeds as affected by priming. *Seed Science, 10,* 35–42.
5. Bray, C. M., Davison, P. A., Ashraf, M., & Taylor, R. M. (2009). Biochemical changes during osmopriming of *leek* seeds. *Annals of Botany, 63,* 185–193.

Practices in Seed Priming

6. Bray, C. M. (2003). Biochemical processes during the osmopriming of seeds (Chapter 28). In: Kigel, J., & Galili, G. (eds.). *Seed Development and Germination.* Marcel Dekker Inc., NY, pp. 767–790.
7. Burgass, R. W., & Powell, A. A. (2005). Evidence for repair processes in the invigoration of seed by hydration. *Annals of Botany, 53,* 753–757.
8. Cheung, C. P., Wu, J., & Suhadolnik, R. J. (2009). Dependence of protein synthesis on RNA synthesis during the early hours of germination of wheat embryos. *Nature, 277,* 66–67.
9. Chiu, K. Y., Chuang, S. J., & Sung, J. M. (2006). Both anti-oxidant and lipid-carbohydrate conversion enhancements are involved in priming improved emergence of *Echinacea purpurea* seeds that differ in size. *Scientia Horticulturae, 108,* 220–226.
10. Coolbear, P., Slater, R. J., & Bryant, J. A. (2004). Changes in nucleic acid levels associated with improved germination performance of tomato seeds after low-temperature pre-sowing treatments. *Annals of Botany, 65*(2), 187–195.
11. De Castro, R. D., Zheng, X. Y., Bergervoet, J. H. W., De Vos, C. H. R., & Bino, R. J. (2010). β-tubulin accumulation and DNA replication in imbibing tomato seeds. *Plant Physiology, 109,* 499–504.
12. Dearman, J., Brocklehurst, P. A., & Drew, R. L. K. (2006). Effects of osmotic priming and aging on onion seed germination. *Annals of Applied Biology, 108,* 639–648.
13. Dell'Aquila, A. (2009). Development of novel techniques in conditioning, testing and sorting seed physiological quality. *Seed Science and Technology, 37,* 608–624.
14. Eleiwa, M. E. (1989). Effect of prolonged seed soaking on the organic and mineral components of immature pods of soybeans. *Egyptian Journal of Botany, 32,* 149–160.
15. Georgiou, K., Thanos, C. A., & Passam, H. C. (2007). Osmo-conditioning as a means of counteracting the aging of pepper seeds during high temperature storage. *Annals of Botany, 60,* 279–285.
16. Gupta, A. K., Singh, J., Kaur, N., & Singh, R. (2003). Effect of polyethylene glycol induced water stress on uptake and transport of sugars in chickpea seedling. *Plant Physiology and Biochemistry, 31,* 743–747.
17. Hamed, A. N., & Somayeh, F. (2012). Evaluating the potential of seed priming techniques in improving germination and early seedling growth of *Aeluropus macrostachys* under salinity stress conditions. *Annals of Biological Research, 3,* 5099–5105.
18. Iqbal, M., Ashraf. M., Jamil, A., & Rehman, S. (2006). Does seed priming induce changes in the levels of some endogenous plant hormones in hexaploid wheat plants under salt stress. *Journal of Integrative Plant Biology, 48,* 181–189.
19. Job, D., Capron, I., Job, C., Dacher, F., & Corbineau, F. C. D. (2009). Identification of germination-specific protein markers and their use in seed priming technology. *Seed Biology Advances and Application, 1,* 449–459.
20. Kaur, S., Gupta, K., & Kaur, N. (2000). Effect of GA3, Kinetine and Indole acetic acid on carbohydrate metabolism in chickpea seedling germination under water stress. *Plant Growth Regulator, 30,* 61–70.
21. Kausar, M., Mahmood, T., Basra, S. M. A., & Arshad, M. (2009). Invigoration of low vigor sunflower hybrids by seed priming. *International Journal of Agriculture and Biology. 11,* 521–528.

22. Kaya, M. D., Ipek, A., & Ozturk, A. (2007). Effects of different soil salinity levels on germination and seedling growth of safflower (*Carthamus tinctorius* L.). *Turkish Journal of Agriculture and Forestry, 27,* 221–227.
23. Khajeh, H. M., Powell, A. A., & Bingham, I. J. (2008). The interaction between salinity stress and seed vigor during germination of soybean seeds. *Seed Science and Technology, 31,* 715–725.
24. Khan, H. A., Ayub, C. M., Pervez, M. A., Bilal, R. M., Shahid, M. A., & Ziaf, K. (2009). Effect of seed priming with NaCl on salinity tolerance of hot pepper (*Capsicum annuum* L.) at seedling stage. *Soil and Environment. 28,* 81–87.
25. McDonald, M. B. (2000). Seed priming. In: Black, M., & Bewley, J. D. (eds.), *Seed Technology and Its Biological Basis.* Sheffield Academic Press, Sheffield, UK, pp. 287–325.
26. Moghanibashi, M., Karimmojeni, H., Nikneshan, P., & Delavar, B. (2012). Effect of hydropriming on seed germination indices of sunflower (*Helianthus annuus* L.) under salt and drought conditions. *Plant Knowledge Journal, 1*(1), 10–15.
27. Nannipieri, P., Ascher, J., Ceccherini, M. T., Landi, L., Pietramellara, G., & Renella, G. (2003). Microbial diversity and soil functions. *European Journal of Soil Science, 54,* 655–670.
28. Ozbingol, N., Corbineau, F., Groot, S. P. C., Bino, R. J., & Come, D. (2009). Activation of the cell cycle in tomato (*Lycopersicon esculentum* Mill.) seeds during osmoconditioning as related to temperature and oxygen. *Annals of Botany, 84,* 245–251.
29. Parera, C. A., & Cantliffe, D. J. (2002). Enhanced emergence and seedling vigor in shrunken-2 sweet corn via seed disinfection and solid matrix priming. *J. Am. Soc. Hortic. Sci., 117,* 400–403.
30. Pietruszewski, S., & Kania, K. (2010). Effect of magnetic field on germination and yield of wheat. *International Agrophysics, 24,* 297–302.
31. Posmyk, M. M., Corbinau, F., Vinel, D., Bailly, C., & Come, D. (2001). Osmo conditioning reduces physiological and biochemical damage induced by chilling in soybean seeds. *Physiologia Plantarum, 111,* 473–482.
32. Rukui, H., Sutevee, S., Thongket, T., & Sunanta, J. (2002). Effect of Hydro priming and Redrying on the Germination of Triploid Watermelon Seeds. *Journal of Natural Sciences, 36,* 219–224.
33. Sallam, H. A. (1999). Effect of some seed-soaking treatments on growth and chemical components on faba bean plants under saline conditions. *Annals of Agricultural Sciences Journal, 44,* 159–171.
34. Sedghi, M., Ali, N., & Esmaielpour, B. (2010). Effect of seed priming on germination and seedling growth of two medicinal plants under salinity. *Emirates Journal of Food and Agriculture 22,* 130–139.
35. Sturz, A. V., & Christie, B. R. (2003). Beneficial micro bialallelopathies in the root zone: the management of soil quality and plant disease with rhizobacteria. *Soil and Tillage Research, 72,* 107–123.
36. Sung, F. J. M., & Chang, Y. H. (2003). Biochemical activities associated with priming of sweet corn seed to improve vigor. *Seed Science and Technology, 21,* 97–105.
37. Varier, A., Vari, A. K., & Dadlani, M. (2010). The subcellular basis of seed priming. *Current Science, 99*(4), 450–456.

CHAPTER 9

MODIFIED PEARL MILLET STARCH: A REVIEW ON CHEMICAL MODIFICATION, CHARACTERIZATION AND FUNCTIONAL PROPERTIES

MANDIRA KAPRI, DEEPAK KUMAR VERMA, AJESH KUMAR, SUDHANSHI BILLORIA, DIPENDRA KUMAR MAHATO, BALJEET SINGH YADAV, and PREM PRAKASH SRIVASTAV

CONTENTS

9.1   Introduction 192

9.2   Plant Description 192

9.3   Nutritional Description 194

9.4   Modification of Pearl Millet Starch 194

9.5   Characterization of Native and Modified Pearl Millet Starch 204

9.6   Functional Properties of Native and Modified
       Pearl Millet Starch 207

9.7   Summary 216

Keywords 217

References 218

9.1 INTRODUCTION

The term "millet" is used to define various grass crops whose seeds are yielded for both human and animal consumption. In the worldwide, Pearl millet (*Pennisetum glaucum*) is accepted as the most important species among the millet crops which are grown for food to feed the hungery population. This millet crop is also known as bulrush or cattail millet [35]. Pearl millet belongs to the Genus *Pennisetum* and species *P. glaucum,* which have their various vernacular names varying from country to country such as in India as "bajra" (Hindi language), in Nigeria as "gero" (Hausa language), in Niger as "hegni" (Djerma language), in Mali as "sanyo", in Sudan as "dukhon" (Arabic language), and in Namibia as "mahangu" [110]. It is generally accepted that this crop was domesticated since 3,000 to 5,000 years ago in Africa, probably in Sahara (the southern edge) and west Nile; and subsequently, spread to the southern Asia [11]. Millets can easily be cultivated in less fertile soils and in wired conditions, such as intense heat and low rainfall and additionally requires shorter growing seasons compared to other cereal. It is one of the drought-tolerant crops grown in semi-arid tropic regions in the world [22].

This chapter reviews chemical modification, characterization and function properties of modified pearl millet starch.

9.2 PLANT DESCRIPTION

The extent of the growth of pearl millet plant (Figure 9.1) limits within 0.5 to 4 m and kernel color (Figure 9.2) may range from white, pale yellow, brown, gray, slate blue to purple (http://www.fao.org/). The length of grains may vary from 3 to 4 mm, which is comparatively greater than those of other millets, and the 1000-kernel weight varies from 2.5 to 14 g having a mean value of 8 g. The size of a sorghum kernel is about thrice the size of a pearl millet kernel (http://www.fao.org/). The presence of the ratio between the germs to endosperm within a kernel of pearl millet is comparatively higher than that of sorghum [34]. The starch, soluble sugar and amylose content (AC) of in different pearl millet genotypes vary from 62.8 to 70.5%, 1.2 to 2.6% and 21.9 to 28.8%, respectively [54].

Modified Pearl Millet Starch

FIGURE 9.1 Standing field crop of *Pennisetum glaucum* with the grains.

FIGURE 9.2 Kernel color of pearl millet: (A) and (B) grain in ear; and (C) grain in petri fish. (Source: http://www.icrisat.org/)

9.3 NUTRITIONAL DESCRIPTION

Pearl millet, being a cereal crop for human being, nourishes the life of the poverty-stricken population of Africa and Asia [8], and is being considered to be highly palatable and serves a good source of protein, minerals and energy as depicted in Table 9.1 [32].

Pearl millet is a good source of well-balanced protein, containing high concentration of threonine, but having deficiency in lysine and lower leucine concentration than in protein obtained from sorghum. Tryptophan levels are found to be present in ample quantity in pearl millet compared to other cereal crops [25]. The nutritional aspects of pearl millet have drawn higher concentration with compare to other common millet crops (Table 9.2) [133] and its nutritional qualities are much more affluent than most other cereal crops (Table 9.3); having altitudinous levels of calcium, iron, zinc, lipids and high quality proteins [65].

9.4 MODIFICATION OF PEARL MILLET STARCH

Millet starches typically have polygonal and spherical starch granules [41]. The polygonal starch granules tend to be larger than the spherical ones. In rare cases, large spherical granules are observed [69]. Hoover et al. [49] reported starch granule sizes of 2–22 μm for three varieties of pearl millets. Millets starches have not been studied as extensively as other cereal starches. Likewise, studies on millet starches over the years have focused on the physicochemical characteristics, with nominal information on its structural characteristics. Under cooking, the native starch has been studied to loose its clarity, form gel and endure syneresis while storing due to the property of retrogradation which results in deteriorated quality of food products. Since, it has been reported that starch in its native state shows low shear stress resistance, thermal decomposition, high retrogradation and syneresis, its applications are restricted [2] and to surmount this problem starch is being modified.

The process of starch modification is recasting and remodeling the basic structure of starch by creating an impact on the hydrogen bond in a regulatory manner [131]. High tendency for modifications is being

Modified Pearl Millet Starch

TABLE 9.1 Nutritional Value of Pearl Millet

Nutrients Components	Value per 100 g (3.5 oz)
Energy	1,582 kJ (378 kcal)
Carbohydrates	72.8 g
Dietary fiber	8.5 g
Fat	4.2 g
Saturated	0.7 g
Monounsaturated	0.8 g
Polyunsaturated	2.1 g
Omega (ω)3	0.1 g
Omega (ω)-6	2.0 g
Protein	11.0 g
Vitamins	
Thiamine (B_1)	0.42 mg
Riboflavin (B_2)	0.29 mg
Niacin (B_3)	4.72 mg
Pantothenic acid (B_5)	0.85 mg
Vitamin B_6	0.38 mg
Folate (B_9)	85 µg
Vitamin K	0.9 µg
Minerals	
Calcium (Ca)	8 mg
Iron (Fe)	3.0 mg
Magnesium (Mg)	114 mg
Manganese (Mn)	1.6 mg
Phosphorus (P)	285 mg
Potassium (K)	195 mg
Sodium (Na)	5 mg
Zinc (Zn)	1.7 mg
Other constituents	
Copper (Cu)	0.8 mg
Selenium (S)	2.7 µg

Source: USDA Nutrient Database [http://ndb.nal.usda.gov/ndb/search/list].

196   Engineering Interventions in Agricultural Processing

TABLE 9.2   Nutrient Content of Pearl Millet with Comparison to Various Millets Crops

Millet crop	Protein (g)	Fiber (g)	Minerals (g)	Iron (mg)	Calcium (mg)
Barnyard millet	11.2	10.1	4.4	15.2	11
Finger millet	7.3	3.6	2.7	3.9	344
Foxtail millet	12.3	8	3.3	2.8	31
Kodo millet	8.3	9	2.6	0.5	27
Little millet	7.7	7.6	1.5	9.3	17
Pearl millet	10.6	1.3	2.3	16.9	38
Proso millet	12.5	2.2	1.9	0.8	14

Source: Millet Network of India [http://www.milletindia.org].

TABLE 9.3   Nutrient Profile of Other Cereal Crops (per 100 g Portion, Raw Grain)

Nutrients components	Wheat*	Rice**	Sweet corn***
Energy (kJ)	1368	1527	360
Carbohydrates (g)	71.2	79	19
Fiber (g)	12.2	1	3
Fat (g)	1.5	1	1
Calcium (Ca)	29	28	2
Folate (B_9) (µg)	38	8	42
Iron (Fe)	3.2	0.8	0.5
Magnesium (Mg)	126	25	37
Manganese (Mn)	3.9	1.1	0.2
Niacin (B_3) (mg)	5.5	1.6	1.8
Pantothenic acid (B_5) (mg)	0.9	1.0	0.7
Phosphorus (P)	288	115	89
Potassium (K)	363	115	270
Protein (g)	12.6	7	3
Riboflavin (B_2) (mg)	0.1	>0.1	0.1
Thiamine (B_1) (mg)	0.38	0.1	0.2
Vitamin B_6 (mg)	0.3	0.2	0.1
Zinc (Zn)	2.6	1.1	0.5

Source: USDA Nutrient Database (http://ndb.nal.usda.gov/ndb/search/list);

Legend: *Hard red winter;

** White, long grain, raw;

*** Sweet, yellow, raw.

Modified Pearl Millet Starch

observed due to the technological development and the evolution on food industry process to preserve desirable natural characteristics of starches. Determining that how differently the starches respond to these modifications should be the first and foremost function for an industrial classification and control [31]. The techniques for starch modification are of different types (Table 9.4) but broadly, these have been classified into four categories with a purpose of producing various unaccustomed derivatives with improvised physicochemical aspects along with their useful structural properties [89]: (i) physical, (ii) chemical, (iii) enzymatic, and (iv) genetic modification.

9.4.1 CHEMICAL MODIFICATION OF PEARL MILLET STARCH

Chemical modification is commercially carried out to overcome the inherent deficiencies of native starch, i.e., higher retrogradation, which results in developing an extensive cohesive gel with elevated syneresis, [79, 111, 114, 125, 129].

Chemical modifications of starch include oxidation [86], cross-linking [102]; and dual modification are commonly practiced [113]. Dilute conditions (aqueous slurry or solution) or semi-dry conditions (extrusion) are best forms that are used for chemical modification of starches [40]. Common process for starch modifications is reactive extrusion [128]. Normal maize, waxy maize, high-amylose maize, tapioca, potato and wheat starch are the most popular and obtainable starches, but varieties of rice (including waxy rice), rye, barley, oat, sago, amaranth, sweet potato, pea (smooth and wrinkled) and certain other exotic starches are few to be used as localized commercial sources. Potentially, more options have been provided by conventional hybrid breeding and genetic engineering [17, 23, 38].

9.4.1.1 Cross Linking of Starch

Cross-linking reagents reported to be in general use are: phosphorus oxychloride ($POCl_3$), sodium trimetaphosphate (STMP) ($Na_3P_3O_9$) and epichlorohydrin (C_3H_5ClO) [80, 106, 109, 111, 115, 124, 129, 130]. Also

TABLE 9.4 Types of Modifications Along with their Objectives

Modification	Objectives	Benefit to the user	Typical uses
Acid Thinning	• Lower viscosity and increase gel strength	• Enhance textural properties at higher usage concentration of starch	• Gums, pastilles, jellies
Cross-linking	• Strengthen starch granule • Delay viscosity development by retarding granule swelling	• Improved process tolerance to heat, acid and sheer • Production efficiency: Increased heat penetration allowing shorter process time	• Ambient stable products • Bottled sauces
Dextrinisation	• Breakdown and rearrange starch molecule providing lower viscosity, increased solubility and a range of viscosity stability from liquid to gel	• Easily handled or applied at higher dosage than parent native starch for desired effect • Create film-forming properties	• Fat replacers • Bakery glazes
Enzyme conversion (Biochemical modification)	• Produce varied viscosity, gel strength, with thermo reversibility and sweetness	• Contributes texture and rheology • Economic dispersant	• Fat mimetic • Flavor carriers • Dry mix filters
Lipophilic Substitution	• Introduce lipophilic groups	• Emulsion stabilizer which improves quality of any fat/oil- containing product • Reduces rancidity by preventing oxidation	• Beverage and salad dressings • Flavor encapsulating agents

Modified Pearl Millet Starch

TABLE 9.4 (Continued)

Modification	Objectives	Benefit to the user	Typical uses
Oxidation	• Introduce carbonyl and carboxyl groups which increases clarity and reduces retrogradation of cooked starch pastes	• Improves adhesion of coatings • Creates soft stable gels at higher dosage than parent native starch	• Battered meat, poultry and fish • Confectionery
Pregelatinization	• Pre-cooked starch to give cold water thickening properties	• Cold water thickening eliminates need to cook, offers convenience and energy savings	• Instant soups, sauces, dressings, desserts, bakery mixes
Stabilization	• Prevent shrinkage of starch granule and provide stability at low temperatures • Lower gelatinization temperature	• Excellent chill and freeze/thaw stability to extend self-life • Easy to cook in high solid system	• Chilled and frozen processed foods • High brix fillings and toppings
Thermal treatment	• Strengthen starch granule • Delay viscosity development by retarding granule swelling	• Unique functional native starch with starch with "Starch" on label declaration • Improved process tolerance to heat, acid and shear • Production efficiency: increased heat penetration allowing shorter process time	• Ambient stable products • Bottled sauces • Sterilized soups and sauces

STMP, sodium tripolyphosphate (STPP) ($Na_5P_3O_{10}$), C_3H_5ClO and $POCl_3$ are used for commercial cross-linking of starch [13, 18, 103, 124, 125, 129,130]. Lim and Seib [77] reported that the preparation of starch phosphates using a combination of phosphate salts such as STMP and STPP gives superior results than using STMP alone for the preparation of di-starch phosphate (cross-linked starch). Cross-linking occurs when bonding channels are acquainted within abutting molecules [64]. The cross-linkers should have the ability to react with at least two hydroxyl groups within single polymer molecule or in abutting molecules for the formation of cross-linking polymers, such as starches, having backbones made up of hydroxyl groups [108]. During starch cross-linking, for the formation of ether, ester or other linkages within starch molecules along with the hydroxyl groups, bi-functional and multifunctional reagents are normally used which result in restricted swelling of the granules during gelatinization and minimize granule rupture, and at the same time it strengthens the starch granules to be highly opposed towards acidic media, heat and shearing [13, 33, 91, 121]. Starch cross-linking can be accomplished either by covalent cross-linking or non-covalent interactions but among them. Non-covalent cross-linked bondings are highly pliant, such as hydrogen bonding, and in some cases can be reversible for obtaining polymer network which can be controllable and cross-linking extent can be controlled by varying cross-linker dosages, starch concentration, pH, and temperatures [13].

Depending on cross-linking reagents, conditions of the reaction vary and can be conducted either in granular state or in semi-liquid state under neutral to alkaline conditions. In some cases where aldehydes are used, the modification should be conducted under acidic conditions and also can be achieved by irradiation without any additives, which is normally used in grafting processes [13]. Restricted swelling of the granules during gelatinization and minimize granule rupture occurs when intermolecular bridges are introduced by multifunctional reagents used during cross-linking [33, 91, 121]. Highly viscous bodies are more stable to heat, shear and low pH, reduced stringy paste formation under cooking and ruined viscosity are the few common characters of paste made from cross-linked starches [70, 121, 122].

According to Liu et al. [78] in waxy and non-waxy starches, different effects of cross-linking were observed. The height of cross-linking may be influenced by some susceptible factors: the size distribution population of starch granule [52] and smaller starch granules however reported earlier that the larger size starch granules are negatively influenced compared to smaller granules [19]. Jane et al. [55] observed that mainly in amylopectin region of the starch, cross-linking occurred, which introduces covalent bonds which are much stronger than hydrogen bonds and makes the granules more difficult to disrupt by chemical or mechanical means and thus are less susceptible to acid attack or shear, and are widely used as thickeners in food, specially where extended and stable viscosity is required in the food industry. Slight decrease in level of hydrolysis of cross-linked starch was observed as compared to unmodified starch [26. 46] and are also used to produce starch which acts as resistant against amylose digestion and resistance to attack by enzyme with elevated degree of cross-linking [121]. Studies revealed that lower degree of cross-linking showed lesser effect in digestion pattern compared to the starch which are unmodified, though the starch resists pasting property [121]. The confirmation of cross-linking of starch can be done by reduced paste clarity [61, 77, 124].

9.4.1.2 Oxidation of Starch

Due to oxidation of starch, different alternations within the structure of starch molecule results in modification of properties of starch in various aspects. Botanical origin of the native starch, the type of agent used for oxidation of starch and the surrounding conditions for the reaction shows a great effect on the magnitude of changes over structure and physicochemical aspects of oxidized starch. Various studies on oxidation of starch have been conducted by many workers in different cereal and tuber crops [36. 51, 67, 68, 87, 93, 94, 95, 99, 100, 116, 119]. It has been found that compared to cereal starches, tuber starches get more willingly oxidized [36, 67]. The ability of hypochlorite oxidation is significantly maintained by AC of the native starch [68]. Starch oxidized with varying levels of sodium hypochlorite (NaClO) results in yielding oxidized products with varying pasting properties [93, 116]. It has been reported that during oxidation by hypochlorite, the rate at which the oxidation reaction taking place and the type and the quantity of production of functional groups within the starch molecules directly depends on the pH of the solution during the reaction [51, 87, 94, 99, 100, 119].

Native starches are heated and their cooled-off gels provide viscous, cohesive and sticky pastes [6] and generally, show low shear stress resistance and thermal decomposition, in addition to high retrogradation and syneresis. To overcome this problem, the molecular weight of native starches has to be reduced via oxidation by using some oxidizing agents, such as: NaClO, bromine (Br_2), per-iodate (IO_4), permanganate ($KMnO_4$), hydrogen peroxide (H_2O_2) and ammonium sulfate [$(NH_4)_2SO_4$].

According to Rutenberg and Solarek [92], oxidation causes de-polymerization, which leads to low dispersion viscosity and incorporates carbonyl and carboxyl groups which results in impeded re-crystallization. The hydroxyl groups present within the starch molecules are primarily oxidized to carbonyl groups, and then finally to carboxyl groups and the degree of substitution (DS) depends on the average number of carboxyl and carbonyl groups introduced per glucose unit. For the oxidized solutions, due to the steric effect of introduced carboxyl and carbonyl groups, the associative tendencies of starch molecules are hindered [57] and thus, with compared to native starches, oxidized starches have a much lower viscosity and higher slurry stability even at a high concentration, which makes it useful for many industry applications without run ability problems.

By decreasing degree of oxidation, a reduction of the peak viscosity (PV), hot paste viscosity and set-back property has be observed and the post-cooking quality of noodles were found to be negatively influenced, when prepared from oxidized starch by hypochlorite oxidation [125]. Adebowale and Lawal [5] reported that starch obtained from velvet bean (also renowned as mucuna bean) showed a tendency of reduced gelatinization and retrogradation when oxidized with NaClO and starch obtained after modification results lower viscosity due to the degradation of starch molecules. The above properties promote the utilization and application of starch oxidized by different oxidizing agents, where higher concentrated solid particles are required [125]. Hypochlorite being an efficient oxidizing agent but during the oxidation reaction may result in formation of chlorinated by-products that are toxic in nature, whereas H_2O_2 creates no harmful by-products, which can be commercially used and can be easily decomposed by oxygen and water [53] and therefore considered to be more eco-friendly and are utilized when there is a requirement of chlorine free process [62].

The botanical origin of the native starch, type of agent used for oxidation of the starch and reaction circumstances have a great effect on extent of structural and physic-chemical changes in the native starch after oxidation and thereby causes molecular alterations of the starch structures and thus different characters are incorporated in the modified starches. Many studies have revealed that tuber starch gets more readily oxidized than cereal starches when the oxidizing agent is NaClO [36, 67]. The AC of native starch was suggested to play a significant role in controlling the efficiency of hypochlorite oxidation [68] and with varying levels of NaClO, pasting characteristics of the oxidized starch alters [93, 116]. Studies concluded that pH is the parameter on which during oxidation reaction with hypochlorite as an oxidizing agent, depends rate of reaction and the type and quantity of functional groups formed within the starch molecule [51, 87, 94, 99, 100, 119].

During oxidation reaction with H_2O_2 as the oxidizing agent, Whistler and Schweiger [120] reported that amylopectin present within the starch molecules gets extensively degraded when pH was above 7, on the other hand a decreased rate of oxidation reaction was observed when pH level was below 5. Parovuori et al. [96] demonstrated that the presence of different metal catalysts influences the oxidation of starch obtained from potato by peroxide as oxidizing agent. Infrequent studies of starches that are oxidized using various oxidizing agents have been found and very few studies on structural aspects and substitution pattern of starch extracted from potato and further oxidized by NaClO and H_2O_2 have been reported [82, 134, 135].

9.4.1.3 Dual Modification (Oxidation Followed by Cross-Linking) of Starch

Cross-linking shows the higher retrogradation tendency; and on the other hand, oxidation significantly diminishes the viscosity of the starch paste and the capability to resist shear stress and thus dual modification (oxidation followed by cross-linking) can be applied for further improvement of the unwanted characters of starch which results in reduced degree of hydrolysis due to steric hindrance imposed by the bulky substituent and the electron withdrawing effect of carbonyl oxygen of the acetyl group [45]. The reaction conditions for preparing dual-modified starches with STMP and STPP in several starchy crops such as rice, wheat, corn, waxy corn, waxy barley and tapioca for starch have been studied [109, 111, 124, 129, 130]. Common structural features of dual modified starch are amylose-amylose re-crystallization, which implies that amylose re-crystallization is most effective parameter in formation of nutritionally beneficial starch fractions with reduced digestibility [29]; and amylose-amylose interactions, which are much stronger than amylose-amylopectin or amylopectin-amylopectin interaction may have continued to exist after gelatinization and thereby partly restricts the accessibility of starch chains to the hydrolyzing enzymes in case of dual modified starch [28].

Oxidized starch in its prime form or gelatinized form diminishes the ultimate effect of hydrolysis of starch by enzymes. Whereas the increased primary period digestion of starch granules were observed, because of enhanced swelling of starch granules. Similar results were obtained by Chung et al. [26] in corn starch. Locating the substrate properly against the active site of amylose would sterically be hindered when bulky substituent gets added on C_2 within the glucose unit and restricts effectively the attack of enzyme on adjacent glycosidic bonds of substituted glucose residues [46, 47, 85]. Studies by Wattanachant et al. [118] revealed that clarity of starch gets decreased after dual modification.

9.5 CHARACTERIZATION OF NATIVE AND MODIFIED PEARL MILLET STARCH

9.5.1 MOISTURE CONTENT (MC)

Moisture content (MC) of 10.8, 13.7, 14.1, 14.0 and 13.0% at different levels were reported by Beleia et al. [15] in different varieties of pearl millet (viz. in HMP 1700, Serere 3A, HMP 550, RMP' 78 and RMP' 76, respectively), which were in agreement with the reports of Freeman and Bocan [37]. The MC of pearl millet starch has been reported to vary from 10.8 to 14.1% for different African cultivars ([15].

Abdalla et al. [2] reported significantly high MC of 7.67% and 8.87% for Ashana and Dembi, respectively which can be characterized to the hydrophilic properties of the pure starch [42]. Bhupender et al. [21] also reported different moisture contents of 8.2, 7.94, 9.37, 7.47, 8.87% for HHB 67, Varun 666, Pioneer 86M86, Roagro9444, Sri Ram 8494, respectively. Bhardwaj et al. [20] reported starch extracted from pearl millet with a MC of 4.5% (db). The MC of starch was reported to be 11.4% [59]. The MC of pearl millet starch below 13% has been recommended to be safe for storage [21].

9.5.2 AMYLOSE CONTENT (AC)

The amylose content (AC) of starch significantly influences its functional properties. Native millet (species not specified) starch has been reported to contain 12 to 19% AC. Badi et al. [14] reported AC range from 17 to 29%. A narrow range of AC was also reported by Beleia et al. [15], which ranged from 22–24%. However, Jambunathan et al. [54] also reported high AC of about 21.9–28.8% for pearl millet starch. However, higher AC has also been reported by further studies, which concluded AC ranges from 15.64 to 19.46 g/100 g of starch for cultivar Roagro-9444 and HHB-67, respectively [21]. Native and modified form of starch contained amylose ranging from 12.04 to 14.1%. The AC of native starch sample was reported to be lowest (12.04%) and highest in oxidized cross-linked modified starch (14.1%) [59].

An increase in AC after oxidation and cross-linking of starch has been observed, which was 13.55% and 12.41%, respectively [59]. Increase in AC might be due to de-polymerization of starch molecules into the polymer chain with shorter chain lengths [68]. Besides introduction of carbonyl and carboxyl groups while oxidation reaction, degrades the starch molecules by forming cleavage between amylose and amylopectin within the starch molecules at α-1,4-glucosidic linkages [116]. The AC also gets increased in cross-linked starch, because within the starch molecule, in the amorphous region, amylose is present and this region becomes more accessible for cross-linking than amylopectin side chains, during the modification of starch [60].

A significant increase in AC of cross-linked and dual modified starches was observed compared to native sample of bambarra groundnut starch [4]. The reason for such observation may be, amylose might have willingly reacted with NaClO, and thus insufficient accessibility of NaClO was reported for the amylopectin oxidation. It was earlier reported that increased susceptibility to oxidative deformation of amylose may result due to randomly arranged, linear structure of amylose. In case of cross-linked starch, cross-linking mainly takes place within the amorphous region of starch granule [46]. Though amylose and amylopectin plays a role in retrogadation, but retrogadation of amylose are renowned to be a quick process which takes only few hours generally due to its linear structure and this makes easier and faster reassotiation, while retrogradation process of amylopectin takes over a period of days [72].

An increase in AC has also been observed with respect to the native form of pearl millet starch and was reported to be 14.10% by Kapri [59]. Hence, the AC can increase after the modification of cross-linking because the amorphous region of the starch granule are acquired by amylose and during the process of modification, this amorphous region becomes more susceptible than amylopectin side chains and thus structural and modification based differences occurs within amylose and amylopectin during modification.

9.5.3 DEGREE OF OXIDATION (CARBOXYL AND CARBONYL CONTENT) AND CROSS-LINKING

Studies concluded that carboxyl and carbonyl content of the oxidized pearl millet starch was 4.25% and 0.07%, respectively [59]. Earlier reported that hydroxyl groups present within starch molecules are primarily oxidized to carbonyl groups and further to carboxyl groups [68, 119]. Studies have showed that oxidation with hypochlorite as an oxidizing agent produces functional groups, of which carboxyl groups are produced in majority and its amount increases with increase in reaction time along with minor production of carbonyl groups [95]. Earlier studies reported that when oxidation of starch was done using hypochlorite as an oxidizing agent under alkaline conditions, the reaction favors majorly the production of carboxyl groups [125]. The paternal formation of the functional group, i.e., the

Modified Pearl Millet Starch

carbonyl and carboxyl group during the oxidation of starch with hypochlorite is: successively carbonyl groups are formed as intermediate during the reaction, which again gets oxidized into carboxyl groups and obtained as primary final product. The results suggest that the successive conversion of functional groups from carbonyl to carboxyl groups was very rapid.

Wattanachant et al. [118] reported that starch cross-linked with POCl$_3$ and the mixture of phosphate salts help in formation of di-starch phosphate derivatives by permitting the phosphate groups to replace the bonds, which results higher phosphorus content (P < 0.05) compared to the native starch. Koo et al. [66] reported that an increase in concentration of phosphate salts will successively increase the degree of cross-linking; and these results were supported by Kapri [59], who indicated that degree of cross-linking was 68.8%. The modified starch granules showed higher cross-linking in amylose chains. Wattanachant et al. [118] reported that, to obtain a very appreciable effect, the very low degree of cross-linking reaction with C$_3$H$_5$ClO made it difficult to determine the DS directly and earlier reported the characterization of cross-linked starch and its quality control are dependent on the measurement of physical properties [90, 92].

9.6 FUNCTIONAL PROPERTIES OF NATIVE AND MODIFIED PEARL MILLET STARCH

9.6.1 SWELLING AND SOLUBILITY ASPECTS

The pattern of swelling and solubility aspects of starch describes the associative binding force inside the starch granules [75]. Leach et al. [75] and Zeleznak and Hoseney [133] concluded that swelling property of starch reflects the amylopectin content within the starch, whereas amylose acting as diluents and inhibits swelling activity specially by forming insoluble complex with lipids during swelling and gelatinization. According to Schoch and Maywald [101], the swelling behavior of pearl millet starches can be considered as highly restricted and starches showing restricted swelling behavior are relatively stable against shearing action during cooking in water [39]. The swelling behavior of starch was reported to be a functional property of starch with respect to temperature and the average value of swelling behavior significantly increases with increase in temperature.

Beleia et al. [15] reported that at 65°C and 70°C starches of RMP categories showed significantly less swollen properties, whereas at 80°C to 90°C they showed greatest swelling behavior but the values were not significant and the suspected reason was, the method was not as precise as that of followed by Leach et al. [75]. Malleshi et al. [81] reported a decline in the swelling factor of millet starches when germinated. On the other hand, Abd Allah et al. [1] also concluded higher swelling factor for pearl millet. Swelling behavior of starch depends on AC, structure of amylose and amylopectin and it provides evidence of magnitude of interaction between starch chains within amorphous and crystalline domains [112].

Hoover et al. [49] studied the swelling behavior and leaching of amylose from starches extracted from pearl millet which was observed having an increasing effect in swelling factor and amylose leaching with increasing temperature for the millet starches and was found to be predominant between 60°C to 70°C. They also observed that the swelling factor of the starches extracted from pearl millet was much higher than those of wheat and rice. Swelling power was concluded to be achieved due to amylopectin content and also the chemical aspects (distribution across the granule, amylose/amylopectin distribution) [71]. Swelling power of pearl millet starch has also been observed to range from 2.03 to 14.5 g/g with change in temperature from 60–90°C [21].

According to Bhupender et al. [21], swelling power was observed to be highly affected by temperature and was found to be significantly increased with temperature for Shri Ram 8494 cultivar, whereas opposite results were observed for Pioneer 86M86 cultivar. Damilola et al. [30] reported that swelling index of millet starch was 2.2. Swelling power of native starch was found to be varied from 2.03 to 8.76 g/g with change in temperature from 50 to 90°C [59]. Therefore, it can be concluded that swelling behavior of starch mainly depends and can vary with AC, structural property of amylose and amylopectin, presence of non-carbohydrate substances and especially the lipids present, acting as barrier of swelling [112].

Studies revealed that a reduced swelling power was due to the formation of inter-molecular channels by phosphorous residues after the reaction of cross-linking [27, 56, 125]. Hoover and Manuel [44] concluded that after the modification of starch, increase in starch crystallinity might have been limited the penetration of water within starch matrices causing restricted

Modified Pearl Millet Starch

swelling behavior. However, it was reported earlier that higher level of cross-linking caused a reduced swelling power [58]. A decrease in swelling behavior of cross-linked starch was also observed and it ranged from 1.81 to 8.62 g/g with the change in temperature from 50 to 90°C [59]. An increase in swelling behavior was observed in case of oxidized starch, which ranged from 2.14 to 11.4 g/g as the temperature was increased from 50 to 90°C. The rise in swelling power with increase in temperature may be due to gelatinization of starch which unfolds the structure of starch granule and an insertion of functional group. Increase in swelling behavior of starches with increase in temperature has been reported in some studies [10, 72]. A similar increase in swelling behavior with temperature was observed in the case of oxidized cross-linked modified starch in comparison with the native starch and the range was 2.16 to 9.56 g/g with change in temperature. Increase in swelling behavior of starch with increase in temperature has been reported [126]. Due to the presence of elongated chains of amylopectin structure within the pearl millet starch, increased swelling behavior may have observed [97].

Beleia et al. [15] reported that cultivars of HMP550 and HMP1700 showed low solubility at 75°C, whereas, starches isolated from RMP cultivars showed an opposite behavior. Significantly more soluble than other starches and at 95°C starch from RMP78 cultivar was most soluble. They also reported that starch having low swelling and solubility behavior at low temperature possesses high swelling and solubility behavior at high temperature and this phenomenon can be related to two stage relaxation of bonding forces within the starch granules during swelling. Beleia et al. [15] also observed a two-stage solubility pattern in millet starches.

Malleshi et al. [81] reported an elevation in the solubility of millet starches when germinated. Hoover et al. [49] studied that the leaching of pearl millet starch at temperatures between 50 and 95°C and found that amylose leaching with increasing temperature for the millet starch was most predominant between 60 and 70°C. Difference in solubility pattern with varying temperature has been observed [16], which was characterized due to the presence of different chain length within the starch molecule. During the process of swelling, the quantity of amylose leaching out from the starch granule is referred to as solubility behavior of the starch at that particular temperature; therefore, the swelling behavior will get increased with elevated leaching of amylose [107]. Lawal [73] reported that solubility corresponds to hydrophilicity and AC, i.e., more the dissociation of inter- and intra-hydrogen bonds, more will be the amylose leaching and hence solubility will increase.

Bhupender et al. [21] also reported that the increase in solubility for cv. Shri Ram 8494 was recorded with increasing temperature, whereas decrease in solubility with increasing temperature in case of cultivar Pioneer 86M86 was also reported. Lower swelling factor in one of the millet starch was due to its excess content of lipid-complexes amylose and/or the availability of more crystallites within the granule [12]. Solubility of starch reflects the degree of starch granules dispersed after cooking and solubility behavior of native pearl millet starch at various temperature (50 to 90°C) was reported to be a function of temperature and the average value of swelling capacity gets elevated significantly with elevation in temperature and the native sample of starch isolated from pearl millet showed higher solubility [59]. And this might be due to the elevation in temperature, the starch molecules imbibe water and the amylose and linear branches of amylopectin split in suspension and the solubility of starch gets elevated [20]. Thus it can be concluded that solubility patterns does not readily follow the swelling behavior pattern [15].

Damilola et al. [30] reported a solubility of 12.5% for millet starch. A lower solubility compared to native starch was observed in oxidized starch. According to Wang and Wang [116], aldehyde groups were initially generated due to the minimization of formation of oxidant hemiacetal cross-links in oxidized starch, which stabilized the swelling behavior of starch granules and also decrease its solubility as the cross-links hinders the amylose molecules from leaching out. Cross-linked and dual starches showed lower solubility than native starch. This may be due to the fact that the process of cross-linking of starch strengthens the granule structure and hinders the starch molecules from being leached out. Thus minimizes disintegration of starch granules during gelatinization [126]. Oxidized cross-linked starch also showed a lower solubility pattern than the native starch. Beleia et al. [15] reported that HMP550 and HMP1700 had low solubility at temperature 75°C and on the other hand RMP78 cv. was significantly more soluble and, most soluble were leached from RMP starches at 95°C. Therefore, it can be concluded that starches have low swelling and solubility behavior at 70°C compared to high swelling and solubility behavior at 80–95°C.

9.6.2 LEAST GELATION CONCENTRATION (LGC)

Least gelation concentration (LGC) can be referred to as gelation index. The process of starch gelation is considered as a complex process that involves steps like swelling, water adsorption or imbibition to form a three-dimensional structure, which offers structural support and rigidity to different foods. Variations among LGC within different varieties may be corelated to the relative ratio of various constituents like protein, carbohydrates and lipids [98] and due to high content of starch which restricts the process of gelation due to starch-starch or starch-lipid interaction creates a decrease in LGC [104]. Oshodi et al. [84] observed starch extracted from pearl millet having a GLC of 12% (w/v). Kapri [59] reported that native starch formed strong gel at 8%, weak gel at 6% and 4% concentrations and no gel was obtained at 2%.

Similar results were obtained by Abdalla et al. [2] for Khadim and Rigeiba varieties of pearl millet starch, and they reported a strong gel at 8%, a week gel at 6% and 4% concentrations and no gel was observed at 2%. In contrast, starch extracted from both Ashana and Dembi, varieties formed a week gel at 2%, a strong gel at 4% and a very strong gel at 6%, 8% and 10% concentration.

Oxidized and cross-linked starches formed strong gel at 10%, weak gel at 8% and 6%. The high LGC for oxidized starch may be because of building of the structural network that involves the bridging of the intergranular binding forces within the molecules of the starch, which mainly includes hydrogen bonding. Insertion of functional groups by oxidation, like carbonyl and carboxyl groups, might inhibit this interaction to cause electrostatic repulsion within the starch molecules. Similar results have been reported by Lawal [72] and Yusuf et al. [132] in their studies on hybrid maize starch and jack bean starch, respectively. Activity of cross-linking agents forms a hard crust that restricts gel formation [50]. Thus, there is an increased LGC after cross-linking. Oxidized cross-linked starch formed gel at a higher concentration of 1.2%, and weak gel at 10% and 8%.

9.6.3 SWELLING VOLUME

The cold water binding capacity (WBC) or swelling volume (SV) of HMP550 and HMP1700 pearl millet cultivars ranged from 83.6% and

99.5%, respectively as reported by Beleia et al. [15] and the variation of water binding may be due to inherent difference in proportion of amorphous and crystalline region in starch granules. Starch containing higher amorphous regions will imbibe more water. Beleia et al. [15] reported a range of 83.6–99.5% WBC for starches obtain from African cultivars. Lawal [72] reported that variations in WBC may be caused by inherent differences in proportion to crystalline and amorphous areas in granules. Starch containing elevated proportion of amorphous material would have more water binding sites thus absorbing more water.

Abdalla et al. [2] reported that WBC values were 2.10 and 2.24 ml/g for Khadim and Rigeiba, respectively; 1.72 ml/g for Ashana and 1.69 ml/g for Dembi and thus resulted water absorption capacities of Khadim and Rigeiba starch varieties were higher compared to Ashana and Dembi pearl millet starch varieties. The elevated levels of protein might be a reason for high bonding of hydrogen, and both facilitates binding and imbibitions or entrapment of water [9]. Bhupender et al. [21] reported that WBC of starches from pearl millet was highest and ranged from 83.27 to 88.84% for Roagro 9444 and lowest for Pioneer 86M86. Bhardwaj et al. [20] reported that SV of pearl millet starch to be 291.8% of unknown variety.

Damilola et al. [30] reported that water absorption capacity of millet starch was 50.64% and the low values for the native starch than modified one implies firmness cf the natural granules [7]. The SV is an indicator of WBC of the starch. All modified starches showed decrease in SV compared to the native starch and the SV was 2.2 ml/g, 2.0 ml/g, 2.0 ml/g and 1.7 ml/g for native, cross-linked, oxidized and oxidized cross-linked starch, respectively. This is because of tempering or rupturing of granules [59].

The process of cross-linking has been reported to decrease the SV of starches which depend on type of agents used for cross-linking and degree of cross-linking occurred [46]. Singh et al. [105] also reported that the cross-linked starches suffer a reduction in SV and the resistance towards swelling was increased by increasing concentration of cross-linking agent. Wang and Wang [116] also reported that due to the process of cross-linking movement of starch molecules, there is a reduction in water imbibition. Maximum formation of hydrophilic groups (-COOH) may increase the SV property in the oxidized starch [76]. The availability of water binding sites within the starch molecules of modified starch leads to the variation in WBC [3, 123].

9.6.4 X-RAY DIFFRACTION (XRD) STUDIES

Starch, being a semi-crystalline material, can be identified structurally by light microscope and through X-ray diffraction (XRD) patterns. Four renowned types of XRD found in native starches (Figure 9.3) are A, B, C and V [136]:

- Type A represents to most starches originated from the cereals.
- Type B represent starches originated from potato and other root and tubers and also the retrograded starch.
- Type C represent starches obtained from smooth pea and various beans and the final type.
- Type V re represents only those starches having amylose helical complex starches after the process of gelatinization associated with lipid or related compounds.

Figure 9.3. X-ray diffraction (XRD) studies of different starches: Labeling (A) refers to: A-type from cereal starches, (B) refers to: B-type from tuber starches, (C) refers to: C-type from seed and legume starches, (V) refers to: V-type from helical amylose complexes.

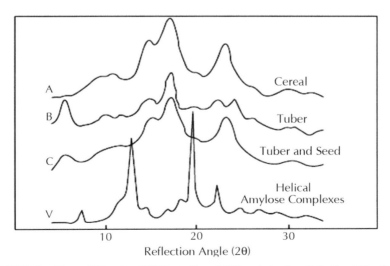

FIGURE 9.3 X-ray diffraction (XRD) studies of different starches: Labeling (A) refers to: A-type from cereal starches, (B) refers to: B-type from tuber starches, (C) refers to: C-type from seed and legume starches, (V) refers to: V-type from helical amylose complexes.

Differences in crystallinity within different starches are generally influenced by the factors like: crystallite size, number of crystallites that are arranged in a crystalline array, MC and polymorphic content [48]. The millet starches possessed a typical A-type XRD pattern. Perez et al. [88] reported that starch showing elevated crystallinity level contains more amylose, which is contrary to the fact that native starch having higher AC shows a lower degree of crystallinity [83]. Studies revealed that Pearl millet starch possess A-type crystallinity and has a crystallinity of about 24.145% [59].

Kuakpetoon and Wang [68] reported that after oxidative treatment of corn starch with 0.8% NaClO as an oxidative agent, the relative crystallinity was found to get elevated, whereas a slight reduction in relative crystallinity was observed when hypochlorite concentration was increased to 2% and 5%; and this may be due to the degradation of the crystalline region. They indicated that there was no noticeable variation between relative crystallinity of the native and cross linked pearl millet starches and these results are in agreement with those of Hoover and Sosulski [46], who reported that cross-linking mainly takes place in the amorphous regions of starch granules and does not change the crystalline patterns of starch. It has been observed that cross-linked and oxidized starches have a relatively lower crystallinity percentage as compared to the native pearl millet starch [59].

Relative crystallinity of oxidized cross-linked was observed to be lower than the native pearl millet starch because of the crystallinity of starch [59]. Jane et al. [55] reported that amylopectin molecules seem to be more cross-linked than amylose. Hood and Mercier [43] speculated that cross-linked sites are located at the amorphous region of the amylopectin.

9.6.5 PASTING PROPERTIES

A paste is defined as a viscous mass, which consists of a continuous phase of solubilized amylose or amylopectin and a discontinuous phase of granule ghosts and fragments. After holding at 95°C for one hour and also during the cooling, however larger differences were observed among the pearl millet starches. Hoover et al. [49] also reported that differences in the pasting properties of pearl millet varieties and viscosity differences at 95°C may be due to degree of crystallinity, extent of amylose leaching and amount of amylose lipids. Whereas, they also observed an increase in viscosity during the holding period at 95°C. However, decrease in viscosity during the same period has been reported by other investigators [15. 37, 117]. Abd Allah et al. [1] reported a temperature of gelatinization of 90.8°C for pearl millet. Choi et al. [24] reported an initial temperature of gelatinization of 75.2°C for waxy millet, which was higher than that (69.0 to 69.3°C) observed for waxy millet [63].

The pasting temperature, PV, setback viscosity and breakdown viscosity of native and modified pearl millet starch differed significantly; and the peak and final viscosities of starch (that are oxidized) were observed to be lower than native starch [59]. This was in agreement with reduced PV and final viscosity of oxidized rice, corn and potato starches [127]. This reduction in paste viscosity and pasting temperature of oxidized pearl millet starch might be caused by partial breakdown of the glycosidic linkages due to oxidation, which results in reduced molecular weight of starch molecules and forms a lower viscosity due to unresisting shearing action and unable to maintain starch granule integrity because of partially degraded network [92]. Due to oxidative reaction, starch chains under goes cleavage and diminishes the viscosity, which leads to low molecular size of starch granules [67, 92]. After the application of oxidation reaction on starch obtained from mucuna bean, reduced PV has been observed [5].

Wattanachant et al. [118] reported a reduction in pasting temperature and delayed PV temperature when cross-linking of starch was done using mixture of phosphate salts as a cross-linking agent, after hydroxyl-propylation with 8% propylene oxide. A restricted PV for cross-linked starch with compared to native starch was reported, and the lower PV and high pasting temperature (89.6°C) of cross-linked starches could be because of strengthening of bonds between starch chains resulting in opposing in swelling of starch granules, leading to elevated pasting temperature and the decrease in PV and cross-linking for pearl millet starch also resist the setback viscosity, and the molecules might have less mobility as a result of cross-linking along with reduced interactions of starch molecules with water molecules due to cross-linking. Therefore cross-linked starch shows lower viscosity as compared to the native starch [59]. Correspondingly, an elevation of granular resistance to temperature and heating time was studied after the cross-linking of starch [74, 105].

Cross-linking gave a significant reduction in the breakdown. The breakdown viscosities of cross-linked pearl millet starch were however lower than native starch samples. After dual modification (oxidized cross-linked) of pearl millet starch, compared to native and cross-linked starch samples, the dual modified starch sample suffers a lower setback value due to the introduction of carbonyl and carboxyl functional groups [96].

Wattanachant et al. [118] reported that PV did not show any significant difference when different cross-linking reagents were used. However, the mixture of phosphate salts affected highest PV among the samples. It has been earlier reported that cross-linking can decrease the viscosity breakdown of starch [92]. They also confirmed that the mixture of phosphate salts acts effectively as cross-linking agents and showed lowest viscosity breakdown. The retaining granular integrity under heat and shearing conditions, elevated setback properties due to elevated developed viscosity during cooling period that have been developed in cross-linked starches.

9.7 SUMMARY

Starch contributes a major portion of human diet. Among the various sources of starch, millet has been one of promising crops that have been neglected despite its great potential to fulfill the increasing demand for food and feed. As the crop requires short duration to grow even in drought conditions, it can be the prominent and reliable source for starch. Besides the use of starch for food and feed purpose, there are many industrial applications where starch can be utilized but with slight modifications. In order to achieve the same goal, chemical modifications are undergone which includes from cross-linking, oxidation or oxidation followed by cross-linking. These oxidized and/or cross-linked starch has better functional properties like swelling power, solubility as well as gelling and pasting properties appropriate for various applications viz., formation of edible films and coatings to protect and enhance the quality and safety of different food matrices. Oxidized starch is used as a surface sizing agents and coating binders in the paper industry; for the encapsulation of flavor oils or for uses as stabilizer, emulsifier, thickening agent, dusting agent, drying aids, binder, clouding agent, suspending agent and for freeze-thaw stability. Oxidized starch is commonly produced by reaction of starch with an oxidizing agent under

Modified Pearl Millet Starch

controlled temperature and pH. During oxidation process, hydroxyl groups on starch molecules are oxidized to carbonyl and carboxyl groups, contributing improved stability to starch paste. The reaction also causes degradation of starch molecules resulting in a modified starch with low viscosity. This allows the use of oxidized starch in application where high solid concentration is needed. Future prospect of modified starch should be explored in more details as the demands for healthy foods are increasing. Besides this, new methods for developing genetically modified starches should be looked into as an interesting area as compared to the conventional starch modifications by using enzyme, physical and chemical treatment.

KEYWORDS

- acid thinning
- acidic conditions
- aldehyde group
- amylopectin
- amylose
- bi-functional reagent
- carboxyl group
- covalent cross-linking
- cross-linked starch
- electrostatic repulsion
- enzymatic modification
- enzyme conversion
- epichlorohydrin
- genetic engineering
- glycosidic linkage
- hydrogen bonding
- least gelation concentration
- lipophilic substitution
- non-covalent interactions
- non-waxy starch
- polygonal starch granules

- **retrogradation**
- **spherical starch granules**
- **starch-lipid interaction**
- **starch-starch interaction**
- **swelling volume**
- **syneresis**
- **water binding capacity**
- **x-ray diffraction**

REFERENCES

1. Abd Allah, M. A., Foda, V. H., Mahmoud, R. M., & Abou Arab, A. A. (1987). X-ray diffraction of starches isolated from yellow corn, sorghum, sordan and pearl millet. *Starch, 39*, 40–42.
2. Abdalla, A. A., Ahmed, U. M., Ahmed, A. R., Tinay, A. H. E., & Ibrahim, K. A. (2009). Chemical properties of native and malted finger millet, pearl millet and foxtail millet starches. *Journal of Applied Science Research, 11*, 2016–2027.
3. Abraham, T. E. (1993). Stabilization of paste viscosity of cassava starch by heat moisture treatment. *Starch/Staerke, 45*, 131–135.
4. Adebowale, K. O., & Lawal, O. S. (2002). Effect of annealing and heat moisture conditioning on the physicochemical characteristics of Bambara groundnut (*Voandzeia subterranean*) starch. *Nahrung/Food, 46*, 311–316.
5. Adebowale, K. O., & Lawal, O. S. (2003). Functional properties and retrogradation behavior of native and chemically modified starch of mucuna bean (*Mucuna pruriens*). *Journal of the Science of Food and Agriculture, 83*, 1541–1546.
6. Adebowale, K. O., Olu-Owolabi, B. I., Olawumi, E. K., & Lawal, O. S. (2005). Functional properties of native, physically and chemically modified bread fruit (*Artocarpus artilis*) starch. *Industrial Crops and Products, 21*, 343–351.
7. Akin-Ajani, O. D, Itiola, O. A., & Odeku, O. A. (2014). Effect of acid modification on the material and compaction properties of fonio and sweet potato starches. *Starch/Starke, 66*, 749–759.
8. Ali, M. A. M., El Tinay, A. H., & Abdalla, A. H. (2003). Effect of fermentation on the *In vitro* protein digestibility of pearl millet. *Food Chemistry, 80*, 51–54.
9. Altschul, A. M., & Wilcks, H. L. (1985). *New Proteins Foods: Food Science and Technology*. Academic Press, Orlando, Florida.
10. Ancana, D. B., Guerrero, L. C., & Harnandez, E. C. (1997). Acetylation and characterization of *Canavalia ensiformis* starch. *Agricultural and Food Chemistry, 45*, 375–382.
11. Andrews, D. J., & Anand Kumar, K. (1992). Pearl millet for food, feed, and forage. *Advance Agronomy, 48*, 89–139.

Modified Pearl Millet Starch

12. Annor, G. A., Marcone, M., Bertoft, E., & Seetharaman, A. (2013). *In vitro* digestibility and expected glycemic index of Kodo millet (*Paspalum scrobiculatum*) as affected by starch–protein–lipid interactions. *Cereal Chemistry, 90*, 211–217.
13. Ayoub, A. S., & Rizvi, S. S. H. (2009). An overview on the technology of crosslinking of starch for nonfood applications. *Journal of Plastic Film and Sheeting, 25*(1), 25–45.
14. Badi, S. M., Hoseney, R. C., & Finney, P. L. (1976). Pearl millet characterization by amino acid analysis, lipid composition and prolamine solubility. *Cereal Chemistry, 53*, 478–487.
15. Beleia, A., Varriano–Marston, E., & Hoseney, R. C. (1980). Characterization of starch from pearl millets. *Cereal Chemistry, 57*(5), 300–303.
16. Bello–Perez, L. A., Contreras Ramos, S. M., Jimennez-Afarican, A., & Paredese Lopez, O. (2000). Acetylation and characterization of banana starch. *Acta Ceintifica Venezolona, 51*, 143–149.
17. BeMiller, J. N. (1997). Starch modification: Challenges and prospects. *Starch/Starke, 49*, 127–131.
18. Bergthaller, W. (2004). Starch world markets and isolation of starch. In: Tomasik, P. (Ed.), *Chemical and Functional Properties of Food Saccharides*. CRC Press, Boca Raton, pp. 103–122.
19. Bertolini, A. C., Souza, E., Nelson, J. E., & Huber, K. C. (2003). Composition and reactivity of A- and B-type starch granules of normal, partial waxy and waxy wheat. *Cereal Chemistry, 80*, 544–549.
20. Bhardwaj, N., Balasubramanian, S., Sharma, R., & Kaur, J. (2014). Characterization of modified pearl millet (*Pennisetum typhoides*) starch. *Journal of Food Science and Technology, 51*(2), 294–300.
21. Bhupender, S. K., Rajneesh, B., & Baljeet, S. Y. (2013). Physicochemical, functional, thermal and pasting properties of starches isolated from pearl millet cultivars. *International Food Research Journal, 20*(4), 1555–1561.
22. Burton, G. W., Wallance, A. T., & Radice, K. O. (1972). Chemical composition during maturation and nutritive value of pearl millet. *Crop Science, 12*, 187–192.
23. Chiu, C., & Solarek, D. (2009). Modification of starches. In: *Starch: Chemistry and Technology*, 3rd Ed. Elsevier Inc., pp. 629–648.
24. Choi, H., Kim, W., & Shin, M. (2004). Properties of Korean amaranth starch compared to waxy sorghum starches. *Food Chemistry, 56*, 469–477.
25. Chung, O. K., & Pomeranz, Y. (1985). Aminoacids in cereal protein and protein fractions. In: Finley, H. (Ed.), *Digestibility and Amino Acid Availability in Cereals and Oilseeds. American Association of Cereal Chemists*. St. Paul, MN, USA. pp. 65–109.
26. Chung, H. J., Shin, D. H., & Lim, S. T. (2008). *In vitro* starch digestibility and estimated glycemic index of chemically modified corn starches. *Food Research International, 41*, 579–585.
27. Chung, H. J., Woo, K. S., & Lim, S. T. (2004). Glass Transition and enthalpy relaxation of cross-linked corn starches. *Carbohydrate Polymers, 55*, 9–15.
28. Chung. H. J., Liu, Q., & Hoover, R. (2009). The impact of annealing and heat- moisture treatments on rapidly digestible, slowly digestible and resistant starch levels in native and gelatinized corn, pea and lentil starches. *Carbohydrate Polymers, 75*, 436–447.

29. Chung, H. J., Liu, Q , & Hoover, R. (2010). Effect of single and dual hydrothermal treatments on the crystalline structure, thermal properties and nutritional fractions of pea, lentil and navy bean starches. *Food Research International, 43*(2), 501–508.

30. Damilola, B. A., Ajala, T. O., & Oyewo, M. N. F. (2015). The compaction, mechanical and disintegration properties of modified *Pennisetum glaucaum* (Poaceae) starch in directly compressed chloroquine tablet formulations. *Journal of Applied Pharmaceutical Science, 5*(2), 43–50.

31. Dolmatova, L., Ruckebush, C., Dupuy, N., Huvenne, J. P., & Legrand, P. (1998). Identification of modified starches using infrared spectroscopy and artificial neural network processing. *Society for Applied Spectroscopy, 52*, 329–338.

32. Durojaiye, A. A., Falade, K. O., & Akingbala, J. O. (2010). Chemical composition and storage properties of fura from pearl millet (*Pennisetum americanum*). *Journal of Food Processing and Preservation, 34*(5), 820–830.

33. Eliasson, A. C. (1996). *Carbohydrates in food, volume II. Series: Food science and technology.* Marcel Dekker, Inc. Madison Avenue, New York. pp. 355–357.

34. FAO, (1995). *Sorghum and millets in human nutrition 198.* FAOSTAT, 1999. Database Results.

35. FAO. (1986). *Production Yearbook.* Volume 40. FAO, United Nations, Rome.

36. Forssell, P., Hamunen, A., Autio, K., Suortti, T., & Poutanan, K. (1995). Hypochlorite oxidation of barley and potato starch. *Starch/Stärke, 47*, 371–377.

37. Freeman, J. E., & Bozan, B. J. (1973). Pearl millet a potential crop for wet milling. *Cereal Science, 18*, 69–74.

38. Friedman, R. B., Hauber R. J., & Katz, P. R. (1993). Behavior of starches derived from varieties of maize containing different genetic mutations, III: Effects of biopolymer source on starch characteristics including paste clarity. *Journal of Carbohydrate Chemistry, 12*(4/5), 611–624.

39. Galvez, F. C. F., & Resurreccion, A. V. A. (1992). Reliability of the focus group technique in determining the quality characteristics of navy bean noodles. *Journal of Sensory Study, 7*, 315.

40. Gotlieb, K. F., & Capelle, A. (2005). *Starch Derivatization–Fascinating and Unique Industrial Opportunities.* Wageningen Academic Publishers.

41. Hadimani, N. A., Muralikrishna, G., Tharanathan, R. N., & Malleshi, N. G. (2001). Nature of carbohydrates and proteins in three pearl millet varieties varying in processing characteristics and kernel texture. *Journal of Cereal Science, 33*, 17–25.

42. Hahn, R. R. (1969). Dry milling of grain sorghum. *Cereal Science Today, 14*(7), 10–13.

43. Hood, L. F., & Mercier, C. (1978). Molecular structure of unmodified and chemically modified manioc starches. *Carbohydrate Research, 61*, 53–66.

44. Hoover, R., & Manuel, H. (1996). The effect of heat-moisture treatment on the structure and physicochemical properties of normal maize, waxy maize, dull waxy maize and amylomaize V starches. *Journal of Cereal Science, 23*, 153–162.

45. Hoover, R., & Sosulski, F. W. (1985). Studies on the functional characteristics and digestibility of starches from phaseolus vulgaris biotypes. *Starch, 37*, 181–191.

46. Hoover, R., & Sosulski, F. W. (1986). Effect of crosslinking on functional properties of legume starches. *European Journal, 38*, 149–155.

Modified Pearl Millet Starch

47. Hoover, R., & Zhou, Y. (2003). *In vitro* and in vivo hydrolysis of legume starch by α- amylase and resistance starch formation in legumes. *Carbohydrate Polymers, 54,* 401–417.
48. Hoover, R., Hughes, T., Chung, H. J., & Lui, Q. (2010). Composition, molecular structure, properties and modification of pulse starches a review. *Food Research International, 43,* 399–413.
49. Hoover, R., Swamidas, G., Kok, L. S., & Vasanthan, T. (1996). Composition and physicochemical properties of starch from pearl millet grains. *Food Chemistry, 56,* 355–367.
50. Huber, K. C., & BeMiller, J. N. (2001). Location of sites of reaction within starch granules. *Cereal Chemistry, 78,* 173–180.
51. Hullinger, C. H., & Whistler, R. L. (1951). Oxidation of amylose with hypochlorite and hypochlorous acid. *Cereal Chemistry, 28,* 153–157.
52. Hung, P. V., & Morita, N. (2005). Physiochemical properties of hydroxypropylated and crosslinked starches from A-type and B-type wheat starch granules. *Carbohydrate Polymers, 59,* 239–246.
53. Isbell, H. S., & Frush, H. L. (1987). Mechanisms for hydroperoxide degradation of disaccharides and related compounds. *Carbohydrate Research, 161,* 181–193.
54. Jambunathan, R., Singh, U., & Subramanian, V. (1988). Proceedings of a workshop on interface between agriculture, grain quality of sorghum, pearl millet, pigeon-pea and chick pea. *Nutrition and Food Science, 15,* 227–243.
55. Jane, J., Xu, A., Radosavljevlc, M., & Selb, P. A. (1992). Location of amylose in normal starch granules. Susceptibility of amylose and amylopectin to crosslinking reagents. *Cereal Chemistry, 69*(4), 405–409.
56. Janzen, G. L. (1969). Digestibility of starches and phosphatized starches by means of pancreatin. *Starch/Starke, 38,* 149–155.
57. Jonhed, A. (2006). *Properties of Modified Starches and their Use in the Surface Treatment of Paper.* PhD Dissertation, Faculty of Technology and Science, Chemical Engineering, Karlstad University, Karlstad, Sweden.
58. Jyothi, A. N., Moorthy, S. N., & Rajasekharan, K. N. (2006). Effect of crosslinking with epichlorohydrin on the properties of cassava (*Manihotesculenta crantz*) starch. *Food Chemistry, 58,* 292–299.
59. Kapri, M. (2014). *Characterization of Native and Chemically Modified Pearl Millets Starch for their Physicochemical Properties and In Vitro Digestively.* MSc Dissertation, Department of Food Technology, Maharshi Dayanand University, Rohtak, Haryana, India.
60. Karmakar, S., Raj, D., & Ganguly, S. (2013). Identification of the N-terminal glycine – arginine rich (GAR) domain in gairdia lamblia fibrillarin and evidence of its essentiality for RNA binding. *International Journal of Tropical Disease & Health, 4,* 318–327.
61. Kerr, R. W., & Cleveland, F. C. (1959). Orthophosphate esters of starch. US Patent. 2884413.
62. Ketola, H., & Hagberg, P. (2003). *Modified Starch.* US Patent Office, Patent No. 6,670,470.
63. Kim, N., Seog, H., & Nam, Y. (1987). Physiochemical properties of domestic millet starches. *Korean Journal Food Science and Technology, 9,* 245–249.

64. Kling, M. (2001). *Stiffening of Cellulose Fibers: A Comparison Between Cross-Linking the Fiber Wall and Lumen Loading.* Department of Chemical and Metallurgical Engineering, Division of Chemical Technology, Luleå University of Technology, Luleå, Sweden.
65. Klopfenstein, C. F., & Hoseney, R. C. (1995). Nutritional properties of sorghum and the millets. In: Dendy, D. A. V. (Ed.), *Sorghum and Millets: Chemistry and Technology.* St. Paul, MN: American Association of Cereal Chemists. pp. 125–168.
66. Koo, S. H., & Lee, H. G. (2010). Effect of crosslinking on the Physiochemical and physiological properties of corn starch. *Food Hydrocolloids, 24,* 619–625.
67. Kuakpetoon, D., & Wang, Y. J. (2001). Characterization of different starches oxidized by hypochlorite. *Starch/Stärke, 53,* 211–218.
68. Kuakpetoon, D., & Wang, Y. J. (2006). Structural characteristics and physicochemical properties of oxidized corn starches varying in amylose content. *Carbohydrate Research, 341,* 1896–1915.
69. Kumari, S. K., & Thayumanavan, B. (1998). Characterization of starches of proso, foxtail, barnyard, kodo and little millets. *Plant Food Human Nutrition, 53,* 47–56.
70. Langan, R. T. (1986). Uses and properties of modified starches in the food industry. *Food Chemistry, 12,* 427–434.
71. Larsson, C. T., Khoshnoodi, J., Ek, B., Rask, L., & Larsson, H. (1998). Molecular cloning and characterization of starch-branching enzyme II from potato. *Plant Molecular Biology, 37,* 505–511.
72. Lawal, O. S. (2004). Composition, Phisicochemical properties and retrogradation characteristics of native, oxidized and acetylated and acid-thinning new cocoyam starch. *Food Chemistry, 87,* 205–218.
73. Lawal, O. S. (2009). Starch hydroxyalkylation: physicochemical properties and enzymatic digestibility of native and hydroxypropylated finger millet (*Eleusine coracana*) starch. *Food Hydrocolloids, 23,* 415–425.
74. Lawal, O. S., Adebowale, K. O., Ogunsanwo, B. M., Barba, L. L., & Ilo, N. S. (2005). Oxidized and acid thinned starch derivatives of hybrid maize: functional characteristics, wide-angle X-ray diffractometry and thermal properties. *International Journal Biological Macromolecules, 35,* 71–79.
75. Leach, H. W., Mccowen, L. D., & Schoch, T. J. (1959). Structure of the starch granules. I. Swelling and solubility patterns of various starches. *Cereal Chemistry, 36,* 534–541.
76. Lee, J. W., Lee, D. S., Bhoo, S. H., Jeon, J. S., Lee, Y. H., & Hahn, T. R. (2005). Transgenic Arabidopsis plants expressing Escherichia coli pyrophosphatase display both altered carbon partitioning in their source leaves and reduced photosynthetic activity. *Plant Cell Report, 24,* 374 – 382.
77. Lim, S., & Sieb, P. A (1993). Preparation and pasting properties of wheat and waxy corn starch phosphates. *Cereal Chemistry, 70*(2), 137–144.
78. Liu, H. J., Ramsden, L., & Corke, H. (1999). Physical properties and enzymatic digestibility of hydroxypropylated and normal maize starches. *Carbohydrate Polymer, 40,* 175–182.
79. Lopez, A. (1987). *Packaging; aseptic processing; ingredients, volume II.* A complete course in canning (12th Ed.). Maryland: The Canning Trade Inc.
80. Luallen, T. E. (1985). Starch as a functional ingredient. *Food Technology, 39*(1), 59–63.

Modified Pearl Millet Starch

81. Malleshi, N., Desikachar, H., & Tharanathan, R. (1986). Physico-chemical properties of native and malted finger millet, pearl millet and foxtail millet starches. *Nutrition and Food Science, 38,* 202–205.

82. Manelius, R., Buleon, A., Nurmi, K., & Bertoft, E. (2000). The substitution pattern in cationized and oxidized potato starch granules. *Carbohydrate Research, 329,* 621–633.

83. Myllärinena, P., Partanena, R., Seppäläb, J., & Forssella, P. (2002). Effect of glycerol on behavior of amylase and amylopectin films. *Carbohydrate Polymer, 50*(4), 355–361.

84. Oshodi, A. A., Ogungbenle H. N.and Oladimeji, M. O. (1999). Chemical composition, nutritionally valuable minerals and functional properties of binniseed (*Sesamum radiatum*), pearl millet (*Pennisetum typhoids*) and qinoa (*Chenopodium quinoa*) flours. *International Journal Food Science Nutrition, 50,* 325–331.

85. Ostergard, K., Bjorck, I., & Gunnarsson, A. (1988). A study of native and chemically modified potato starch, Part I: Analysis and enzyme avilibility in vitro. *Food Chemistry, 40,* 58–66.

86. Parovuori, P., Hamunen, A., Forssell, P., Autio, K., & Poutanen, K. (1995). Oxidation of potato starch by hydrogen peroxide. *Starch/Stärke, 47,* 19–23.

87. Patel, K. F. Mehta, H. U., & Srivastava, H. C. (1974). Kinetic and mechanism of oxidation of starch with sodium hypochlorite. *Journal of Applied Polymer Science, 18,* 389–399.

88. Perez, E., Segovia, X., Tapia, M. S., & Schroeder, M. (2012). Native and cross-linked modified *Dioscorea trifida* (cush-cush yam) starches as bio-matrices for edible films. *Journal of Cellular Plastics, 48*(6), 545–556.

89. Phillips, G. O., & Williams, P. A. (2009). *Handbook of Hydrocolloids.* 2nd Ed., CRC Press.

90. Radley, J. A. (1976). *Industrial Uses of Starch and Its Derivatives.* London: Applied Science Publishers Ltd.

91. Reddy, I., & Seib, P. A. (1999). Paste properties of modified starches from partial waxt wheats. *Cereal Chemistry, 76,* 341–349.

92. Rutenberg, M. W., & Solarek, D. (1984). Starch derivatives: Production and Uses. In: Whistler, R. L., BeMiller, J. N., & Paschall, E. F. (Eds.), *Starch Chemistry and Technology.* 2nd Ed., pp. 311–388.

93. Sangseethong, K., & Sriroth, K. (2002). Effect of hypochlorite levels on the modification of cassava starch. *Zywnosc Technologia Jakosc, 9,* 191–197.

94. Sangseethong, K., Lertphanich, S., & Sriroth, K. (2009). Physicochemical properties of oxidized cassava starch prepared under various alkalinity levels. *Starch/Stärke, 61,* 92–100.

95. Sangseethong, K., Termvejsayanon, N., & Sriroth, K. (2010). Characterization of physiochemical properties of hypochlorite and peroxide- oxidized cassava starches. *Carbohydrate Polymers, 82,* 446–453.

96. Sarmento, S. B. S., Polesi, L. F., & Franco, C. M. L. (2011). Production and physiochemical properties of resistant starch from hydrolysed winkled pea starch. *International Journal of Food Science and Technology, 46*(11), 2257–2264.

97. Sasaki, T., & Matsuki, J. (1998). Effect of wheat structure on swelling power. *Cereal Chemistry, 74,* 525–529.

98. Sathe, S. K., Deshpande, S. S., & Salunkhe, D. K. (1982). Functional properties of winged bean (psophocarpus tetragonolobus) proteins. *Journal of Food Science*, *47*, 503–508.

99. Schmorak, J., & Lewin, M. (1963). The chemical and physicochemical properties of wheat starch mildly oxidized with alkaline sodium hypochlorite. *Journal of Polymer Science: Part A*, *1*, 2601–2620.

100. Schmorak, J., Mejzler, D., & Lewin, M. (1961). A kinetic study of the mild oxidation of wheat starch by sodium hypochlorite in the alkaline pH range. *Journal of Polymer Science*, *49*, 203–216.

101. Schoch, T. J., & Maywald, E. C. (1968). Preparation and properties of various legume starches. *Cereal Chemistry*, *45*, 564–573.

102. Seidel, C., Kulicke, W. M., He, X. C., Hartmann, B., Lechner, M. D., & Lazik, W. (2004). Synthesis and characterization of cross-linked carboxymethyl potato starch ether gels. *Starch-Starke*, *56*, 157–166.

103. Silva, M. C., Ibezim, E. C., Riberio, T. A. A., Carvalho, C. W. P., & Andrade, C. T. (2006). Reactive Processing and mechanical properties of cross-linked maize starch. *Industrial Crops and Products*, *24*(1), 46–51.

104. Singh, J., & Singh, N. (2001). Studies on the morphological, thermal and rheological properties of starch separated from some Indian Potato cultivars. *Food Chemistry*, *75*, 67–77.

105. Singh, J., Kaur, L., & Mc Carthy, O. J. (2007). Factors influencing the physio-chemical, morphological, thermal and rheological properties of some chemically modified starches for food applications. *Food Hydrocolloids*, *21*, 1–22.

106. Smolka, G. E., & Alexander, R. J. (1985). *Modified Starch, Its Method of Manufacture and the Salad Dressings Produced Therewith.* US Patent 4562086.

107. Srichuwong, S., Sunarti, T. C., Mishima, T., Isono, N., & Hisamatsu, M. (2005). Starches from different botanical sources II contribution of starch structure to swelling and pasting properties. *Carbohydrate Polymers*, *62*, 25–34.

108. Steiger, F. H. (1966). *Surgical dressing.* US Patent 3241553.

109. Takahashi, S., Maningat, C. C., & Seib, P. A. (1989). Acetylated and hydroxypropylated wheat starch: paste and gel properties compared with modified maize and tapioca starches. *Cereal Chemistry*, *66*, 499–506.

110. Taylor, J. R. N. (2004). Grain production and consumption: Africa. In: Wrigley, C., Corke, H., & Walker, C. E. (Eds.), *Encyclopedia of Grain Science*. Elsevier, London, pp. 70–78.

111. Tessler, M. M. (1975). *Hydroxypropylated, Inhibited High Amylase Retort Starches.* US Patent 3904601.

112. Tester, R. F., & Morrison, W. R. (1990). Swelling and gelatinization of Cereal starches, II: Waxy rice Starches. *Cereal Chemistry*, *67*(6), 551–557.

113. Tharanathan, R. N. (2002). Food-derived carbohydrates—structural complexity and functional diversity. *Critical Reviews in Biotechnology*, *22*(1), 65–84.

114. Tuschhoff, J. V. (1986). Hydroxypropylated starches. In: Wurzburg, O. B. (Ed.), *Modified Starches: Properties and Uses*. Florida: CRC Press, Inc. pp. 89–96.

115. Valle, F., Tuschoff, J. V., & Streaty, C. E. (1978). *Hydroxypropylated, Epichlorohydrin Crosslinked Tapioca and Corn Starch Derivatives.* US Patent 4120983.

Modified Pearl Millet Starch

116. Wang, Y. J., & Wang, L. (2003). Physicochemical properties of common and waxy corn starches oxidized by different levels of sodium hypochlorite. *Carbohydrate Polymers, 52*, 207–217.

117. Wankhede, D. B., Rathi, S. S., Gunjal, B. B., Patil, H. B., Walde, S. G., Rodge, A. B., & Sawate, A. R. (1990). Studies on isolation and characterization of starch from pearl millet (*Pennisetum americanum* leeke) grains. *Carbohydrate Polymer, 13*(1), 17–28.

118. Wattanachant. S., Muhammad, K., Hashim, D. M., & Rahman, R. A. (2003). Effect of crosslinking reagents and hydroxypropylation levels on dual modified sago starch properties. *Food Chemistry, 80*, 463–471.

119. Whistler, R. L., & Schweiger, R. (1957). Oxidation of amylopectin with hypochlorite at different hydrogen ion concentrations. *Journal of the American Chemical Society, 79*, 6460–6464.

120. Whistler, R. L., & Schweiger, R. (1959). Oxidation of amylopectin with hydrogen peroxide at different hydrogen ion concentrations. *Journal of the American Chemical Society, 81*, 3136–3139.

121. Woo, K. S., & Seib, P. A. (2002). Cross-linked resistant starch: preparation and properties 1. *Cereal Chemistry, 79*(6), 819–825.

122. Woo, K. S., & Sieb, P. A. (1997). Crosslinking of hydroxypopylated wheat starch in an alkaline slurry with sodium trimetaphosphate. *Carbohydrate Polymers, 32*, 239–244.

123. Wotton, M., & Bamunuarachchi, A. (1978). Water binding capacity of commercial produced native and modified starches. Starch/Starke, *33*, 159–161.

124. Wu, Y., & Sieb, P. A. (1990). Acetylated and hydroxypropylated distarch phosphates from waxy barley: paste properties and free-thaw stability. *Cereal Chemistry, 67*(2), 202–208.

125. Wurzburg, O. B. (1986). Converted starches. In: Wurzburg, O. B. (Ed.), *Modified Starches: Properties and Uses*. Boca Raton, FL: CRC Press, pp. 17–40.

126. Xiao, H., Lin, Q., & Liu, G. Q. (2012). Effect of cross-linking and enzymatic hydrolysis composite modification on the properties of rice starches. *Molecules, 17*, 8136–8146.

127. Xiao, H., Lin, Q., Liu, G. O., Wu, Y., Tian, Q., Wu, W., & Fu, X. (2011). Physicochemical properties of chemically modified starches from different botanical origin. *Scientific Research and Essays, 6*(21), 4517–4525.

128. Xie, F., Yu, L., Liu, H., & Chen, L. (2006). Starch modification using reactive extrusion. *Starch/Stärke, 58*(3–4), 131–139.

129. Yeh, A.-I., & Yeh, S.- L. (1993). Some characteristics of hydroxypropylated and cross-linked rice starch. *Cereal Chemistry, 70*(5), 596–601.

130. Yook, C., Pek, U. H., & Park, K. H. (1993). Gelatinization and retrogradation characteristics of hydroxypropylated and cross-linked rice. *Journal of Food Science, 58*(2), 405–407.

131. Yui, P. H., Loh, S. L., Rajan, S. C., & Wong, C. F. J. (2008). Bong physiochemical properties of sago starch modified by acid treatment in alcohol. *American Journal Apply Science, 5*, 307–311.

132. Yusuf, A. A., Ayedu, H., & Logunleko, G. B. (2007). Functional properties of unmodified and modified Jack bean (*Canavalia ensiformis*) starches. *Nigerian Food Journal, 25*, 141–149.

133. Zeleznak, K. J., & Hoseney, R. C. (1987). The glass transition in starch. *Cereal Chemistry, 64*, 121–124.
134. Zhu, Q., & Bertoft, E. (1997). Enzymatic analysis of the structure of oxidized potato starches. *International Journal of Biological Macromolecules, 21*, 131–135.
135. Zhu, Q., Sjoholm, R., Nurmi, K., & Bertoft, E. (1998). Structural characterization of oxidized potato starch. *Carbohydrate Research, 309*, 213–218.
136. Zobel, H. F. (1988). Molecules to granules: a comprehensive starch review. *Starch/Starke, 40*, 44–50.

PART III

AGRICULTURAL PROCESSING: HEALTH BENEFITS OF MEDICINAL PLANTS

CHAPTER 10

POTENTIAL HEALTH BENEFITS OF TEA POLYPHENOLS—A REVIEW

BRIJ BHUSHAN, DIPENDRA KUMAR MAHATO,
DEEPAK KUMAR VERMA, MANDIRA KAPRI, and
PREM PRAKASH SRIVASTAV

CONTENTS

10.1 Introduction 229
10.2 Tea Polyphenols 232
10.3 Potential Health Benefits of Tea 243
10.4 Conclusions 255
10.5 Summary 256
Keywords 256
References 257

10.1 INTRODUCTION

Tea (*Camellia sinesis*) after water occupies second position in consumption as a beverage in the world due to its beneficial effects on human health [41, 80, 201, 258]. However, consumption of tea per head in the world is 500 ml per day [195]. Tea mainly originated from China and South Asia since past and in India it came in fashion around 1835 [15]. According to the Food and Agriculture Organization (FAO), China is one of the largest tea producer 38.2% while India is second largest producer, contributing 21% of total tea production in world [92]. There are mainly two varieties of tea grown in the world: *C. sinensis var. sinensis* that grows in cold climate with small leaves and Chinese in origin; and *C. sinensis var. assamica (Assan type)* that grows in semitropical climate having large leaves and grows in the Assam region of India [74, 80, 201]. The consumption of various types of tea differs from country to country. For instance, non-fermented tea commonly called green tea (GT) is consumed in China, Japan, Korea, Morocco, North Africa and Taiwan; fermented tea commonly called black tea (BT) is a favorite tea in West; and semi-fermented tea (oolong tea, OT) is widely favored in China and Taiwan [39, 42, 299]. Table 10.1 describes types of tea and place of origin. However, of the total tea production in world about 78% is consumed as black tea, 20% as GT and rest is in form of OT [173].

On the basis of processing patterns, tea is of following types: GT, BT, white tea (WT) and Oolong-type of tea. The polyphenols of tea vary from GT to BT and widely depends on its processing pattern prior to drying [80, 240]. For production of GT, steam is applied on the rolled leaves to deactivate polyphenol oxidases and to minimize oxidation before drying the tea [64]. In the production of BT, the phenolic compound comes in contact with polyphenol oxidases and undergoes oxidation for 4 hours. During the production of OT, fermentation time shorter than BT, for 2–3 hours hence called semi-fermented tea [224]. WT is produced by protecting the buds from sunlight to prevent the formation of chlorophyll in the leaves in order to produce leaves of white color without undergoing oxidation [15, 69, 80]. Figure 10.1 represents processing of different types of tea. GT contain more antioxidants than the others [41, 50, 80].

Now-a-days, people face lot of health issues: related to heart called cardiovascular disease (CVD), aging, cancer, stress, headaches, body aches and pains, digestion, diabetic, and mutation, etc. [23, 76, 201, 331]. Research has proved that consumption of tea reduces the risk of such diseases due to its antioxidant properties [41, 69, 138, 246, 274, 364, 379].

TABLE 10.1 Tea and Place of Origin with Common Name

Native	Color (common name)
Japan, China, Taiwan	Light green (White)
India, East Africa, Argentina	Black (Darjeeling and, Assam)

Potential Health Benefits of Tea Polyphenols—A Review

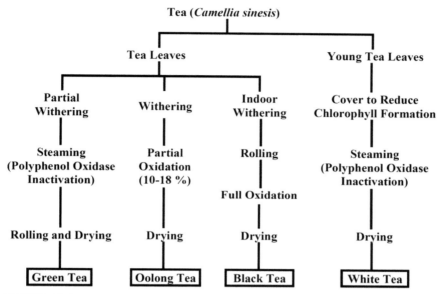

FIGURE 10.1 Processing of different types of tea.

Tea acts as an antioxidant that prevents the undesirable oxidation of biomolecules (i.e., carbohydrates, lipids and proteins). It also prevents oxidative damage of DNA and reduces the risk of cancer [275]. Studies on animals revealed that consumption of GTCs could act as an anti-tumorigenic factor and also can suppress the obesity induced by high fat diet [69]. Catechins, from GT, i.e., epigallocatechin gallate and epicatechin gallate, showed relatively high activity in rat liver microsomes and provided better protection from lipid oxidation and in a study epicatechin gallate is found to be 10 times more effective antioxidant than vitamin E [250, 275]. Studies on animals have shown that GT prevented mice from cancer: lung cancer, liver cancer, prostate cancer, and from mammary cancer in female rats [41, 69]. Flavonoids present in GT and BT have good properties to scavenge the free radical [279]. It scavenges the oxygen-free radicals produced in body and protecting the cells from unwanted oxidative damage thus reducing aging and other chronic diseases. These days, dementia grows rapidly in aged population. However, the exact region of dementia is not known but common causes of Dementia include Alzheimer disease [39, 147]. Consumption of GT, that contains considerable amount

232 Engineering Interventions in Agricultural Processing

of catechins, has preventive action against such diseases [200]. However, some studies have reported that consumption of BT is harmful for health.

10.2 TEA POLYPHENOLS

Flavonoids are low molecular compounds mainly present in plants and are largest class of phenol usually in the form of glycosides [18, 274]. Flavonoid contains two aromatic rings A and B that is attached by a heterocyclic ring C. On the basis of structural variation in substitution pattern to C ring, flavonoids can be subdivided into major two subclasses: flavones and flavanols [18, 66, 274, 388] as shown in Table 10.2.

Tea contains a greater amount of polyphenols that can prevent oxidation [39]. Phenolic compounds have ability to donate their hydrogen to scavenge singlet oxygen. Additionally, it has also metal chelating property [276], due to presence of functional groups (hydroxyl and carbonyl groups) at suitable position in flavanol ring. It is well known that ferrous ion is one of the most important and effective pro-oxidant in the food

TABLE 10.2 Major Flavonoids in Tea

Flavonoid subclass	Structural formula	Examples
Flavone		Apigenin, Chrysin, Luteolin
Flavanols		**Monomers (Catechins):** Catechin, Epicatechin gallate, Epicatechin, Epigallocatechin gallate and Epigallocatechin **Dimers and Polymers:** Isorhamnetin, Kaempferol, Myricetin, Quercetin, Theaflavins and Thearubigins

system and it could be converted into a complex compound by hydroxyl group present in tea [303, 362]. However, the antioxidant property of phenolic compounds depends on its structure and ranges from simple phenolic molecules to highly polymerized compounds [35]. Polyphenols that naturally occurs in BT are: theaflavin (TF1), theaflavin 3-gallate (TF2A), theaflavin 3"-gallate (TF2B), and theaflavin 3,3"-gallate (TF3) [214, 215]. Flavonoids and polyphenols present in GT are [91, 147]:

EC: (-)-epicatechin	ECg: (-)-epicatechin gallat
EGC: (-)-epigallocatechin	EGCg: (-)-epigallocatechin gallat
C: (-)-catechin	Cg: (-)-catechin gallat
GC: (-)-gallocatechin	GCg: (-)-gallocatechin gallat

It has been shown that EGCG constitutes about 40% of the total polyphenols present in the GT [39]. Generally, content of catechins in different types of tea is in the order of GT > pouchong tea > OT > BT, showing that GT have higher catechins [376]. During the processing of tea, catechins react with polyphenol oxidase, leading to formation of flavanol dimmers and polymers called as theaflavins and thearubigens, respectively that is responsible for color and flavor of tea [113]. It has been reported that antioxidant activity of tea can be inhibited by milk protein [209, 301]. However, according to Richell et al. [277], addition of milk in tea does not affect antioxidant properties of tea. Catechins content of the tea is affected by processing time, temperature, and types of tea.

Flavonoid content of tea is in order of brewed hot tea ($541–692$ µg/mL) > instant preparations ($90–100$ µg/mL) > iced tea > ready-to-drink tea [13, 121]. The scavenging capacity of catechins present in tea was increased in order of EC < ECG < EGC < EGCG under active oxygen method (AOM) at $97.8°C$ [147]. In the GT, flavonols were found to be 52% (measured in weight % of extract of solids) while catechins of total dry mass present were 32–40%, and in OT and BT 8–20 and 3–10%, respectively [24, 39]. Table 10.3 provides a list of polyphenoels and the quantity present in different types of tea. Thus GT is more effective antioxidant than other type of tea [39, 66]. It has been found that higher the reduction potential lower is the antioxidant property because higher energy requires for

TABLE 10.3 Tea Polyphenols and Their Quantity (in % dry weight of tea solids)

Polyphenols	Quantity
Theaflavins	2–6
Thearubigins	15–20
Catechins*	32–40
Catechins**	8–20
Catechins***	3–10

*GT; **OT; ***BT.

donating hydrogen or for reduction. EGCG and EGC have lower reduction potential of 0.43 having good antioxidant properties than 0.48 for vitamin E, 0.24 at 7 pH and 20°C [11, 168, 169]. Relative antioxidant activity of tea catechins determined by using troloxequivalent antioxidant activity assay (TEAC) was EGCG ≈ ECG > EGC > EC [137]. Along with these polyphenols, tea also contains Gallic acid (highest amount present in Chinese puerh teas approximately 15 g per kg of dry weight), caffeine (20–40 mg/g), methylxanthines, theobromine and theophylline and amino acid theanine (5-N-ethyl glutamine) [113, 224, 226].

10.2.1 FLAVONES

Flavone, with a molecular formula of $C_{15}H_{10}O_2$, belongs to a class of flavonoids having the backbone of 2-phenylchromen-4-one (2-phenyl-1-benzopyran-4-one). It has three-ring skeletons referred to as A-, C-, and B- rings along with three functional groups, including hydroxy, carbonyl, and conjugated double bond. Flavones are colorless-to-yellow crystallines which are soluble in water and ethanol. Flavones are moderate-to-strong oxygen bases, and are soluble in acids due to formation of oxonium salts having pKa values ranging from 0.8 to 2.45 [71]. Flavones have a planar structure with its C-O-C bond angle of 120.9°. Its bond length between C-O is 1.376 Å and dihedral angle is around 179.2°.

Flavonoid biosynthesis starts with the condensation of one molecule of 4-coumaroyl-CoA (C-CoA) and three molecules of malonyl-CoA (M-CoA) yielding naringenin chalcone, carried out by the enzyme chalcone synthase (CHS). The two immediate precursors of the chalcone originate from two different pathways of primary metabolism. C-CoA is synthesized from the amino acid phenylalanine by three enzymatic steps, collectively called general phenylpropanoid pathway; M-CoA is synthesized by carboxylation of acetyl-CoA (A-CoA) that is a central intermediate in the Kreb's tricarboxylic acid (TCA) cycle. The chalcone is consequently isomerized by the enzyme chalcone flavanone isomerase (CHI) to yield a flavanone. From this central intermediate, the pathway diverges into several different classes of flavonoids [190].

10.2.1.1 Luteolin

Luteolin (3',4',5,7-tetrahydroxyflavone) is widely distributed in fruits, herbs and green vegetables (celery, peppermint, broccoli, cauliflower, green pepper, perilla leaf, camomile tea, cabbage, thyme, honeysuckle and spinach) [125, 305]. Consumption of foods containing luteolin reduces the risk of developing chronic diseases [232]. Luteolin possesses a variety of pharmacological activities, including: antioxidant [239], anti-neoplastic [52], anti-hepatotoxic, anti-allergic, anti-osteoporotic [81], anti-diabetic [384], anti-inflammatory [354], anti-platelet and vasodilatory activity [60]. It induces apoptosis and cell cycle arrest in cancer cells through signaling pathways mediated by TNF-α, JNK, IGF-1, p53, Wnt, and mTOR [337]; also mediates anti-inflammation via NF-κB signaling [163]; glucose disposal via AMPK in adipocytes [361] and adipocyte differentiation [262]. Mesenchymal stem cells (MSCs) upon stimulus differentiate into adipocytes or osteoblasts [29, 263]. Outnumbering of adipocytes over bone-forming cells in the bone marrow leads to age-related osteoporosis [25, 26, 285]. Therefore, a reciprocal regulation of adipogenesis and osteogenesis offer strategies to reduce the incidence of osteoporosis [185]. Several other mechanisms involved in the biological activities of luteolin have been reported like modulation of ROS levels, inhibition of topoisomerases, and inhibition of PI3 K.

10.2.1.2 Apigenin

Apigenin (4',5,7,-trihydroxyflavone) belongs to the flavone subclass of polyphenolic compounds [265], and it is abundantly found in vegetables and fruits such as parsley, onions, oranges, tea, chamomile, wheat sprouts, and in some seasonings [87, 265]. Its biological functions including anti-oxidative, anti-mutagenic, anti-inflammatory, and especially, anti-tumorigenic properties is present in many in vitro systems [183, 265]. It also exhibits anti-proliferative and inhibitory activities against various cancer cells lines, including breast, colon, skin thyroid, leukemia cells, and multiple myeloma cells [38, 45, 349, 387]. Apigenin shows anti-pancreatic cancer property by inhibiting IκB kinase-β (IKK-β) and glycogen synthase kinase-3β (GSK-3β) mediated nuclearfactor-κB (NF-κB) activation, suppressing 4-(methylnitrosamino)-1-(3-pyridyl)-1-butanone (NNK) induced focal adhesion kinase (FAK) and extracellular signal-regulated kinase (ERK) activation, down-regulating the hypoxia response gene HIF-1α, glucose transporter 1 (GLUT-1), and vascular endothelial growth factor (VEGF), and inhibiting the GLUT-1 and phosphoinositide 3-kinase (PI3K)/Akt pathway [358]. It also inhibits the activation of nuclear transcription factor NF-nB, which plays a key role in the regulation of cell growth, cell-cycle regulation, and apoptosis [128]. Apigenin induces growth inhibition, cell cycle arrest and apoptosis human carcinoma cells including breast, colon cancer cells and leukemic cells [241, 347, 349]. Apigenin treatment resulted in G1 cell cycle arrest in synchronized human diploid fibroblasts [212] and G2/M arrest in rat neuronal cells [296]. The p53 regulates the passage through both G1 and G2/M [2, 315]. The over expression of p53 inhibits cellular proliferation [242] and activation of p53 in response to DNA damage led to cell cycle arrest [171]. Apigenin also causes selective cell-cycle arrest and apoptosis of several human prostate carcinoma cells but not of normal cells [119], thus possibility for prevention and therapy of prostate cancer by the use of apigenin.

10.2.1.3 Chrysin

Chrysin (5,7-dihydroxyflavone) is composed of three p–p conjugated rings, which provide strong interaction to most anti-tumor drugs with p–p conjugated structures such as doxorubicin, camptothecin and paclitaxel *via* both hydrophobic and p–p interaction [53, 204, 222]. Chrysin exerts different biological activities like anti-diabetogenic activity [306] and anti-anxiolytic effect [383]. The vaso-relaxing and beneficial effects of chrysin include the release of NO from endothelium which causes aortarelaxation

[85, 342, 343]. It has been reported that chrysin restarts the phorbol ester stimulated superoxide anion formation but has no effect on interleukin-1-induced inducible nitric oxide synthase expression in vascular smooth muscle cells (VSMCs) [130]. Chrysin exhibits tyrosinase inhibitory activity [194], moderate aromatase inhibitory activity and estradiol-induced DNA synthesis inhibitory activity [348]. Different derivatives of chrysin have been synthesized for example: C-isoprenylatedhydrophobic derivatives of chrysin which are potential P-glycoprotein (Pgp) modulators in tumor cells [68].

10.2.2 *FLAVANOLS*

Flavanols and their oligomeric derivatives reduce the risk of CVDs. The consumption of flavanol-containing foods improves arterial function [20, 86, 96, 132, 133], decreases blood pressure [327, 328], modulates hemostasis [97, 140, 248], improves insulin sensitivity [114, 115], reduces risk of cancer [187], stroke [179] and coronary heart disease [14]. Flavanols also exert anti-inflammatory, anti-bronchioconstrictory, anti-arteriosclerotic and anti-carcinogenic properties. In addition to this, these inhibit the activity of pro-oxidative enzymes of the arachidonic pathway [307], counteract the decline of gap-junctional intercellular communication (GJIC) [322], act as vasoprotector [1] and photoprotector [1] and improve microcirculation and quality of the skin [131, 251]. Flavanols even interact with other anti-oxidants (e.g., α-tocopherol, β-carotene or ascorbic acid), thereby protecting the cells from oxidative damage [233].

10.2.2.1 Catechins

Catechins belong to flavan-3-ol type and constitute 25–30% of the dry weight of tea leaves. High amount of catechins develops astringent and bitter taste [221]. Therefore, GT requires a higher content of free amino acids and an appropriate content of catechins [288]. The accumulation of catechins in tea leaves depends on factors like: UV irradiation [389], light intensity [141, 236], soil humidity [49, 62] and even nutrient elements available in the soil [59, 227, 289]. Catechins contain high anti-oxidant activity with anti-oxidative, anti-tumor and anti-inflammatory effects, thus beneficial to human health [34, 102, 120]. They also have anti-obesity effect in animals [247] and humans [124, 333]. It mainly includes (-)-epigallocatechin gallate (EGCG), (-)-epigallocatechin (EGC), (-)-epicatechin gallate (ECg), (-)-epicatechin (EC), and their geometric isomers (-)-gallocatechin gallate (GCg), (-)-gallocatechin (GC), (-)-catechin gallate (Cg), and (-)-catechin (C), among which EGCG is regarded as the most important catechin due to high content in tea leaf and excellent bioactivity [167]. The scavenging abilities of EGCg and GCg are higher than those of non-gallated catechins EGC, GC, EC and C due to the presence of a gallate moiety at C-3 position [118]. Han et al. [122] and Huvaere et al. [151] have reported that catechins affect the composition of curds and cheese due to interactions between milk components and catechins. These catechins interact non-covalently with proteins [381] or with the milk-fat globules [206, 375]. These challenges can be overcome by nano-encapsulation to protect catechins from interacting with milk components [271]. It has been reported that the tea catechins inhibit the activity of α-amylase and α-glucosidase, which are essential for starch digestion inhuman [191, 228, 377]. Among the four types of digestive enzymes (viz., α-amylase, pepsin, trypsin and lipase), α-amylase activity is inhibited the most [129]. Catechins with galloyl group are natural tyrosine kinase inhibitors thus signaling kinases, ERK1/2, protein kinase B (Akt), PI3K and p38 mitogen-activated protein kinase [313]. Phenolic hydroxyl group of catechins can inhibit lipidperoxidation and fat hydrolysis while galloyl group contributes to the prostacyclin production, reduction of vascular cell adhesion molecule-1 expression [16].

10.2.2.2 Theaflavins

Theaflavins (TFs) and thearubigins (TRs) are produced during fermentation process by enzymatic oxidation. The orange red or brown color and astringent taste of BT is due to TFs [127, 372]. There are four major TFs in BT and OT viz., theaflavin (TF1), theaflavin-3-gallate (TF2a), theaflavin-3'-gallate (TF2b), and theaflavin-3,3'-digallate (TF3) [41, 214, 215, 373]. Theaflavins are characterized by a benzotropolone ring structure and they contribute to the unique taste of BT [367]. The consumption of BT has several beneficial health benefits, such as prevention of cancer, CVD, anti-mutagenicity, anti-inflammatory action, neurological disorders, suppression of extracellular signals to inhibit cell proliferation, anti-obesity, lowering hepatic lipid accumulation, insulin sensitization as well as repair DNA oxidative damage [109, 146, 220, 223, 253, 339, 340]. Theaflavins contribute importantly to properties of BT including its color, mouth feel and extent of tea cream formation [370]. The oxidation process also forms larger and less well-characterized polymers, referred to as thearubigins. Theaflavins are major polyphenols in BT [370]. It was previously demonstrated that TF3 prevents adipocyte-triggered inflammatory response (metaflammation) by promoting the switch of inflammatory M1 macrophage toward anti-inflammatory M2 phenotype [189].

10.2.2.3 Thearubigins

The briskness, brightness, strength, body and total color of BT is due to levels of theaflavins and thearubigins [284]. Thearubigins contribute to the mouth feel (thickness) and color of the tea [30]. The most abundant and heterogeneous substances (TRs), first described by Roberts et al. [283], are found in BT infusion [205]. Thearubigins have shown to influence the sensorial quality and bring several health benefits through their anti-oxidant activity [103, 259]. Theaflavins are formed by an oxidation catalyzed by the endogenous enzymes polyphenol oxidase [297], part of which is converted to thearubigins [316]. Thearubigins inhibit human DNA methyltransferases and even bind to quadruplex DNA selectively [243, 270]. Though, thearubigins were introduced fifty years ago, their chemical nature still remains mystery and many efforts are being done to understand their composition [127, 281, 282]. Researchers have shown that these consist of a highly complex mixture of phenolic coupling products comprising of 5000 to 30,000 individual chemical constituents [84, 195, 196, 197]. EGCG, the major constituent of GT leaves, acts as a precursor for thearubigins formation. A combination of electrochemical oxidation of EGCG solutions, LC–MS, and direct infusion-tandem mass spectrometry experiments were performed to reveal the formation of thearubigin [323]. The H^1 and C^{13} NMR spectra of thearubigins displayed line broadening and poor signal-to-noise ratios, indicating to presence of heterogeneous, polymeric and paramagnetic metals. According to Bailey et al. [17], thea-fulvins contain flavan-3-ol species linked via their 'B' rings (C-2 to C-20), which was similar to the findings of Davies et al. [70], who suggested the presence of flavan-3-ol molecules, benztropolone rings and C-2 to C-20 linkages based on the NMR signals.

10.2.2.4 Quercetin

Quercetin (3,5,7,30,40-pentahydroxyflavone, a bioflavonoid), is often found in foods like apples, berries, onions, tea and Brassica vegetables. It has many health benefits, including cardiovascular protection, anti-cancer activity, cataract prevention, anti-viral activity and anti-inflammatory effects, antioxidant, anti-inflammatory and hepato-protective [4, 309]. It has even been proposed as a therapeutic drug which can be used against aluminum induced oxidative stress as it increases transcriptional regulation which results in increased expression of PGC-1a [72]. Quercetin also augments free radical scavenging, aluminum chelating and anti-inflammatory activities [123]. It decreases plasma cholesterol and hepatic lipids, ameliorates diabetes-induced oxidative stress and preserves pancreatic beta cell integrity [378]. Administration of quercetin in rats improved dyslipidemia, hypertension and hyper insulinemia, and reduced the body weight gain [280]. It has also been seen that the quercetin–iron complexes exhibit high lipophilicity and anti-cancer activities [325]. Quercetin is used in the range of 0.008–0.5% or 10–125 mg/serving for the purpose of a nutraceutical for functional foods [126]. Quercetin exhibits a wide range of biological effects including inhibition of low-density lipoprotein oxidation and platelet aggregation as well as improvement of endothelial function [75, 149, 230]. Quercetin metabolism may affect its anti-inflammatory properties [234, 321, 335, 352]. It protects the myocardium from ischemic heart injury when administered before ischemia [165]. Besides this, a significant reduction of the myocardial infarct size in both normal and diabetic animals has been observed [12]. However, the molecular mechanism underlying Quecetin-mediated cardioprotection is yet to be elucidated. Other investigations have shown the potential of quercetin against cognitive deficit in various animal models [6, 198, 229, 268, 308, 334, 374].

10.2.2.5 Myricetin

Myricetin (3, 5, 7, 3', 4', 5'-hexahydroxyflavone) have several therapeutic uses as an anti-carcinogenic, anti-inflammatory, anti-atherosclerotic, anti-thrombotic, anti-carcinogen, anti-thrombotic, anti-diabetic, and anti-viral properties [256]. It has been shown to exert both anti- and pro-oxidant effects, along with mutagenic and anti-mutagenic potential, suggesting a vital role in mutagenesis and carcinogenesis [10, 46, 77, 88, 89, 256. 291]. It is one of the most potent inhibitors of MMP-2 enzyme activity in COLO 205 cells and the growth of Gram-positive and Gram-negative bacterial species [188]. It also inhibits BACE1 [48, 304] which, along with inhibition of fibril growth [257], proves it as a therapeutical drug in Alzheimer's disease. Myricetin arrests the G2/M cell-cycle phase in HepG2 cells by two distinct mechanisms. Western blotting has shown that myricetin inactivates the cyclin B/CDK1 (cyclin-dependentkinase 1, which promotes mitosis) complex by increasing the phosphorylation of the CDK1 protein at Thr14/Tyr15. The phosphorylation of CDK1[Thr14/Tyr15] alters the orientation of adenosine triphosphate (ATP) and prevents the efficient kinase activity of CDK1 [386]. In addition to this, myricetin protects against oxidative injury in neuro-degenerative disorders at physiological concentration [203, 304]. Though, myricetin has neuro-protective effect, the mechanism underlying myricetin-mediated neuro-protection is still unclear [238]. Additionally, myricetin has been found to exhibit anti-hyperlipidemic and anti-obesity activities in high-fat diet–fed rats [51]. It also inhibits adipogenesis during differentiation of 3T3-L1 preadipocytes [9, 51, 369] by decreasing mRNA levels of PPARγ, CEBP/α, and ap2 [9].

10.2.2.6 Kaempferol

Kaempferol is well known for its cancer fighting properties and prevents breast cancer, ovarian cancer and prostate cancer along with other pharmacological functions like anti-oxidant, anti-microbial, cardio-protective, anti-inflammatory, neuro-protective, anti-diabetic, anti-osteoporotic, estrogenic/anti-estrogenic, anxiolytic, analgesic anti-oxidant, anti-atherogenic and anti-allergic activities [44, 55]. It also reduces the risk of pancreatic and lung cancer. Besides this, it prevents arteriosclerosis and heart disease. Additionally, kaempferol is a powerful antioxidant and phytestrogen (negative menopausal symptoms). Kaempferol has anti-Alzheimer property and might be used for the prevention and therapeutics of Alzheimer's disease. *In vitro* studies have suggested that kaempferol protects PC12 neuro-blastoma and T47D human breast cancer cells against the toxic effects and aggregation famyloid-β that may cause the progressive neuronal loss in Alzheimer disease [257, 287]. Kaempferol significantly improves the learning and memory capability, ameliorate the oxidative stress, enhance the activity of Na^+, K^+-ATPase, regulate the expression of extracellular signal-regulated kinases-cyclic AMP response element binding protein pathway and offer the neuro-protective effect in a D-galactose-induced cognitive impairment model rat [210]. It has also been reported that Kaempferol inhibits cancer cell growth by inducing apoptosis in various cancer cell lines like breast cancer (MDA-MB-453), lung carcinoma (H460), chronic myelogenous leukemia (K562) and pro-myelocitic leukemia (U937) [63, 213, 252]. It also reduces the risk of ovarian cancer [110]. The protective effect of kaempferol on CVD is due to its ability to attenuate cellular ROS levels [336, 360]. The inhibition of NOX has also been demonstrated as a mechanism to prevent ROS production [160, 184, 314]. On a molecular level, it inactivates the serine/threonine protein kinase Akt and in turn activates the proapoptotic protein Bax and the apoptosis [148], while in oral cavity cancer cells, a caspase-3-dependent apoptosis was observed [170]. Though kaempferol acts as a natural anti-cancer agent and the identification of several different molecular mechanisms, details of kaempferol-induced cell cycle arrest and apoptosis in cancer cells still remains unclear. For instance, Kaem inhibited cancer cell growth and induced cell death in A549 lung cancer cells through mitogen-activated protein kinase (MEK)-MAPK pathway and inhibited the phosphorylation of MAPK pathway and the activation of c-Fosand NFATc in bone marrow cells [208]. Kaem also down-regulated the expression of ERK, JNK and p38 in U-2 OS human osteosarcoma cells [57]. Kaem suppressed PI3K/Akt signaling pathways by directly binding with PI3K and the subsequent inhibition of nuclear factor kappa B (NF-κB) and AP-1 activities, which impact a number of cellular processes, including proliferation, angiogenesis and apoptosis [207].

10.3 POTENTIAL HEALTH BENEFITS OF TEA

The beneficial effect of tea has been known for many years ago when it was produced in China. Free radicals, i.e., reactive oxygen species called as ROS (superoxide, hydroxyl, and peroxyl radicals and hydrogen peroxide) and nitrogen species are capable for producing oxidative stress that results onset of chronic diseases including coronary heart disease, cancer, and aging [23, 76]. Now-a-days, number of research studies has proved that tea consumption is related to preventing and treating cancer and CVDs due to its antioxidant properties [27, 101, 139, 330]. It has been reported that GT delays the lipid oxidation in different types of foods along with dry-fermented sausage, soybean oil, corn oil and marine oil [346, 371]. The onset of CVD and cancer is due to oxidation of biomolecules such as DNA. Tea acts as an antioxidant, reduces oxidation of biomolecules [43, 341]. Tea reduces the risk of CVD by decreasing the serum cholesterol and triacylglycerides level [274]. Compounds present in GT, mainly EGCG, act as anti-carcinogenic and constrain carcinogenic activity in lungs, breast, colon and melanoma cancer by blocking nitrosamines [39]. Green tea polyphenols (GTPs) can also enhance the immunity thus preventing body from diseases like cancer by increasing humoral and cell-mediated immunity [186]. Quercetin in tea prevents damage of pancreatic ß-cell thus reducing risk of diabetes. Consumption of 2 or more cups of GT daily reduces neurodegenerative disorders such as Alzheimer's and Parkinson's diseases [54]. One of the major problems related to human especially to the elderly women is hip fracture, and this can be prevented by drinking GT [79]. Tea also acts as anti-inflammatory and immunomodulatory [19, 98, 150, 244, 261]. On one hand it has several beneficial effect but another hand it has some demerits. Tea catechins not only act as an antioxidant but also as a pro-oxidant, because in the presence of bleomycin-iron complex it accelerates DNA damage [147].

10.3.1 EFFECTS OF TEA ON DIFFERENT TYPES CANCER

Cancer is generally defined as the uncontrolled cell division resulting in proliferation of cells to form a disease state called tumor. Carcinogenic agents interfere with specific codons present in DNA strands that result in somatic mutation of oncogens and leads to formation of a tumor [231]. In the United States, cancer is the second reason of mortality after heart disease and is increasing constantly and continuously during past half century [8, 39]. According to the report of World Health Organization (WHO) in 2020, cancer data would increase by 50% to 15 million annually [353]. Cancer may be due to genetic mutation, heavy metal ingestion, smoking, pollution and improper diet, etc. Cancer is also characterized as aging related disease and dietary polyphenols have ability to reduce cellular aging, thus inhibiting the onset of cancer in body [65, 245]. EGCG, a major catechin present in GT, was able to inhibit cancer in rodent cell culture system and was non-toxic [3, 104. 154, 366]. Due to non-toxic nature of tea catechins and its anti-carcinogenic effect on wide range of organs like digestive tract, liver, lung, pancreas, mammary gland, skin brain, kidney, uterus and ovary or testes, glandular stomach, duodenum, colon in mice and rats, an increasing interest has been observed on the consumption of tea throughout the world [318]. An enzyme urokinase (u-plasminogen activator), which is responsible for the cancer, was inhibited by the polyphenols present in GT [161]. However, some of researchers show harmful aspects of tea, mostly due to BT consumption associated with cancer of different organs.

10.3.1.1 Effects on Lung Cancer

Lung cancer, called as bronchogenic cancer, is a fatal disease and can cause death due to uncontrolled division and growth of cells in the lungs. Parkin et al. [264] reported that in 2002 nearly 1.8 million people died due to lung cancer. Studies on EGCG proved that when small cell lung carcinoma (SCLC) was incubated with EGCG for 24 hrs, reduced telomerase activity of caspases 3 and 9 activity by 50–60%, 50% and 70% respectively, without defecting DNA and caspase 8 [290]. Lu et al. [235] reported that angiogenesis was inhibited by GT and showed preventive action against pulmonary cancer. Cancer preventive action of GT is due to its antioxidant activity, inhibition of kinases protein, inhibition of cell proliferation, and induction of apoptosis, a programmed type of cell death that differs from necrotic cell death and is known as a normal process of cell elimination

[67, 95, 105, 362]. Lung tumorigenesis induced by AA-Nitrosodiethylamine (NDEA) in A/J Mice was inhibited by GT. Dreosti et al. [83] reported that more than 90% of female A/J mice developed adenomas in lung with an average of 2.5 lung adenomas per mouse, when they were treated with NDEA at a dose of 10 mg/kg/week for 8 weeks. They were dying 16 weeks after from the last dose of NDEA given to them. However, before 2 weeks of first dose of NDEA 1.25%, GT infusions were given to mouse and continued until 1 week after last dose of NDEA. They observed that the tumor incidence and its multiplicity was decreased by 39% and 56%, respectively. Ohno et al. [255] studied Okinawan tea (partially fermented and similar to GT) that prevented lung cancer especially in women. However, researchers did not ask how much GT they drank, so it was not clear that anti-carcinogenic effect were from one or both tea. Higher GT consumption prevented the risk of lung cancer while taking 2 cups of GT per day could reduce 18% risk of the lung cancer [325]. Zhong et al. [390] reported that consumption of GT reduces the risk of lung cancer among Chinese (Table 10.4). However, Tewes et al. [329] reported an inverse relationship between tea and lung cancer. Their studies in Hong-Kong showed flavonoid in tea to be mutagenic (Table 10.4).

TABLE 10.4 Tea and Cancer

Country	Type of Study/ Study population	Cancer type	Health effects	Ref.
China	Population-based case-control study 649 cases of primary lung cancer in women and 675 population controls	Lung cancer	Regular consumption of GT reduced risk of lung cancer in non-smoker women (OR = 0.65; 95% CI = 0.45–0.93); and risk decreased with increasing doses of tea.	Zhong et al. [390]
	Case-control study, 1009 breast cancer women and 1009 population controls	Breast cancer	The onset of breast cancer was reduced among women, drinking GT (1–249 g of dried GT leaves yearly), regularly (OR = 0.87; 95% CI = 0.73–1.04)	Zhang et al. [385]

246 Engineering Interventions in Agricultural Processing

TABLE 10.4 (Continued)

Country	Type of Study/ Study population	Cancer type	Health effects	Ref.
	Population based case-control study, 734 cases, 1,552 population controls	Esophageal cancer	High consumption of GT reduced the risk of cancer in men and women who did not smoke cigarette/drink alcohol, High use: (OR = 0.83 95% CI = 0.59–1.16 for men) and (OR = 0.29; 95% CI = 0.13–0.65 for women)	Gao et al. [108]
	Case-control study, 130 cases of adenocarcinoma of prostate and 127 population controls	Prostate cancer	Reduction in development of prostate cancer in men who drank 3 cups of GT/ day (OR = 0.27; 95% CI = 0.15–0.48)	Jian et al. [164]
Hong-Kong	Population based case-control study, 200 lung cancer female case and 200 Population controls (matched by age and area of residence)	Lung cancer	ORs adjusted for age, no. of live births, schooling, smoking, alcohol, and frequency of consumption of vegetables, fruit and BT; 23 cases (11.5%) and 13 controls (6.5%) claimed to be regular GT drinkers (OR = 2.74; 95% CI = 1.10–6.80).	Tewes et al. [329]
Italy	Case-control study, 199 cases and 6147 hospital controls	Oral cancer	BT associated with reduction of oral cancer (OR = 0.6; 95% CI = 0.3–1.1) while BT consumption was ≥1 cup/day	La Vecchia et al. [202]
Japan	Case-control study, 139 stomach cancer cases and 278 population controls	Stomach cancer	Higher consumption of GT, reduce the risk of stomach cancer (OR = 0.36; 95% CI = 0.16–0.80, for highest vs. lowest level), lowest and highest consumption level of GT ≤ 4 cups/ day and ≥10 cups/day respectively.	Kono et al. [193]

Potential Health Benefits of Tea Polyphenols—A Review

TABLE 10.4 (Continued)

Country	Type of Study/ Study population	Cancer type	Health effects	Ref.
	Case-control study, 294 cases having stomach cancer and 294 population controls	Stomach cancer	High intake of GT reduced stomach cancer (OR = 0.8; 95% CI = 0.5–1.3, for highest vs. lowest level), lowest and highest consumption of GT, ≤4 cups/day ≥8 cups/day respectively.	Hoshiyama and Sasaba [143]
	Cohort study, 314 cases and 39,290 population controls	Stomach cancer	Consumption of GT associated with stomach cancer (OR = 1.19; 95% CI 0.89–1.59, for highest vs. lowest level), lowest and highest consumption level of GT <1 cup/day and ≥5 cups/day respectively.	Koizumi et al. [192]
	Case-control study, 140 cases and 140 hospital controls	Prostate cancer	Consumption of GT ≥ 5 cups/day reduced risk of prostate cancer (OR = 0.27; 95% CI = 0.27–1.64)	Sonoda et al. [310]
Taiwan	Case-control study, 343 cases and 755 population controls	Esophageal cancer	Tea consumption was inversely related with esophageal cancer (OR = 0.4; 95% CI = 0.2–0.6) while tea consumption was ≥7 cups/week	Chen et al. [58]
USA	Population based case-control study, 501 breast cancer women Chinese, Japanese, and Filipino in Los Angeles County 594 population control	Breast cancer	There was a significant reduction in breast cancer among women who drank GT than did not. ORs adjusted for age and potential confiding factors. The risk of breast cancer further decreased with increasing dose of GT (OR = 1.00, 0.71- 95% CI = 0.51–0.99) and (OR = 0.53; 95% CI = 0.35–0.78), respectively in association with number of GT cups/day (none, 0 85.7 mL, and > 85.7).	Wu et al. [355, 357]

248 Engineering Interventions in Agricultural Processing

TABLE 10.4 (Continued)

Country	Type of Study/ Study population	Cancer type	Health effects	Ref.
USA (Hawaii)	Cohort study, 108 cases (64 men, 44 women), 11799 population controls	Stomach cancer	Consumption of GT associated with stomach cancer (OR = 1.5; 95% CI = 0.9–2.3, for highest vs. lowest level), lowest and highest consumption of GT, 0 cups/day ≥2 cups/day.	Galanis et al. [106]

OR = odds ratio; CI = 95% confidence interval.

10.3.1.2 Effects on Breast Cancer

Today breast cancer is one of the common and fatal diseases in women worldwide. In the USA cases of breast cancer are 126.1/100,000 women/ year followed by Japan and China 32.7 and 18.7 cases/100,000 women/ year respectively [94]. Many factors are responsible for breast cancer including circulation of sex hormones, i.e., estrogens and androgens, high mammographic density, pre-diagnostic circulating levels of insulin-like growth factor-1 (IGF-1) and IGF-binding protein-3 (IGFBP-3) may influence risk of breast cancer development [32, 33, 82, 180, 181, 266].

Heterocyclic amines (HCAs) present in cooked meat are carcinogens and associated with breast cancer in western world that is inhibited by GTPs [135, 319, 351]. *In vivo* study on rodents indicates that GTCs reduce the proliferation of breast cancer cells and reduce breast tumor growth in rodents. Nagata et al. [249] performed cross section study in Japan and reported that GT consumption was associated with lowering circulating estrogen level in premenopausal women. Wu et al. [359] also reported that GT consumption was associated with lowering circulating estrogen level in postmenopausal Chinese women in Singapore and found higher consumption of GT reduced 13% estrogen level as compared to non-drinker while consumption of BT increased 19% level of estrogen in women. This study was performed by selecting 130 women out of which 84 were non or irregular, 27 were GT and 19 were BT drinkers. Wu et al. [355, 357] performed a case control study among Chinese, Japanese and Filipino women in Los Angeles Country, California, USA, and they reported that there was a significant relationship between GT consumption and reduction of breast cancer. Higher was the consumption of GT, lower was the risk of breast cancer. They found that risk of breast cancer was reduced among GT women drinkers (64 cases and 99 controls) and increased among non-drinkers (Table 10.4). Zhang et al. [385] concluded that consumption of GT reduced the risk of breast cancer in women (Table 10.4).

10.3.1.3   Effects on Stomach Cancer

There are several factors responsible for development of stomach cancer, such as: genetic factors, smoking, improper dietary intake and infection with bacteria like *Helicobactor pylori* [100, 134]. GT shows a chemo-preventive action against different types of cancer including stomach cancer. Two studies in Japan reported that higher consumption of GT reduced onset of stomach cancer [142, 143, 193]. Hoshiyama and Sasaba [143] reported that there was a significant reduction in stomach cancer in people taking higher doses of GT (OR = 0.8; 95% CI = 0.5–1.3, for highest vs. lowest level), while lowest and highest consumption of GT were ≤4 cups/day ≥8 cups/day respectively (Table 10.4). However, two another cohort studies in USA and Japan showed inverse association between GT consumption and stomach cancer (Table 10.4); and results were not statistically significant [106, 192].

10.3.1.4   Effects on Esophageal Cancer

There were 462,117 cases of esophageal cancer reported worldwide in 2002 out of which 315,394 cases were males and 146,723 cases were females [107]. Gao et al. [108] reported that consumption of two or three cups of GT daily reduced the risk of esophageal cancer among non-smokers and people who did not drank alcohol in Shanghai (Table 10.4). In South America, there was 38% reduction of esophageal cancer in men and women who consumed more than 500 ml of tea per day than those who did not drank tea [47]. It was found that there was no association between drinking of normal GT and esophageal cancer, however esophageal cancer was increased two- to three-folds. Taking very hot tea means higher the tea temperature, higher the risk of esophageal cancer [2, 61, 177]. According to Chen et al. [58], consumption of tea is associated with risk of esophageal cancer. In their study, 343 cases of squamous cell carcinoma of esophagus and 755 control reported that consumption of ≥ 7 cups of tea/week, was related to esophageal cancer (OR = 0.4; 95% CI = 0.2–0.6) as compared to <1 cup of tea/week. Notani and Jayant [254] reported that Indian who drank BT more than 2 cups/day was more prone to risk of esophageal cancer than those drinking 2 or less. Another Indian study in Mumbai reported 4-fold excess risk of esophageal cancer among tea drinkers [107].

10.3.1.5 Effects on Oral Cancer

Major risk factor, that is responsible for oral cancer, includes smoking tobacco and alcohol drinking [155, 156]. Other factors may be poor oral hygiene, ill-fitting dentures and poor dentition [152]. GTPs showed protective action against the oral cancer [39]. La Vecchia et al. [202] reported reduction in oral cancer in non-alcohol drinkers with consuming BT $\geq$ 1 cup/day. However, result was statistically not-significant (Table 10.4). GT and the constituents present in it induced apoptosis in oral carcinoma cell while EGCG of GT suppressed the growth of oral carcinoma cell and preventing from oral cancer [145]. Oral cancer was induced by cigarette smoking and was inhibited by intake of mixed tea due to its preventing action on DNA damage in oral leukoplakias [219]. Recently, Ren et al. [273] reported an inverse relationship between drinking of hot tea and oral cancer (OR = 0.75; 95% CI = 0.53–1.06).

10.3.1.6 Effects on Skin Cancer

The human skin is an effective barrier against physical, chemical and environmental pollutants [172, 174]. Skin cancer is one of the major problems in Australia and USA: about 2.5% men and 1.7% women in Australia and 55% population of the USA are suffering from skin cancer [7]. In the USA, there are one million cases of new skin cancer recognized per year [144]. Ultraviolet rays from sun play an important role in skin cancer development and are of three types on the basis of wavelength: UVA (320–400 nm), UVB (290–320 nm) and UVC (200–290 nm) [73]. Exposure of skin to UVB or a combination of UVA and UVB light can initiate skin cancer [272]. Several studies reported that GTPs especially EGCG provided protection against UV light and showed preventive action against UV induced skin cancer [175]. The *in vitro* and *in vivo* studies on animals and human showed that GTPs have ability to prevent skin disorders including skin photo aging, skin cancer (melanoma and non-melanoma) [240, 355–357]. According to Wang et al. [350], oral administration of GTP (0.1% in drinking water) to mice, which were exposed in UVB light, showed significant reduction in skin tumor as compared to non-GTP drinkers. Bushman [39] reported 62% and 29% reduction in skin cancer in mice treated with 10 or 50 g EGCG three times per week for three weeks while mice were exposed to UV light and developed 95% skin cancer at 28 weeks after first UV treatment.

10.3.1.7   Effects on Prostate Cancer

In USA, nearly 200,000 men suffer from prostate cancer out of which 29,000 die per year [166]. According to Jian et al. [164], consumption of GT reduced the development of prostate cancer in men (Table 10.4). They reported that taking three cups of GT frequently reduced prostate cancer among the men however not in those who did not took GT (Table 10.4). Another case-control study in Japan on 140 men with prostate cancer and same numbers of hospital controls indicated that higher consumption of GT, i.e., ≥ 5 cups/day reduced prostate cancer in men than those who consumed < cup of GT/day in Japan [310] (Table 10.4). A report in Japan indicated that a dose dependent inverse relationship among GT drinkers of Japan and risk of prostate cancer, for higher consumption of GT, i.e., ≥ 5 cups/day, OR was 0.52(95% CI = 0.28–0.96) in men as compared to <1 cup of GT/day. Kikuchi et al. [182] in cohort study of 110 prostate cancer cases, found that there was no significant association among GT drinkers and prostate cancer who consumed ≥ 5 cups of GT/day when compared to < 1 cup of GT/day. Another cohort study of Japanese ancestry in Hawaii (USA) reported that consumption of GT associated with statistically non-significant increased risk of prostate cancer in men OR was 1.47(95% 0.99–2.19) [302].

10.3.2 EFFECTS ON CARDIOVASCULAR DISEASES (CVD)

CVDs are related to heart and blood vessels. There are many types of CVDs in which coronary heart disease (ischemic heart disease) and cerebrovascular disease are major cause of death in world [331]. One of the major factors that play an important role in the development of CVD is oxidative stress [28]. Other factors are high blood pressure, cigarette smoking, and high cholesterol level in blood, diabetes, obesity, physical activity, and occupation [159, 211]. In coronary heart disease due to deposition of plaques of fatty minerals in the inner wall of arteries, reduction of blood flow occurs in arteries and condition is known as atherosclerosis [159]. EGCG, a GTCs prevented the onset of CVD by reducing oxidative stress produced by reactive oxygen species and nitrogen species [40]. High plasma cholesterol level can be reduced by consuming high levels of GT [117, 274]. Asia and Japan have highest consumption of tobacco; show lowest incident of atherosclerosis and cancer due to high consumption of GT nearly 1.2 liter per day show most beneficial effect in these regions [320]. Hypertension is a common problem in today's lifestyle and a major factor that increases high blood pressure in the arteries and leads to CVDs. High blood pressure is defined as systolic pressure greater than 140 mm Hg and diastolic pressure greater than 90 mm Hg [382]. According to Yang et al. [365], moderate amount of consumption of GT and OT reduced hypertension. GTCs act as an anti-inflammatory by inhibiting transcriptional factor kappa B (NF-κB) that is responsible for heart inflammation [28, 225]. Sasazuki et al. [294] in their cross section study in Japan on 512 coronary heart patients (302 men and 210 women) reported that higher consumption of GT, i.e., ≥4 cups per day reduced atherosclerosis in men (OR = 0.4; 95% CI = 0.2–0.9) as compared those with one or less than one cup per day.

10.3.3 EFFECTS ON NEURODEGENERATIVE DISEASES

Alzheimer's disease (AD) and Parkinson's disease (PD) are common neurodegenerative disease in aged population. Major factor, that is associated with onset of AD and PD, is oxidative damage of neural biomolecules and accumulation of higher amount of iron in specific brain areas [111, 278].

AD is an age related disease and it is found that half of the populations over 80 years of old are suffering from this disease [36, 237]. AD can be characterized by dementia (memory failure and personality disorder due to memory loss) [267, 286]. It is well known that PD and AD are due to deposition of iron in brain tissue that leads to neuronal death [380]. Iron regulatory protein IRP 1 and 2 and divalent metal transporter DMT 1 are mutated iron metabolism genes which also play an important role in development of neurodegenerative disease such as AD, Huntington disease, amyotrophic lateral sclerosis, Friedreich ataxia, and multiple sclerosis [93, 238, 380]. GTPs have ability to prevent neurodegenerative diseases due to its antioxidant property, because GTPs control the homeostasis of calcium, inhibit dopamine (DA) presynaptic transporters and catechol-O-methyltransferase (COMT), amyloid-β fibril formation (Aβ fibrils); activate mitogen-activated protein kinase (MAPK), protein kinase C (PKC), antioxidant and phase II detoxifying enzymes and survival genes; process amyloidprecursor protein (APP) for secretion of nontoxic soluble amyloid precursor protein (sAPPα) and destabilize of preformed of β-amyloid-β peptide as a primary compound of senile plaques [56, 112, 157, 216, 217, 218, 238, 257, 260, 298]. There are beneficial actions of GTPs to prevent neurodegenerative diseases. Due to presence of 3, 4-dihydroxil group in B ring, GTCs neutralize ferric ion to form redox-inactive ion and protecting the cells from oxidative damage [116, 136].

One of the crucial factor which plays an important role in neuro-degeneration is induction of mitochondrial membrane permeabilization (MMP) and voltage-dependant anion channel (VDAC), a constituent of outer mitochondrial membrane that regulates MPP, combines with hexokinase to prevent the opening of mitochondrial transition pore complex while adenine nucleotide translocase (ANT) locates at inner mitochondrial membrane converters in to non-specific pore that causes neuron death [238]. GTPs interact with ANT to prevent neural death. GT consumption prevents the neuro-degeneration and provides protection against PD and AD. The neuro-protective activity of EC and EGCG is higher as compare to catechin or resveratrol [293]. EC have simple structure and good blood-brain barrier penetration properties; it might be a good alternative for combating neurodegenerative diseases including HIV-related neurocognitive disorder that is created due to oxidative stress.

254   Engineering Interventions in Agricultural Processing

10.3.4   EFFECTS ON ANTI-MICROBIAL ACTIVITY

Tea also plays an important role to prevent microbial diseases. Tiwari et al. [332] reported that both GT and BT extract inhibited the growth of *Salmonela typhimurium* 1402/84, *S. typhi*, *S typhi* Ty2a, *Shigella dysenterie*, *Yersinia enterocolitica* C770, and *Escherichia coli* (EPEC P2 1265). However, it was found that BT extracts showed higher inhibitory activity as compared to GT extract. The synergistic effect of tea was reported in combination with chloramphenicol and other antibiotics like gentamycin, methicillin, nalidixic acid [332]. Tea polyphenols also inhibited the growth of influenza-A virus that brought about worldwide outbreak in human and animals. In 1918–1919 the outbreak of Spanish flu occurred that causes about 20 million human deaths [326]. Recently in year of 2009 worldwide mortality came in light due to outbreak of H1N1, a type virus responsible for Human influenza-A. The inhibitory activity of tea polyphenols is in order of Theaflavin > procynadin B2 > procynadin B2 diagallate > (-)-epigallocatachingallate [368]. In 2006, the Food and Drug Administration of the United States has approved a drug called Veregene™ that is prepared by GT catechin and is used for the treatment of genital and anal warts caused by human papilloma viruses in adults [338]. GTPs also act as an anti-fungal since it is effective against tenia pedis [153]. Tenia pedis is a toe web spaces infection caused by *Trichophyton rubrum*, *Trichophyton mentagrophytes* or *Epidemophyton floccosum* [31]. Lukewarm water bathing with GTP effectively reduced the symptoms created by tenia pedis [176, 199]. The anti-microbial activity of tea was increased by decreasing the fermentation time [21].

10.3.5   ANTI-DIABETIC EFFECTS

From time immemorial, GT has been used as therapeutic agent for the treatment of diabetes. Diabetes mellitus is characterized by insulin deficiency due to which the body glucose level raise. Common symptoms of diabetes mellitus type 1 include: polyuria, polydipsia, glucosuria, slow wound healing, and weakness. However, indirect symptoms are atherosclerosis, hypertension, hypertriglyceridemia, hypercholestoremia, myocardial infarction, ischemic attack, retinopathy and neuropathy [37, 269,

Potential Health Benefits of Tea Polyphenols—A Review 255

312, 345]. Al-Attar and Zari [5] reported that tea leave extract acts as anti-hyperglycemic, anti-hyperlipidemic, and anti-hyperproteinemic. Type 2 diabetes and insulin resistance can also be controlled with GT that is associated with obesity. Type 2 diabetes is due to interruption of insulin signaling that leads to hyperinsulenemia [78, 90, 99]. Islam and Choi [158] reported that GT significantly reduced the blood glucose level in diabetic rat treated with low dose, i.e., 0.5% GT extract. However, high dose (2.0%) increased the blood glucose level. Recently, Shri Lankan BTs have proved to be as anti-hyperglycemic, anti-diabetic and hypoglycemic [162]. Both GT and BT showed high anti-glycation activity equally [344]. The diabetes was induced in mice by injecting streptozotocin (STZ) and it was reported that after 15 and 30 days the body weights were significantly decreased [178, 292, 295, 300, 311, 317]. However, when these diabetic mice are treated with tea leave extract, it decreased their body weight loss [5]. Tea is a good alternative source to reduce the diabetes and diabetes related complications.

10.4 CONCLUSIONS

Tea as a beverage that is a very popular drink in the world and a good source of antioxidant at low cost. Tea contains phenolic compounds that act as an antioxidant. There are so many types of teas in which GT is good source of antioxidant. GT contains 40% EGCG that protect the cells from oxidative damage by scavenging free radicals. Quercetin in the tea controls diabetes by preventing pancreatic β-cells damage. EGCG also prevents cancer of different organs like lungs, esophagus, breast, stomach, and skin, oral and prostrate. GTPs inhibit the activity of urokinase enzyme, kinase protein and cell proliferation. GT reduces circulation of estrogen levels in women thus reduces breast cancer. GT reduced 13% estrogen level while BT increased it 19%. Drinking very hot tea increased the risk of cancer by two to three folds. EGCG act as a barrier against UV rays and prevent skin cancer. High plasma cholesterol and triacylglycerides level can also be controlled by GTPs and provided protection to the heart. Tea also acts as anti-inflammatory and reduces heart inflammation. Neurodegenerative diseases like PD and AD can be controlled with GT by reducing oxidative stress. Tea polyphenols inhibit

256 Engineering Interventions in Agricultural Processing

the growth of microorganisms and its anti-microbial activity widely depends on fermentation time.

However, there are many conflicting results between case control study and cohort study conducted in different countries. For instance, two of the case-control studies claimed to reduce the stomach cancer by drinking GT while two of the cohorts claimed to increase the risk of stomach cancer with GT. These studies show that there is more need of research on tea to clear the confusion and determines the daily intake of tea.

10.5 SUMMARY

Tea is consumed worldwide next to water due to its beneficial effect. Recent studies on GT prove that it has ability to prevent cancer, CVDs, AD, and PD, Obesity, diabetes. Epigallocatechin-3-gallate present in GT reduces the oxidative stress. However, there are some studies on BT that does not show beneficial effects. It has been proved by researchers that GT is a good antioxidant and protects the cells from oxidative damage. This chapter provides an overview of beneficial effect of tea mainly GT. Besides many of beneficial effects, some of harmful aspects of tea have also been discussed.

KEYWORDS

- **anti-anxiolytic effect**
- **anti-arteriosclerotic**
- **anti-atherogenic**
- **anti-atherosclerotic**
- **anti-estrogenic**
- **anti-inflammatory**
- **anti-microbial activity**
- **antioxidant activity**
- **anti-tumorigenic**
- **catechins**

- **epigallocatechin**
- **epigallocatechin gallate**
- **flavonoid**
- **kaempferol**
- **oxidative damage**
- **tea**
- **tea polyphenols**
- **theaflavins**
- **thearubigins**

REFERENCES

1. Actis-Goretta, L., Ottaviani, J. I., Keen, C. L., & Fraga, C. G. (2003). Inhibition of angiotensin converting enzyme (ACE) activity by flavan-3-ols and procyanidins. *FEBS Letters, 555*, 597–600.
2. Agarwal, M. L., Agarwal, A., Taylor, W. R., & Stark, G. R. (1995). p53 controls both the G2/M and the G1 cell cycle checkpoints and mediates reversible growth arrest in human fibroblasts. *Proceedings of the National Academy of Sciences of the United States of America, 92*, 8493–8497.
3. Ahmad, N., Feyes, D. K., Nieminen, A. L., Agarwal, R., & Mukhtar, H. (1997). Green tea constituent epigallocatechin-3-gallate and induction of apoptosis and cell cycle arrest in human carcinoma cells. *Journal of the National Cancer Institute, 89*, 1881–1886.
4. Ahn, J., Lee, H., Kim, S., Park, J., & Ha, T. (2008). The anti-obesity effect of quercetin ismediated by the AMPK and MAPK signaling pathways. *Biochemical and Biophysical Research Communications, 373*, 545–549.
5. Al-Attar, A. M., & Zari, T. A. (2010). Influences of crude extract of tea leaves, *Camellia sinensis*, on streptozotocin diabetic male albino mice. *Saudi Journal of Biological Sciences, 17*, 295–301.
6. Ali, H. A., Afifi, M., Abdelazim, A. M., & Mosleh, Y. Y. (2014). Quercetin and omega 3 ameliorate oxidative stress induced by aluminum chloride in the brain. *Journal of Molecular Neuroscience, 53*(4), 654–660.
7. ACS (1991). American Cancer Society: Cancer facts and data. *American Cancer Society,* Washington. DC.
8. ACS (1997). *American Cancer Society: Cancer Facts and data.* http://www.cancer.org/97facts.html
9. An, Y., Zhang, Y., Li, C., Qian, Q., He, W., & Wang, T. (2011). Inhibitory effects of flavonoids from Abelmoschus manihot flowers on triglycerideaccumulation in 3 T3-L1 adipocytes. *Fitoterapia, 82*, 595–600.

10. Anderson, D., Dobrzynska, M. M., Basaran, N., Basaran, A., & Yu, T. W. (1998). Flavonoids modulate comet assay responses to food mutagens in human lymphocytes and sperm. *Mutation Research, 402*, 269–277.

11. Anderson, R. F., Fisher, L. J., & Hara, Y. (2001). Green tea catechins partially protect DNA from OH radical induced strand breaks and base damage through fast chemical repair of DNA radicals. *Carcinogenesis, 22*, 1189–1193.

12. Annapurna, A., Reddy, C. S., Akondi, R. B., & Rao, S. R. (2009). Cardioprotective actions of two bioflavonoids, quercetin and rutin, in experimental myocardial infarction in both normal and streptozotocin-induced type I diabetic rats. *Journal of Pharmacy and Pharmacology, 61*, 1365–1374.

13. Arts, I. C. W., Van, D. P. B., & Hollman, P. C. H. (2000). Catechin contents of foods commonly consumed in The Netherlands. 1. Fruits, vegetables, staple foods, and processed foods. *Journal of Agricultural and Food Chemistry, 48*(5), 1746–1751.

14. Arts, I. C., Hollman, P. C., Feskens, E. J., Bueno de Mesquita, H. B., & Kromhout, D. (2001). Catechin intake might explain the inverse relation between tea consumption and ischemic heart disease: the Zutphen Elderly Study. *The American Journal of Clinical Nutrition, 74*, 227–232.

15. Awasom, I. (2011). Commodity of the Quarter Tea. *Journal of Agricultural and Food Information, 12*, 12–22.

16. Babu, P. V., & Liu, D. (2008). Green tea catechins and cardiovascular health: anupdate. *Current Medicinal Chemistry, 15*(18), 1840–1850.

17. Bailey, R. G., Nursten, H. E., & McDowell, I. (1992). The isolation of a polymeric thearubigin fraction from black tea. *Journal of the Science of Food and Agriculture, 59*, 365–375.

18. Balasundram, N., Sundram, K., & Samman, S. (2006). Phenolic compounds in plants and agri industrial by-products: Antioxidant activity, occurrence, and potential uses. *Food chemistry, 99*, 191–203.

19. Baliga, M. S., & Katiyar, S. K. (2006). Chemoprevention of photocarcinogenesis by selected dietary botanicals photochem. *Photochemical and Photobiological Sciences, 5*, 243–253.

20. Balzer, J., Rassaf, T., Heiss, C., Kleinbongard, P., Lauer, T., Merx, M., Heussen, N., Gross, H. B., Keen, C. L., Schroeter, H., & Kelm, M. (2008). Sustained benefits in vascularfunction through flavanol-containing cocoa in medicated diabetic patients: a double-masked, randomized, controlled trial. *Journal of the American College of Cardiology, 51*, 2141–2149.

21. Bancirova, M. (2010). Comparison of the antioxidant capacity and the antimicrobial activity of black and green tea. *Food Research International, 43*, 1379–1382.

22. Bashirov, M. S., Nugmanov, S. N., Kolycheva, N. I. (1968). Epidemiological study of esophageal cancer in the Akhtubinsk region of the Kazakh Socialist Republic (Russian). *Voprosy Onkologii, 14*, 3.

23. Benzie, I. F. (2000). Evolution of antioxidant defense mechanisms. *European Journal of Nutrition, 39*, 53–61.

24. Benzie, I. F. F., & Szeto, Y. T. (1999). Total Antioxidant Capacity of Teas by the Ferric Reducing/Antioxidant Power Assay. *Journal of Agricultural and Food Chemistry, 47*, 633–636.

25. Beresford, J. N. (1989). Osteogenic stem cells and the stromal system of bone and marrow. *Clinical Orthopedics and Related Research, 240*, 270–280.
26. Beresford, J. N., Bennett, J. H., Devlin, C., Leboy, P. S., & Owen, M. E. (1992). Evidence for an inverse relationship between the differentiation of adipocytic and osteogenic cells in rat marrow stromal cell cultures. *Journal of Cell Science, 102*, 341–351.
27. Berger, M. M. (2005). Can oxidative damage be treated nutritionally? *ClinicalNutrition, 24*, 172–183.
28. Bhardwaj, P., & Khanna, D. (2013). Green tea catechins: defensive role in cardiovascular disorders. *Chinese Journal of Natural Medicines, 11*(4), 0345−0353.
29. Bianco, P., Riminucci, M., Gronthos, S., & Robey, P. G. (2001). Bone marrow stromal stem cells: nature, biology, and potential applications. *Stem Cells, 19*(3), 180–192.
30. Biswas, A. K., Biswas, A. K., & Sarkar, A. (1973). Biological and chemical factors affecting the valuations of North-East Indian plain teas. *Journal of the Science of Food and Agriculture, 24*, 1457–1477.
31. Bolognia, J. L., Jorizzo, J. L., & Rapini, R. P. (2003). *Dermatology.* New York, NY, USA: Mosby.
32. Boyd, N. F., Lockwood, G. A., Byng, J. W., Tritchler, D. L., & Yaffe, M. J. (1998). Mammographic densities and breast cancer risk. *Cancer Epidemiology, Biomarkers and Prevention, 7*, 1133–1144.
33. Boyd, N. F., Rommens, J. M., Vogt, K., Lee V Hopper, J. L., Yaffe, M. J., & Paterson, A. D. (2005). Mammographic breast density as an intermediate phenotype for breast cancer. *The Lancet Oncology, 6*, 798–808.
34. Braicu, C., Ladomery, M. R., Chedea, V. S., Irimie, A., & Berindan-Neagoe, I. (2013). The relationship between the structure and biological actions of green tea catechins. *Food Chemistry, 141*(3), 3282–3289.
35. Bravo, L. (1998). Polyphenols: chemistry, dietary sources, metabolism, and nutritional significance. *Nutriton Research, 56*, 317–333.
36. Brookmeyer, R., Gray, S., & Kawas, C. (1998). Projections of Alzheimer's disease in the United States and the public health impact of delaying disease onset. *American Journal of Public Health, 88*(9), 1337–1342.
37. Brownlee, M. (2001). Biochemistry and molecular cell biology of diabetic complications. *Nature, 414*, 813–820.
38. Budhraja, A., Gao, N., Zhang, Z., Son, Y. O., Cheng, S., Wang, X., & Shi, X. (2012). Apigenin induces apoptosis in human leukemia cells and exhibits anti-leukemic activity in vivo. *Molecular Cancer Therapeutics, 11*(1), 132–142.
39. Bushman, J. L. (1998). Green tea and cancer in humans: a review of the literature. *Nutrition and Cancer, 31*, 151–159.
40. Butt, M. S., & Sultan, M. T. (2009). Green Tea: Nature's Defense against Malignancies. *Critical Review in Food Science & Nutrition, 49*, 463–473.
41. Cabrera, C., Artacho, R., & Gimenez, R. (2006). Beneficial effects of green tea— a review. *The Journal of the American College of Nutrition, 25*, 79–99.
42. Cabrera, C., Gimenez, R., & Lopez, M. C. (2003). Determination of tea compounds with antioxidant activity. *Journal of Agricultural and Food Chemistry, 51*, 4427–4435.
43. Cai, Q., Rahn, R. O., & Zhang, R. (1997). Dietary flavonoids, quercetin, luteolin and genistein, reduce oxidative DNA damage and lipid peroxidation and quench free radicals. *Cancer Letters, 119*, 99–107.

44. Calderon-Montano, J. M., Burgos-Moron, E., Perez-Guerrero, C., & Lopez-Lazaro, M. (2011). A review on the dietary flavonoid Kaempferol. *Medicinal Chemistry, 11*, 298–344.

45. Caltagirone, S., Rossi, C., & Poggi, A. (2000). Flavonoids apigenin and quercetin inhibit melanoma growth and metastatic potential. *International Journal of Cancer, 87*(4), 595–600.

46. Camoirano, A., Balansky, R. M., Bennicelli, C., Izzotti, A., D'Agostini, F., & De Flora, S. (1994). Experimental databases on inhibition of the bacterial mutagenicity of 4-nitroquinoline 1-oxide and cigarette smoke. *Mutation Research, 317*, 89–109.

47. Castellsague, X., Munoz, N., De Stefani, E., Victora, C. G., Castelletto, R., & Rolon, P. A. (2000). Influence of mate drinking, hot beverages and diet on esophageal cancer risk in South America. *International Journal of Cancer, 88*, 658–664.

48. Chakraborty, S., Kumar, S., & Basu, S. (2011). Conformational transition in the substrate binding domain of secretase exploited by NMA and its implication in inhibitor recognition: BACE1-myricetin a case study. *Neurochemistry International, 58*, 914–923.

49. Chakraborty, U., Dutta, S., & Chakraborty, B. N. (2002). Response of tea plants to water stress. *Biologic Plantarum, 45*, 557–562.

50. Chan, E. W. C., Lim, Y. Y., & Chew, Y. L. (2007). Antioxidant activity of *Camellia sinensis* leaves and tea from a low land plantation in Malaysia. *Food Chemistry, 102*, 1214–1222.

51. Chang, C. J., Tzeng, T. F., Liou, S. S., Chang, Y. S., & Liu, I. M. (2012a). Myricetinincreases hepatic peroxisome proliferator-activated receptorα protein expression and decreases plasma lipids and adiposity in rats. *Evid Based Complement Alternat Med 2012*. http://dx.doi.org/10.1155/2012/787152 [Article ID787152].

52. Chang, J., Hsu, Y., Kuo, P., Kuo, Y., Chiang, L., & Lin, C. (2005). Increase of Bax/Bcl-XL ratio and arrest of cell cycle by luteolin in immortalized human hepatoma cell line. *Life Sciences, 76*, 1883–1893.

53. Chang, J., Li, Y., Wang, G., He, B., & Gu, Z. (2012b). Fabrication of novel coumarin derivative functionalized polypseudorotaxane micelles for drug delivery. *Nanoscale, 5*, 813–820.

54. Checkoway, H., Powers, K., Smith-Weller, T., Franklin, G. M., Longstreth, Jr. W. T., & Swanson, P. D. (2002). Parkinson's disease risks associated with cigarette smoking, alcohol consumption, and caffeine intake. *American Journal of Epidemiology, 155*, 732–738.

55. Chen, A. Y., & Chen, Y. C. (2013). A review of the dietary flavonoid, kaempferol on human health and cancer chemoprevention. *Food Chemistry, 138*, 2099–2107.

56. Chen, C., Yu, R., Owuor, E. D., & Kong, A. N. (2000). Activation of antioxidant-response element (ARE), mitogen-activated protein kinases (MAPKs) and caspases by major green tea polyphenol components during cell survival and death. *Archives of Pharmacal Research, 23*, 605–612.

57. Chen, H. J., Lin, C. M., Lee, C. Y., Shih, N. C., Peng, S. F., & Tsuzuki, M. (2013). Kaempferol suppresses cell metastasis via inhibition of the ERK-p38-JNK and AP-1 signaling pathways in U-2 OS human osteosarcoma cells. *Oncology Reports, 30*, 925–932.

58. Chen, Y. K., Lee, C. H., Wu, I. C., Liu, J. S., Wu, D. C., Lee, J. M., Goan, Y. G., Chou, S. H., Huang, C. T., Lee, C. Y., Hung, H. C., Yang, J. F., & Wu, M. T. (2009).

Food intake and the occurrence of squamous cell carcinoma in different sections of the esophagus in Taiwanese men. *Nutrition, 25*, 753–761.

59. Chen, Y. M., Tsao, T. M., Liu, C. C., Lin, K. C., & Wang, M. K. (2011). Aluminium and nutrients induce changes in the profiles of phenolic substances in tea plants (*Camellia sinensis* CV TTES, No. 12 (TTE)). *Journal of the Science of Food and Agriculture, 91*, 1111–1117.

60. Chen, Y. T., Zheng, R. L., Jia, Z. J., & Ju, Y. (1990). Flavonoids as superoxide scavengers and antioxidants. *Free Radical Biology & Medicine, 9*, 19–21.

61. Cheng, K. K., & Day, N. E. (1996). Nutrition and esophageal cancer. *Cancer Causes and Control, 7*, 33–40.

62. Cheruiyot, E. K., Mumera, L. M., Ngetich, W. K., Hassanali, A., Wachira, F., & Wanyoko, J. K. (2008). Shoot epicatechin and epigallocatechin contents respond to water stress in tea (*Camellia sinensis* L.). *Bioscience, Biotechnology, and Biochemistry, 72*, 1219–1226.

63. Choi, E. J., & Ahn, W. S. (2008). Kaempferol induced the apoptosis via cell cycle arrest inhuman breast cancer MDA-MB-453 cells. *Nutrition Research and Practice, 2*(4), 322–325.

64. Chow, K., & Kramer, I. (1990). All the tea in China. San Francisco: *China Books and Periodicals.*

65. Chung, F. L., Schwartz, J., Herzog, C. R., Yang, Y. M. (2003). Tea and cancer prevention: Studies in animals and humans. *Journal of Nutrition, 133*, 3268–3274.

66. Chung, S. Y., Joshua, D. L., Jihyeung, J., Gang, L., & Shengmin, S. (2007). Tea and cancer prevention: Molecular mechanisms and human relevance. *Toxicology and Applied Pharmacology, 224*, 265–273.

67. Clark, J., & You, M. (2006). Chemoprevention of lung cancer by tea. *Molecular Nutrition & Food Research, 50*(2), 144–151.

68. Comte, G., Daskiewicz, J. B., Bayet, C., Conseil, G., Viornery-Vanier, A., Dumontet, C., Pietro, A., & Barron, D. (2001). C-Isoprenylation of flavonoids enhances binding affinity toward P-glycoprotein and modulation of cancer cell chemoresistance. *Journal of Medicinal Chemistry, 44*, 763–768.

69. Crespy, V., & Williamson, G. (2004). A review of the health effect of green tea catechins in *in vivo* animal models. *Journal of Nutrition, 134*, 3431S–3440S.

70. Davies, A. P., Goodsall, C., Cai, Y., Davis, A. L., Lewis, J. R., Wilkins, J., Wan, X., Clifford, M. N., Powell, C., Parry, A., Thiru, A., Safford, R., & Nursten, H. E. (1999). Black tea dimeric and oligomeric pigments structures and formation. In: Gross, C. G., Hemingway, R. W., Yoshida, T. (Eds.), *Plant Polyphenols 2: Chemistry, Biology, Pharmacology, Ecology.* Kluwer Academic/Plenum Press, New York, pp. 697–724.

71. Davis, C. T., & Geissman, T. A. (1954). Basic dissociation constants of some substituted flavones. *Journal of the American Chemical Society, 76*, 3507–3511.

72. Davis, J. M., Murphy, E. A., Carmichael, M. D., & Davis, B. (2009). Quercetin increases brain and muscle mitochondrial biogenesis and exercise tolerance. *American Journal of Physiology, Regulatory, Integrative and Comparative Physiology, 296*, R1071–R1077.

73. De Gruijl, F. R. and van der Leun, J. C. (1994). Estimate of the wavelength dependency of ultraviolet carcinogenesis in humans and its relevance to the risk assessment of stratospheric ozone depletion. *Health Physics, 67*, 319–325.

262   Engineering Interventions in Agricultural Processing

74. De Mejia, E. G., Ramirez-Mares, M. V., & Puangpraphant, S. (2009). Bioactive components of tea: cancer, inflammation and behavior. *Brain, Behavior & Immunity, 23*(6), 721–731.

75. De Whalley, C. V., Rankin, S. M., Hoult, J. R., Jessup, W., Wilkins, G. M., & Collard, J. (1990). Modification of low-density lipoproteins by flavonoids. *Biochemical Society Transactions, 18*, 1172–1173.

76. Dean, R. T., Fu, S., Stocker, R., & Davies, M. J. (1997). Biochemistry and pathology of radical-mediated protein oxidation. *Biochemical Journal, 324*, 1–18.

77. Delgado, M. E., Haza, A. I., Arranz, N., Garcia, A., & Morales, P. (2008). Dietary polyphenols protect against N-nitrosamines and benzo(a)pyrene-induced DNA damage (strand breaks and oxidized purines/pyrimidines) in HepG2 human hepatoma cells. *European Journal of Nutrition, 47*, 479–490.

78. Deshpande, A. D., Harris-Hayes, M., & Schootman, M. (2008).Epidemiology of diabetes and diabetes-related complications. *Physical Therapy, 88*, 1254–1264.

79. Devine, A., Hodgson, J. M., Dic, I. M., & Prince, R. L. (2007). Tea drinking is associated with benefits on bone density in older women. *The American Journal of Clinical Nutrition, 86*, 1243–1247.

80. Dias, T. R., Toms, G., Teixeira, N. F., Alves, M. G., Oliveira, P. F., & Silva, B. M. (2013). White tea (*Camellia Sinensis* (L.)): Antioxidant properties and beneficial health effect. *Intern. Journal of Food Sciences and Nutrition Diet, 2*(2), 1–15.

81. DiCarlo, G., Mascolc, N., Izzo, A. A., & Capasso, F. (1999). Flavonoids: old and new aspects of a class of natural therapeutic drugs. *Life Sciences, 65*, 337–353.

82. Dorgan, J. F., Stanczyk, F. Z., Kahle, L. L., & Brinton, L. A. (2010). Prospective case-control study of premenopausal serum estradiol and testosterone levels and breast cancer risk. *Breast Cancer Research, 12*, R98.

83. Dreosti, I. E., Wargovich, M. J., & Yang, C. S. (1997). Inhibition of carcinogenesis by tea: The evidence from experimental studies. *Critical Reviews in Food Science and Nutrition, 37*(8), 761–770.

84. Drynan, J. W., Clifford, M. N., Obuchowicz, J., & Kuhnert, N. (2012). MALDI-TOF mass spectrometry: Avoidance of artifacts and analysis of caffeine-precipitated sill thearubigins from 15 commercial black teas. *Journal of Agricultural and Food Chemistry, 60*(18), 4514–4525.

85. Duarte, J., Jimenez, R., Villar, I. C., Perez-Vizcaino, F., Jimenez, J., & Tamargo, J. (2001). Vasorelaxant effects of the bioflavonoid chrysin in isolated rat aorta. *Planta Medica, 67*, 567–569.

86. Duffy, S. J., Keaney Jr., J. F., Holbrook, M., Gokce, N., Swerdloff, P. L., Frei, B., & Vita, J. A. (2001). Short- and long-term black tea consumption reverses endothelial dysfunctionin patients with coronary artery disease. *Circulation, 104*, 151–156.

87. Duthie, G., & Crozier, A. (2000). Plant-derived phenolic antioxidants. *Current Opinion in Lipidology, 11*(1), 43–47.

88. Duthie, S. J., Collins, A. R., Duthie, G. G., & Dobson, V. L. (1997a). Quercetin and myricetin protect against hydrogen peroxide-induced DNA damage (strand breaks and oxidized pyrimidines) in human lymphocytes. *Mutation Research, 393*, 223–231.

89. Duthie, S. J., Johnson, W., & Dobson, V. L. (1997b). The effect of dietary flavonoids on DNA damage (strand breaks and oxidized pyrimdines) and growth in human cells. *Mutation Research, 390*, 141–151.

Potential Health Benefits of Tea Polyphenols—A Review 263

90. Eckel, R. H., Grundy, S. M., & Zimmet, P. Z. (2005). The metabolic syndrome. *The Lancet, 365*, 1415–1428.
91. Fan, F. Y., Shi, M., Nie, Y., Zhao, Y., Ye, J. H., & Liang, Y. R. (2016). Differential behaviors of tea catechins under thermal processing: Formation of non-enzymatic oligomers. *Food Chemistry, 196*, 347–354
92. FAO (2008). http://faostat.fao.org/default.aspx; 2 March 2011.
93. Felletschin, B., Bauer, P., Walter, U., Behnke, S., Spiegel, J., Csoti, I., Sommer, U., Zeiler, B., Becker, G., Riess, O., & Berg, D. (2003). Screening for mutations of the ferritin light and heavy genes in Parkinson's disease patients with hyperechogenicity of the substantia nigra. *Neuroscience Letters, 352*, 53–56.
94. Ferlay, J., Bray, F., Pisani, P., & Parkin, D. M. (2004). *GLOBOCAN 2002: cancer incidence, mortality and prevalence worldwide.* IARC CancerBase No. 5. version 2.0. IARC Press, Lyon.
95. Fesus, L., Szondy, Z., & Ura, I. (1995). Probing the molecular program of apoptosis by cancer chemopreventive agents. *Journal of Cellular Biochemistry, 22*, 151–161.
96. Fisher, N. D., Hughes, M., Gerhard-Herman, M., & Hollenberg, N. K. (2003). Flavanol-richcocoa induces nitric-oxide-dependent vasodilation in healthy humans. *Journal of Hypertension, 21*, 2281–2286.
97. Flammer, A. J., Hermann, F., Sudano, I., Spieker, L., Hermann, M., Cooper, K. A., Serafini, M., Luscher, T. F., Ruschitzka, F., Noll, G., & Corti, R. (2007). Dark chocolateimproves coronary vasomotion and reduces platelet reactivity. *Circulation, 116*, 2376–2382.
98. Fong, H. H. (2002). Integration of herbal medicine into modern medical practices: issues and prospects. *Integrative Cancer Therapies, 1*, 287–293.
99. Fonseca, V., Inzucchi, S. E., & Ferrannini, E. (2009). Redefining the diagnosis of diabetes using glycated hemoglobin. *Diabetes Care, 32*, 1344–1345.
100. Forman, D., & Burley, V. J. (2006). Gastric cancer: global pattern of the disease and an overview of environmental risk factors. *Best Practice & Research Clinical Gastroenterology, 20*, 633–649.
101. Frankel, E. N. (1984). Lipid oxidation mechanisms, products and biological significance. *Journal of the American Oil Chemist's Society, 61*, 1908–1915.
102. Frei, B., & Higdon, J. V. (2003). Antioxidant Activity of Tea Polyphenols In Vivo: Evidence from Animal Studies. *Journal of Nutrition, 133*, 3275S–3284S.
103. Friedman, M. (2007). Overview of antibacterial, antitoxin, antiviral, and antifungal activities of tea flavonoids and teas. *Mol. Nutr. Food Res. 51*, 116–134.
104. Fujiki, H., Suganuma, M., Okabe, S., Komori, A., Sueoka, E., Sueoka, N., Kozu, T., & Sakai, Y. (1996). Japanese green tea as a cancer preventive in humans.*Nutrition Reviews, 54*, 67S–70S.
105. Fujiki, H., Suganuma, M., Okabe, S., Kurusu, M., Imai, K., & Nakachi, K. (2002). Involvement of TNF-alpha changes in human cancer development, prevention and palliative care. *Mechanisms of Ageing & Development, 123*(12), 1655–1663.
106. Galanis, D. J., Kolonel, L. N., Lee, J., & Nomura, A. (1998). Intakes of selected foods and beverages and the incidence of gastric cancer among the Japanese residents of Hawaii: a prospective study. *International Journal of Epidemiology, 27*, 173–180.

107. Ganesh, B., Talole, S. D., & Dikshit, R. (2009). Tobacco, alcohol and tea drinking as risk factors for esophageal cancer: A case–control study from Mumbai, India. *Cancer Epidemiology, 33*, 431–434.

108. Gao, Y. T., McLaughlin, J. K., Blot, W. J., Ji, B. T., Dai, Q., & Fraumeni, J. J. (1994). Reduced risk of esophageal cancer associated with green tea consumptioa. *Journal of National Cancer Institute, 86*, 855–858.

109. Gardner, E. J., Ruxton, C. H. S., & Leeds, A. R. (2007). Black tea –helpful or harmful? A review of the evidence. *European Journal of Clinical Nutrition, 61*(1), 3–18.

110. Gates, M. A., Tworoger, S. S., Hecht, J. L., De, V., Ivo, I., Rosner, B., & Hankinson, S. E. (2007). A prospective study of dietary flavonoid intake and incidence of epithelial ovarian cancer. *International Journal of Cancer, 121*(10), 2225–2232.

111. Gerlach, M., Double, K. L., Ben-Shachar, D., Zecca, L., Youdim, M. B., & Riederer, P. (2003). Neuromelanin and its interaction with iron as a potential risk factor for dopaminergic neurodegeneration underlying Parkinson's disease. *Neurotoxicological Research, 5*, 35–44.

112. Glenner, G. G., & Wong, C. W. (1984). Alzheimer's disease: initial report of the purification and characterization of a novel cerebrovascular amyloid protein. *Biochemical and Biophysical Research Communications, 120*, 885–890.

113. Graham, H. N. (1992). Green tea composition, consumption, and polyphenol chemistry. *Pre-Medical, 21*, 334–350.

114. Grassi, D., Desideri, G., Necozione, S., Lippi, C., Casale, R., Properzi, G., Blumberg, J. B., & Ferri, C. (2008). Blood pressure is reduced and insulin sensitivity increased inglucose-intolerant, hypertensive subjects after 15 days of consuming highpolyphenoldark chocolate. *Journal of Nutrition, 138*, 1671–1676.

115. Grassi, D., Lippi, C., Necozione, S., Desideri, G., & Ferri, C. (2005). Short-term administrationof dark chocolate is followed by a significant increase in insulin sensitivity and adecrease in blood pressure in healthy persons. *The American Journal of Clinical Nutrition, 81*, 611–614.

116. Grinberg, L. N., Newmark, H., Kitrossky, N., Rahamim, E., Chevion, M., & Rachmilewitz, E. A. (1997). Protective effects of tea polyphenols against oxidative damage to red blood cells. *Biochemical Pharmacology, 54*, 973–978.

117. Grundy, S. M. (1986). Cholesterol and coronary heart disease. *A New Era of JAMA, 256*, 2849–2858.

118. Guo, Q., Zhao, B. L., Shen, S. R., Hou, J. W., Hu, J. G., & Xin, W. J. (1999). ESR study on the structure–antioxidant activity relationship of tea catechins and their epimers. *Biochimica et Biophysica Acta – General Subjects, 1427*(1), 13–23.

119. Gupta, S., Afaq, F., & Mukhtar, H. (2001). Selective growth-inhibitory, cellcycle deregulatory and apoptotic response of apigenin in normal versus human prostate carcinoma cells. *Biochemical and Biophysical Research Communications, 287*(4), 914–920.

120. Gupta, S., Saha, B., & Giri, A. K. (2002). Comparative antimutagenic and anticlastogenic effects of green tea and black tea: A review. *Mutation Research/Reviews in Mutation Research, 512*(1), 37–65.

121. Hakim, I., Weisgerber, U., Harris, R., Balentine, D., & Van-Mierlo, C. (2000). Preparation, composition and consumption patterns of tea-based beverages in Arizona. *Nutrition Research, 20*(12), 1715–1724.

122. Han, J., Britten, M., St-Gelais, D., Champagne, C. P., Fustier, P., Salmieri, S., & Lacroix, M. (2011). Effect of polyphenolic ingredients on physical characteristics of cheese. *Food Research International, 44*(1), 494–497.
123. Hanasaki, Y., Ogawa, S., & Fukui, S. (1994). The correlation between active oxygens scavenging and antioxidative effects of flavonoids. *Free Radical Biology & Medicine, 16,* 845–850.
124. Harada, U., Chikama, A., & Saito, S. (2005). Effects of the long-term ingestion of teacatechins on energy expenditure and dietary fat oxidation in healthy subjects. *Journal of Health Science, 51,* 248–252.
125. Harnly, J. M., Doherty, R. F., Beecher, G. R., Holden, J. M., Haytowitz, D. B., Bhagwat, S., & Gebhardt, S. (2006). Flavonoid content of U. S. fruits, vegetables, and nuts. *Journal of Agricultural and Food Chemistry, 54,* 9966–9977.
126. Harwood, M., Danielewska-Nikiel, B., Borzelleca, J. F., Flamm, G. W., Williams, G. M., & Lines, T. C. (2007). A critical review of the data related to the safety of quercetin and lack of evidence of in vivo toxicity, including lack of genotoxic/carcinogenic properties. *Food and Chemical Toxicology, 45,* 2179–2205.
127. Haslam, E. (2003). Thoughts on thearubigins. *Phytochemistry, 64,* 61–73.
128. Hastak, K., Gupta, S., Ahmad, N., Agarwal, M. K., Agarwal, M. L., & Mukhtar, H. (2003). Role of p53 and NF-kappaB in epigallocatechin-3-gallate-induced apoptosis of LNCaP cells. *Oncogene, 22,* 4851–4859.
129. He, Q., Lv, Y., & Yao, K. (2007). Effects of tea polyphenols on the activities of aamylase, pepsin, trypsin and lipase. *Food Chemistry, 101*(3), 1178–1182.
130. Hecker, M., Preiss, C., Schini-Kerth, V. B., & Busse, R. (1996). Antioxidants differentially affect nuclear factor kappa B-mediated nitric oxide synthase expression in vascular smooth muscle cells. *FEBS Letters, 380,* 224–8.
131. Heinrich, U., Neukam, K., Tronnier, H., Sies, H., & Stahl, W. (2006). Long-term ingestion of high flavanol cocoa provides photoprotection against UV-induced erythema and improves skin condition in women. *Journal of Nutrition, 136,* 1565–1569.
132. Heiss, C., Dejam, A., Kleinbongard, P., Schewe, T., Sies, H., & Kelm, M. (2003). Vascular effects of cocoa rich in flavan-3-ols. *JAMA, 290,* 1030–1031.
133. Heiss, C., Finis, D., Kleinbongard, P., Hoffmann, A., Rassaf, T., Kelm, M., & Sies, H. (2007). Sustained increase in flow-mediated dilation after daily intake of high-flavanolcocoa drink over 1 week. *Journal of Cardiovascular Pharmacology, 49,* 74–80.
134. HCCG (Helicobacter and Cancer Collaborative Group). (2001). Gastric cancer and Helicobacter pylori: a combined analysis of 12 case control studies nested within prospective cohorts. *Gut, 49,* 347–353.
135. Hernaez, J. F., Xu, M., & Dashwood, R. H. (1998). Antimutagenic activity of tea towards 2-hydroxyamino-3-methylmidazo [4,5-f]quinolone]: effect of tea concentration and brew time on electrophile scavenging. *Mutation Research, 402,* 200–306.
136. Hider, R. C., Liu, Z. D., & Khodr, H. H. (2001). Metal chelation of polyphenols. *Methods in Enzymology, 335,* 190–203.
137. Higdon, J. V., & Frei, B. (2003). Tea Catechins and Polyphenols: Health Effects, Metabolism, and Antioxidant Functions. *Critical Reviews in Food Science and Nutrition, 43*(1), 89–143.

138. Hodgson, J. M., Puddey, I. B., Burke, V., Beilin, L. J., & Jordan, N. (1999). Effects on blood pressure of drinking green and black tea. *Journal of Hypertension, 17*, 457–463.
139. Hollman, P. C., Feskens, E. J., & Katan, M. B. (1999). Tea flavonols in cardiovascular disease and cancer epidemiology. *Proceedings of the Society for Experimental Biology and Medicine, 220*, 198– 202.
140. Holt, R. R., Schramm, D. D., Keen, C. L., Lazarus, S. A., & Schmitz, H. H. (2002). Chocolate consumption and platelet function. *JAMA, 287*, 2212–2213.
141. Hong, G. J., Wang, J., Zhang, Y., Hochstetter, D., Zhang, S. P., Pan, Y., Shi, Y. L., Xu, P., & Wang, Y. F. (2014). Biosynthesis of catechin components is differentially regulated in dark-treated tea (*Camellia sinensis* L.). *Plant Physiology and Biochemistry, 78*, 49–52.
142. Hoshiyama, Y., & Sasaba, T. (1992a). A case-control study of single and multiple stomach cancers in Saitama Prefecture, Japan. *Japanese Journal of Cancer Research, 83*, 937–943.
143. Hoshiyama, Y., & Sasaba, T. A. (1992b). Case-control study of stomach cancer and its relation to diet, cigarettes, and alcohol consumption in Saitama Prefecture, Japan. *Cancer Causes and Control, 3,* 441–448.
144. Housman, T. S., Feldman, S. R., Williford, P. M., Fleischer, Jr., A. B., Goldman, N. D., Acostamadiedo, J. M., & Chen, G. J. (2003). Skin cancer is among the costly of all cancers to treat for the medicare population. *Journal of the American Academy of Dermatology, 48*, 425–429.
145. Hsu, S. D., Singh, B. B., Lewis, J. B., Borke, J. L., Dickinson, D. P., & Drake, L. (2002). Chemoprevention of oral cancer by green tea. *General Dentistry, 50*, 140–146.
146. Huang, M. T., Liu, Y., Ramji, D., Lo, C. Y., Ghai, G., Dushenkov, S., & Ho, C. T. (2006). Inhibitory effects of black tea the aflavin derivatives on 12-O-tetradecanoylphorbol-13-acetate-induced inflammation and arachidonic acid metabolism in mouse ears. *Molecular Nutrition and Food Research, 50*(2), 115–122.
147. Huang, S. W., & Frankel, E. N. (1997). Antioxidant Activity of Tea Catechins in Different Lipid Systems. Journal of Agricultural and Food Chemistry, *45*, 3033–3038.
148. Huang, W. W., Chiu, Y. J., Fan, M. J., Lu, H. F., Yeh, H. F., & Li, K. H. (2010). Kaempferol induced apoptosis via endoplasmic reticulum stress and mitochondria-dependent pathway in human osteosarcoma U-2 OS cells. *Molecular Nutrition and Food Research, 54*, 1585–95.
149. Hubbard, G. P., Wolfram, S., Lovegrove, J. A., & Gibbins, J. M. (2004). Ingestion of quercetin inhibit splatelet aggregation and essential components of the collagen-stimulated platelet activation pathway in humans. *Journal of Thrombosis and Haemostasis, 2*, 2138–2145.
150. Huffman, M. A. (2003). Animal self-medication and ethno-medicine: exploration and exploitation of the medicinal properties of plants. *Proceedings of the Nutrition Society, 62*, 371–381.
151. Huvaere, K., Nielsen, J. H., Bakman, M., Hammershoj, M., Skibsted, L. H., Sorensen, J., & Dalsgaard, T. K. (2011). Antioxidant properties of green tea extract protect reduced fat soft cheese against oxidation induced by light exposure. *Journal of Agricultural and Food Chemistry, 59*(16), 8718–8723.
152. Ide, R., Fujino, Y., Hoshiyama, Y., Mizoue, T., Kubo, T., Pham, T. M., Shirane, K., Tokui, N., Sakata, K., Tamakoshi, A., & Yoshimura, T. (2007). A Prospective Study

of Green Tea Consumption and Oral Cancer Incidence in Japan. *Annals of Epidemiology, 17*, 821–826.

153. Ikedaa, S., Kanoya, K., & Nagatad, S. (2013). Effects of a foot bath containing green tea polyphenols on interdigital tinea pedis. *The Foot, 23*, 58–62.

154. Imai, K., Suga, K., & Nakachi, K. (1997). Cancer-preventive effects of drinking green tea among a Japanese population. *Preventive Medicine, 26*, 769–775.

155. IARC (International Agency for Research on Cancer). (1986). *IARC monograph on the evaluation of the carcinogenic risk of chemicals to humans: Tobacco smoking.* Lyon: IARC.

156. IARC (International Agency for Research on Cancer). (1988). *IARC Monograph on the Evaluation of the Carcinogenic Risk of Chemicals to Humans: Alcohol Drinking.* Lyon: IARC.

157. Ishige, K., Schubert, D., & Sagara, Y. (2001). Flavonoids protect neuronal cells from oxidative stress by three distinct mechanisms. *Free Radical Biology and Medicine, 30*, 433–446.

158. Islam, M. S., & Choi, H. (2007). Green tea, anti-diabetic or diabetogenic: a dose response study. *Biofactors, 29*, 45–53.

159. Iso, H., Jacobs, D. R., Wentworth, D. N., Neaton, J. D., & Cohen, J. D. (1989). Serum cholesterol levels and six-year mortality from stroke in 350,977 men screened for the multiple risk factor intervention trial. *The New England Journal of Medicine, 320*, 904–910.

160. Jang, Y. J., Kim, J., Shim, J., Kim, J., Byun, S., Oak, M. H., Lee, K. W., & Lee, H. J. (2011). Kaempferol attenuates 4-hydroxynonenal-induced apoptosis in PC12 cells by directly inhibiting NADPH oxidase. *Journal of Pharmacology and Experimental Therapeutics, 337*, 747–754.

161. Jankun, J., Selman, S. H., Swiercz, R., & Skrzypczak-Jankun, E. (1997). Why drinking green tea could prevent cancer. *Nature, 5*(6633), 387–561.

162. Jayakody, J. R. A. C., & Ratnasooriya, W. D. (2008). Blood glucose level lowering activity of Sri Lankan black tea brew (*Camellia sinensis*) in rats. *Pharmacognosy Magazine, 4*, 341–349.

163. Jia, Z., Nallasamy, P., Liu, D., Shah, H., Li, J. Z., Chitrakar, R., Si, H., McCormick, J., Zhu, H., Zhen, W., & Li, Y. (2015). Luteolin protects against vascular inflammation in mice and TNF-alpha induced monocyte adhesion to endothelial cells via suppressing I-Kappa B-alpha/NF-kappa-B signaling pathway. *The Journal of Nutritional Biochemistry, 26*(3), 293–302.

164. Jian, L., Xie, L. P., Lee, A. H., & Binns, C. W. (2004). Protective effect of green tea against prostate cancer: a case-control study in southeast China. *International Journal of Cancer, 108*, 130–135.

165. Jin, H. B., Yang, Y. B., Song, Y. L., Zhang, Y. C., & Li, Y. R. (2012). Protective roles of quercetin in acute myocardial ischemia and reperfusion injury in rats. *Molecular Biology Reports, 39*, 11005–11009.

166. Johnson, J. J., Bailey, H. H., & Mukhtar, H. (2010). Green tea polyphenols for prostate cancer chemoprevention: A translational perspective. *Phytomedicine, 17*, 3–13.

167. Johnson, M. K., & Loo, G. (2000). Effects of epigallocatechin gallate and quercetin on oxidative damage to cellular DNA. *Mutation Research/DNA Repair, 459*(3), 211–218.

268   Engineering Interventions in Agricultural Processing

168. Jovanavic, S. V., Hara, Y., Steenken, S., & Simic, M. G. (1995). Antioxidant potential of gallocatechins: A pulsen radiolysis laser photolysis study. *Journal of the American Chemical Society, 117,* 9881–9888.

169. Jovanavic, S. V., Steenken, S., & Simic, M. G. (1996). Reduction potentials of flavonoid and model phenoxyl radicals. *Journal of the Chemical Society, Perkins Transactions, 2,* 2497–2503.

170. Kang, J. W., Kim, J. H., Song, K., Kim, S. H., Yoon, J. H., & Kim, K. S. (2010). Kaempferol and quercetin, components of Ginkgo biloba extract (EGb 761), induce caspase-3-dependent apoptosis in oral cavity cancer cells. *Phytotherapy Research, 1,* 77–82.

171. Kastan, M. B., Onyekwere, O., Sidransky, D., Vogelstein, B., & Craig, R. W. (1991). Participation of p53 protein in the cellular response to DNA damage. *Cancer Research, 51,* 6304–6311.

172. Katiyar, S. K. (2003). Skin photoprotection by green tea: Antioxidant and immunomodulations effects. *Current Drug Targets, 3,* 234–242.

173. Katiyar, S. K., & Mukhtar, H. (1996). Tea in chemoprevention of cancer: epidemiologic and experimental studies. *International Journal of Oncology, 8,* 221–238.

174. Katiyar, S. K., & Mukhtar, H. (2001). Immunotoxicity of environmental agents in the skin. In: Fuchs, J., & Packer, L. (eds.). *Environmental Stressors in Health and Disease.* New York: Marcel Dekker, pp. 345–364.

175. Katiyara, S., Elmets, C. A., & Katiyar, S. K. (2007). Green tea and skin cancer: photoimmunology, angiogenesis and DNA repair. *The Journal of Nutritional Biochemistry, 18,* 287–296.

176. Kato, Y., Ishiuchi, T., Ura, M., & Nakagawa, T. (1996). Clinical study on the bath and bed bath with Green Tea Polyphenols: Focused on the anti-tinea effect. *Journal of Japanese Nursing Association (General Nursing), 27,* 47–49.

177. Kaufman, B. D., Liberman, I. S., & Tyshetsky, V. I. (1965). Some data concerning the incidence of oesophageal cancer in the Gurjev region of the Kazakh SSR (Russian). *Voprosy Onkologii, 11,* 78.

178. Kavalali, G., Tuncel, H., Goksel, S., & Hatemi, H. H. (2007). Hypoglycemic activity of Urtica pilulifera in streptozotocin-diabetic rats. *Journal of Ethnopharmacology, 84,* 241–245.

179. Keli, S. O., Hertog, M. G., Feskens, E. J., & Kromhout, D. (1996). Dietary flavonoids, antioxidant vitamins, and incidence of stroke: the Zutphen study. *Archives of Internal Medicine, 156,* 637–642.

180. Key, T. J., Appleby, P. N., Reeves, G. K., & Roddam, A. (2003). Body mass index, serum sex hormones, and breast cancer risk in postmenopausal women. *Journal of the National Cancer Institute, 95,* 1218–1226.

181. Key, T., Appleby, P., Barnes, I., & Reeves, G. (2002). Endogenous sex hormones and breast cancer in postmenopausal women: reanalysis of nine prospective studies. *Journal of the National Cancer Institute, 94,* 606–616.

182. Kikuchi, N., Ohmori, K., Shimazu, T., Nakaya, N., Kuriyama, S., Nishino, Y., Tsubono, Y., & Tsuji, I. (2006). No association between green tea and prostate cancer risk in Japanese men: the Ohsaki Cohort Study. *British Journal of Cancer, 95,* 371–373.

183. Kim, H. P., Mani, I., Iversen, L., & Ziboh, V. A. (1998). Effects of naturally-occurring flavonoids and biflavonoids on epidermal cyclooxygenase and lipoxygenase from guinea-pigs. *Prostaglandins, Leukotrienes, and Essential Fatty Acids, 58*, 17–24.

184. Kim, J. M., Lee, E. K., Kim, D. H., Yu, B. P., & Chung, H. Y. (2010). Kaempferol modulates pro-inflammatory NF-kappa B activation by suppressing advanced glycation end products-induced NADPH oxidase. *Age (Dordr), 32*, 197–208.

185. Kim, M. J., Jang, W. S., Lee, I. K., Kim, J. K., Seong, K. S., Seo, C. R., Song, N. J., Bang, M. H., Lee, Y. M., Kim, H. R., Park, K. M., & Park, K. W. (2014). Reciprocal regulation of adipocyte and osteoblast differentiation of mesenchymal stem cells by *Eupatorium japonicum* prevents bone loss and adiposity increase in osteoporotic rats. *Journal of Medicinal Food, 17*(7), 772–781.

186. Klein, C., Sato, T., Meguid, M. M., & Miyata, G. (2000). From food to nutritional support to specific nutraceuticals: a journey across time in the treatment of disease. *Journal of Gastroenterology, 35*, 1–6.

187. Knekt, P., Jarvinen, R., Seppanen, R., Hellovaara, M., Teppo, L., Pukkala, E., & Aromaa, A. (1997). Dietary flavonoids and the risk of lung cancer and other malignant neoplasms. *American Journal of Epidemiology, 146*, 223–230.

188. Ko, C. H., Shen, S. C., Lee, T. J., & Chen, Y. C. (2005). Myricetin inhibits matrix metalloproteinase 2 protein expression and enzyme activity in colorectal carcinoma cells. *Molecular Cancer Therapeutics, 4*, 281–290.

189. Ko, H. J., Lo, C. Y., Wang, B. J., Chiou, R. Y. Y., & Lin, S. M. (2014). Theaflavin-3,3'-digallate, a black tea polyphenol, attenuates adipocyte-activated inflammatory response of macrophage associated with the switch of M1/M2-like phenotype. *Journal of Functional Foods, 11*, 36–48.

190. Koes, R. E., Quattrocchio, F., & Mol, J. N. M. (1994). The flavonoid biosynthetic pathway in plants: function and evolution. *Biossays, 16*, 123–132.

191. Koh, L. W., Wong, L. L., Loo, Y. Y., Kasapis, S., & Huang, D. (2010). Evaluation of different teas against starch digestibility by mammalian glycosidases. *Journal of Agriculture and Food Chemistry, 58*(1), 148–154.

192. Koizumi, Y., Tsubono, Y., Nakaya, N., Nishino, Y., Shibuya, D., & Matsuoka, H. (2003). No association between green tea and the risk of gastric cancer: pooled analysis of two prospective studies in Japan. *Cancer Epidemiology, Biomarkers & Prevention, 12*, 472–473.

193. Kono, S., Ikeda, M., Tokudome, S., & Kuratsune, M. (1988). A case-control study of gastric cancer and diet in Northern Kyushu, Japan. *Japanese Journal of Cancer Research, 79*, 1067–1074.

194. Kubo, I., Kinst-Hori, I., Chaudhuri, S. K., Kubo, Y., Sanchez, Y., & Ogura, T. (2000). Flavonols from *Heterotheca inuloides*: tyrosinase inhibitory activity and structural criteria. *Bioorganics & Medicinal Chemistry, 8*, 1749–1755.

195. Kuhnert, N. (2010). Unraveling the structure of the black tea thearubigins. *Archives of Biochemistry and Biophysics, 501*(1), 37–51.

196. Kuhnert, N., Clifford, M. N., & Mueller, A. (2010a). Oxidative cascade reactions yielding polyhydroxy-theaflavins and theacitrins in the formation of black tea thearubigins: Evidence by tandem LC–MS. *Food & Function, 1*(2), 180–199.

197. Kuhnert, N., Drynan, J. W., Obuchowicz, J., Clifford, M. N., & Witt, M. (2010b). Mass spectrometric characterization of black tea thearubigins leading to an oxida-

tive cascade hypothesis for thearubigin formation. *Rapid Communications in Mass Spectrometry, 24*(23), 3387–3404.

198. Kumar, A., Sehgal, N., Kumar, P., Padi, S. S., & Naidu, P. S. (2008). Protective effect of quercetin against ICV colchicine-induced cognitive dysfunctions and oxidative damage in rats. *Phytotherapy Research, 22,* 1563–1569.

199. Kunihiro, K., Sawa, M., & Kaneda, S. (2001). The comparison study of effective care between bed bath with green tea polyphenols and hot water shower for low- birth-weight baby with diaper rash. *Journal of Japanese Nursing Association (Paediatric Nursing), 32,* 187–189.

200. Kuriyama, S., Hozawa, A., Ohmori, K., Shimazu, T., Matsui, T., Ebihara, S., Awata, S., Nagatomi, R., Arai, H., & Tsuji, I. (2006). Green tea consumption and cognitive function: across sectional study from the tsurugaya project. *American Journal of Clinical Nutrition, 83,* 355–361.

201. Kuroda, Y., & Hara, Y. (1999). Antimutagenic and anticarcinogenic activity of tea polypheols. *Mutation Research, 436,* 69–97.

202. La Vecchia, C., Negri, E., Franceschi. S., D'Avanzo, B., & Boyle, P. (1992). Tea consumption and cancer risk. *Nutrition and Cancer, 17,* 27–31.

203. Laabich, A., Manmoto, C. C., Kuksa, V., Leung, D. W., Vissvesvaran, G. P., Karliga, I., Kamat, M., Scott. I. L., Fawzi, A., & Kubota, R. (2007). Protective effects of myricetin and related flavonols against A2E and light mediated-cell death in bovine retinal primary cell culture. *Exp. Eye Research, 85,* 154–165.

204. Lai, Y., Long, Y., Lei, Y., Deng, X., He, B., & Sheng, M. (2012). A novel micelle ofcoumarin derivative monoend-functionalized PEG for anti-tumor drug delivery: in vitro and in vivo study. *Journal of Drug Targeting, 20,* 246–254.

205. Lakenbrink, C., Lapczynski, S., Maiwald, B., & Engelhardt, U. H. (2000). Flavo-noids and other polyphenols in consumer brews of tea and other caffeinated bever-ages. *Journal of Agricultural and Food Chemistry, 48,* 2848–2852.

206. Langley-Evans, S. C. (2000). Consumption of black tea elicits an increase in plasma antioxidant potential in humans. *International Journal of Food Science and Nutrition, 51*(5), 309–315.

207. Lee, K. M., Lee, D. E., Seo, S. K., Hwang, M. K., Heo, Y. S., & Lee, K. W. (2010). Phosphatidylinositol 3-kinase, a novel target molecule for the inhibitory effects of kaempferol on neoplastic cell transformation. *Carcinogenesis, 31,* 1338–43.

208. Lee, W. S., Lee, E. G., Sung, M. S., & Yoo, W. H. (2014). Kaempferol inhibits IL-1beta-stimulated, RANKL-mediated osteoclastogenesis via down regulation of MAPKs, c-Fos, and NFATc1. *Inflammation, 37,* 1221–30.

209. Leenen, R., Roodenburg, A. J., Tijburg, L. B. M., & Wiseman, S. A. (2000). A single dose of tea with or without milk increase plasma antioxidant activity in humans. *European Journal of Clinical Nutrition, 54,* 87–92.

210. Lei, Y. F., Chen, J. L., Zhang, W. T., Fu, W., Wu, G. H., Wei, H., Wang, Q., & Ruan, J. L. (2012). *In vivo* investigation on the potential of galangin, kaempferol and myric-etin for protection of D-galactose-induced cognitive impairment. *Food Chemistry, 135,* 2702–2707.

211. Leino, M., Raitakari, O. T., Porkka, K. V. K., Taimela, S., Viikari, J. S. A. (1999). Asso-ciation of education with cardiovascular risk factors in young adults: the Cardiovascu-lar Risk in Young Finns Study. *International Journal of Epidemiology, 28,* 667–675.

212. Lepley, D. M., & Pelling, J. C. (1997). Induction of p21/WAF1 and G1 cell-cycle arrest by the chemopreventive agent apigenin. *Molecular Carcinogenesis, 19*, 74–82.

213. Leung, H. W., Lin, C. J., Hour, M. J., Yang, W. H., Wang, M. Y., & Lee, H. Z. (2007). Kaempferol induces apoptosis in human lung non-small carcinoma cells accompanied by an induction of antioxidant enzymes. *Food and Chemical Toxicology, 45*, 2005–2013.

214. Leung, L. K., Su, Y., Chen, R., Zhang, Z., Huang, Y., & Chen, Z. Y. (2001). Theaflavins in black tea and catechins in green tea are equally effective antioxidants. *Journal of Nutrition, 131*, 2248–2251.

215. Leung, L., Su, Y., & Chen, R. (2001). Theaflavins in black tea and cate-chins in green tea are equally effective antioxidants. *Journal of Nutrition, 131*, 2248–2251.

216. Levites, Y., Amit, T., Mandel, S., & Youdim, M. B. H. (2003). Neuroprotection and neurorescue against amyloid beta toxicity and PKC-dependent release of non-amyloidogenic soluble precursor protein by green tea polyphenol (-)-epigallocate-chin-3-gallate. *The FASEB Journal, 17*, 952–954.

217. Levites, Y., Amit, T., Youdim, M. B., & Mandel, S. (2002). Involvement of protein kinase C activation and cell survival/cell cycle genes in green tea polyphenol (-)-epigallocatechin-3-gallate neuroprotective action. *The Journal of Biological Chemistry, 277*, 30574–30580.

218. Levites, Y., Weinreb, O., Maor, G., Youdim, M. B. H., & Mandel, S. (2001). Green tea polyphenol (-)-epigallocatechin-3-gallate prevents N-methyl-4-phenyl-1,2,3,6-tetrahydropyridine-induced dopaminergic neurodegeneration. *Journal of Neurochemistry, 78*, 1073–1082.

219. Li, N., Sun, Z., Liu, Z., & Han, C. (1998). Study on the preventive effect of tea on DNA damage of the buccal mucosa cells in oral leukoplakias induced by cigarette smoking. *Wei Sheng Yen Chiu, 27*, 173–174.

220. Liang, Y. C., Chen, Y. C., Lin, Y. L., Lin-Shiau, S. Y., Ho, C. T., & Lin, J. K. (1999). Suppression of extracellular signals and cell proliferation by the black tea polyphenol, theaflavin-3,3′-digallate. *Carcinogenesis, 20*(4), 733–736.

221. Liang, Y. R., Lu, J. L., Zhang, L. Y., Wu, S., & Wu, Y. (2003). Estimation of black tea quality by analysis of chemical composition and color difference of tea infusions. *Food Chemistry, 80*, 283–290.

222. Liang, Y., Lai, Y., Li, D., He, B., & Gu, Z. (2013). Novel polymeric micelles with cinnamic acidas lipophilic moiety for 9-nitro-20 (S)-camptothecin delivery. *Materials Letters, 97*, 4–7.

223. Lin, C. L., Huang, H. C., & Lin, J. K. (2007). Theaflavins attenuate hepatic lipid accumulation through activating AMPK in human HepG2 cells. *Journal of Lipid Research, 48*(11), 2334–2343.

224. Lin, J., Lin, C., Liang, Y., Lin-Shiau, S., & Juans, I. (1998). Survey of catechins, gallic acid, and methylxanthines in green, oolong, pu-erh, and black teas. *Journal of Agricultural and Food Chemistry, 46*, 3635–3642.

225. Lin, Y. L., & Lin, J. K. (1997). (–)-Epigallocatechin-3-gallate blocks the induction of nitric oxide synthase. By down-regulating lipopolysaccharide-induced activity of transcription factor nuclear factor-kappa B. *Molecular Pharmacology, 52*, 465–472.

226. Lin, Y., Tsai, Y., Tsay, J., & Lin, J. (2003). Factors affecting the levels of tea polyphenols and caffeine in tea leaves. *Journal of Agricultural and Food Chemistry, 51*, 1864–1873.

227. Lin, Z. H., Qi, Y. P., Chen, R. B., Zhang, F. Z., & Chen, L. S. (2012). Effects of phosphorus supply on the quality of green tea. *Food Chemistry, 130*, 908–914.
228. Liu, J., Wang, M., Peng, S., & Zhang, G. (2011). Effect of green tea catechins on the postprandial glycemic response to starches differing in amylose content. *Journal of Agriculture and Food Chemistry, 59*(9), 4582–4588.
229. Liu, J., Yu, H., & Ning, X. (2006). Effect of quercetin on chronic enhancement of spatial learning and memory of mice. *Science China Life Sciences, 49*, 583–590.
230. Lodi, F., Jimenez, R., Moreno, L., Kroon, P. A., Needs, P. W., & Hughes, D. A. (2009). Glucuronidated and sulfated metabolites of the flavonoid quercetin preventendothelial dysfunction but lack direct vasorelaxant effects in rat aorta. *Atherosclerosis, 204*, 34–9.
231. Loeb, L. A. (2001). A mutator phenotype in cancer. *Cancer Research, 61*, 3230–3239.
232. Lopez-Lazaro, M. (2009). Distribution and biological activities of the flavonoid luteolin. *Mini Reviews in Medicinal Chemistry, 9*, 31–59.
233. Lotito, S. B., & Fraga, C. G. (2000). Catechins delay lipid oxidation and alpha-tocopherol and beta-carotene depletion following ascorbate depletion in human plasma. *Proceedings of the Society for Experimental Biology and Medicine, 225*, 32–38.
234. Lotito, S. B., & Frei, B. (2006). Dietary flavonoids attenuate tumor necrosis factor alpha-induced adhesion molecule expression in human aortic endothelial cells. Structure function relationships and activity after first pass metabolism. *The Journal of Biological Chemistry, 281*, 37102–37110.
235. Lu, G., Liao, J., Yang, G., Reuhl, K. R., Hao, X., & Yang, C. S. (2006). Inhibition of adenoma progression to adenocarcinoma in a 4-(methylnitrosamino)-1-(3-pyridyl)-1-butanone-induced lung tumorigenesis model in A/J mice by tea polyphenols and caffeine. *Cancer Research, 66*(23), 11494–11501.
236. Lu, Z. W., Liu, Y. J., Zhao, L., Jiang, X. L., Li, M. Z., Wang, Y. S., Xu, Y. J., Gao, L. P., & Xia, T. (2014). Effect of low-intensity white light mediated de-etiolation on the biosynthesis of polyphenols in tea seedlings. *Plant Physiology and Biochemistry, 80*, 328–336
237. Luo, Y. (2006). Alzheimer's disease, the nematode Caenorhabditis elegans, and ginkgo biloba leaf extract. *Life Sciences, 78*, 2066–2072.
238. Mandel, S., & Youdim, M. B. H. (2004). Catechin polyphenols: Neurodegenartion and neuroprotection in neurodegenerative diseases. *Free Radical Biology & Medicine, 37*(3), 304–317.
239. Manju, V., Balasubramaniyan, V., & Nalini, N. (2005). Rat colonic lipid peroxidation and antioxidant status: the effects of dietary luteolin on 1,2-dimethylhydrazine challenge. *Cellelur and Molecular Biology Letters, 10*, 535–555.
240. McKay, D. L., & Blumberg, J. B. (2002). The role of tea in human health: an update. *The Journal of the American College of Nutrition, 21*(1), 1–13.
241. McVean, M., Xiao, H., Isobe, K., & Pelling, J. C. (2000). Increase in wild-type p53 stability and transactivational activity by the chemopreventive agent apigenin in keratinocytes. *Carcinogenesis, 21*, 633–639.
242. Michalovitz, D., Halevy, O., & Oren, M. (1990). Conditional inhibition of transformation and of cell proliferation by a temperaturesensitive mutant of p53. *Cell, 62*, 671–680.

243. Mikutis, G., Karakoese, H., Jaiswal, R., LeGresley, A., Islam, T., & Fernandez-Lahore, M. (2013). Phenolic promiscuity in the cell nucleus – epigallocatechingallate (EGCG) and theaflavin-3,30-digallate from green and black tea bind to model cell nuclear structures including histone proteins, double stranded DNA and telomeric quadruplex DNA. *Food & Function, 4*(2), 328–337.
244. Miller, K. L., Liebowitz, R. S., & Newby, L. K. (2004). Complementary and alternative medicine in cardiovascular disease: a review of biologically based approaches. *American Heart Journal, 147*, 401–411.
245. Mittal, A., Pate, M. S., Wylie, R. C., Tollesfsbol, T. O., & Katiyar, S. K. (2004). EGCG down regulates telomerase in human breast carcinoma MCF-7 cells, leading to suppression of cell viability and induction of apoptosis. *International Journal of Oncology, 24*, 703–710.
246. Muramatsu, K., Fukuyo, M., & Hara, Y. (1986). Effect of green tea catechins on plasma cholestrol level in cholesterol-fed rats. *Journal of Nutritional Science and Vitaminology, 32*, 613–622.
247. Murase, T., Nagasawa, A., & Suzuki, J. (2002). Beneficial effects of tea catechins ondiet-induced obesity: stimulation of lipid catabolism in the liver. *International Journal of Obesity, 26*, 1459–1464.
248. Murphy, K. J., Chronopoulos, A. K., Singh, I., Francis, M. A., Moriarty, H., Pike, M. J., Turner, A. H., Mann, N. J., & Sinclair, A. J. (2003). Dietary flavanols and procyanidinoligomers from cocoa (Theobroma cacao) inhibit platelet function. *The American Journal of Clinical Nutrition, 77*, 1466–1473.
249. Nagata, C., Kabuto, M., & Shimizu, H. (1998). Association of coffee, green tea, and caffeine intakes with serum concentrations of estradiol and sex hormone-binding globulin in premenopausal Japanese women. *Nutrition and Cancer, 30*, 21–24.
250. Namiki, M., & Osawa, T. (1986). Antioxidants/antimutagens in foods. *Basic Life Sciences, 39*, 131–142.
251. Neukam, K., Stahl, W., Tronnier, H., Sies, H., & Heinrich, U. (2007). Consumption of flavanolrichcocoa acutely increases microcirculation in human skin. *European Journal of Nutrition, 46*, 53–56.
252. Nguyen, T. T., Tran, E., Ong, C. K., Lee, S. K., Do, P. T., Huynh, T. T., Nguyen, T. H., Lee, J. J., Tan, Y., Ong, C. S., & Huynh, H. (2003). Kaempferol induced growth inhibition and apoptosis in A549 lung cancer cells is mediated by activation of MEK-MAPK. *Journal of Cellular Physiology, 197*, 110–121.
253. Nishiumi, S., Bessyo, H., Kubo, M., Aoki, Y., Tanaka, A., Yoshida, K., & Ashida, H. (2010). Green and black tea suppress hyperglycemia and insulin resistance by retaining the expression of glucose transporter 4 in muscle of high-fat diet-fed C57BL/6J mice. *Journal of Agricultural and Food Chemistry, 58*(24), 12916–12923.
254. Notani, P. N., & Jayant, K. (1987). Role of diet in upper aerodigestive tract cancers. *Nutrition and Cancer, 10*, 103–113.
255. Ohno, Y., Wakai, K., Genka, K., Ohmine, K., Kawamura, T., Tamakoshi, A., Aoki, R., Senda, M., & Aoki, K. (1995). Tea consumption and lung cancer risk: a case-control study in Okinawa, Japan. *Japanese Journal of Cancer Research, 86*, 1027–1034.
256. Ong, K. C., & Khoo, H. E. (1997). Biological effects of myricetin. *General Pharmacology: The Vascular System, 29*, 121–126.

257. Ono, K., Yoshiike, Y., Takashima, A., Hasegawa, K., Naiki, H., & Yamada, M. (2003). Potent anti-amyloidogenic and fibril-destabilizing effects of polyphenols in vitro: implications for the prevention and therapeutics of Alzheimer's disease. *Journal of Neurochemistry, 87*, 172–181.
258. Otten, D. (2010). Tea—Second most consumed beverage in the world. Available at http://ezinearticles.com/?Tea—The-Second-Most-Consumed-beverage-in-the-World&id=3867563 on 04–03–2010.
259. Owuor, P. O., Obanda, M., Nyirenda, H. E., Mphangwe, N. I. K., Wright, L. P., & Apostolides, Z. (2006). The relationship between some chemical parameters and sensory evaluations for plain black tea (*Camellia sinensis*) produced in Kenya and comparison with similar teas from Malawi and South Africa. *Food Chemistry, 97*, 644–653.
260. Pan, T., Fei, J., Zhou, X., Jankovic, J., & Le, W. (2003). Effects of green tea polyphenols on dopamine uptake and on MPP+-induced dopamine neuron injury. *Life Sciences, 72*, 1073–1083.
261. Parab, S., Kulkarni, R., & Thatte, U. (2003). Heavy metals in herbal medicines. *Indian Journal of Gastroenterology, 22*, 111–112.
262. Park, H. S., Kim, S. H., Kim, Y. S., Ryu, S. Y., Hwang, J. T., Yang, H. J., Kim, G. H., Kwon, D. Y., & Kim, M. S. (2009). Luteolin inhibits adipogenic differentiation by regulating PPAR gamma activation. *Biofactors, 35*(4), 373–9.
263. Park, K. W., Halperin, D. S., & Tontonoz, P. (2008). Before they were fat: adipocyte progenitors. *Cell Metabolism, 8*(6), 454–457.
264. Parkin, D. M., Bray, F., Ferlay, J., & Pisani, P. (2002). Global cancer statistics. *CA: A Cancer Journal for Clinicians, 55*, 74–108.
265. Patel, D., Shukla, S., & Gupta, S. (2007). Apigenin and cancer chemoprevention: Progress, potential and promise (review). *International Journal of Oncology, 30*(1), 233–245.
266. Pollak, M. N., Schernhammer, E. S., & Hankinson, S. E. (2004). Insulin-like growth factors and neoplasia. *Nature Reviews Cancer, 4*, 505–551.
267. Price, D. L., Sisodia, S. S., & Borchelt, D. R. (1998). Alzheimer disease—when and why? *Nature Genetics, 19*(4), 314–316.
268. Pu, F., Mishima, K., Irie, K., Motohashi, K., Tanaka, Y., Orito, K., Egawa, T., Kitamura, Y., Egashira, N., Iwasaki, K., & Fujiwara, M. (2007). Neuroprotective effects of quercetin and rutin on spatial memory impairment in an 8-arm radial maze task and neuronal death induced by repeated cerebral ischemia in rats. *Journal of Pharmacological Sciences, 104*, 329–334.
269. Pushparaj, P. N., Low, H. K., Manikandan, J., Tan, B. K. H., & Tan, C. H. (2007). Antidiabetic effects of Cichorium intybus in streptozotocin induced diabetic rats. *Journal of Ethnopharmacology, 111*, 430–434.
270. Rajavelu, A., Tulyasheva, Z., Jaiswal, R., Jeltsch, A., & Kuhnert, N. (2011). The inhibition of the mammalian DNA methyltransferase 3a (Dnmt3a) by dietary black tea and coffee polyphenols. *BMC Biochemistry, 12*.
271. Rashidinejad, A., Birch, E. J., Sun-Waterhouse, D., & Everett, D. W. (2014). Delivery of green tea catechin and epigallocatechin gallate in liposomes incorporated into low-fat hard cheese. *Food Chemistry, 156*(1), 176–183.

272. Record, I. R., & Dreosti, I. E. (1998). Protection by black tea and green tea against UVB and UVA + B induced skin cancer in hairless mice. *Mutation Research, 422*, 191.
273. Ren, J. S., Freedman, N. D., Kamangar, F., Dawsey, S. M., Hollenbeck, A. R., Schatzkin, A., & Abent, C. C. (2010). Tea, coffee, carbonated soft drinks and upper gastrointestinal tract cancer risk in a large United States prospective cohort study. *European Journal of Cancer, 46*, 1873–1881.
274. Rice-Evans, C. A., Miller, N. J., & Paganga, G. (1997). Antioxidant properties of phenolic compounds. *Trends in Plant Science, 2*, 152–159.
275. Rice-Evans, C. A., Miller, N. J., & Panagga, G. (1996). Structure-antioxidant activity relationships of flavonoid and phenolic acids. *Free Radical Research, 20*(7), 933–956.
276. Rice-Evans, C. A., Miller, N. J., Bolwell, P. G., Bramley, P. M., & Pridham, J. B. (1995). The relative antioxidant activities of plant-derived polyphenolic flavonoids. *Free Radical Research, 22*, 375–383.
277. Richelle, M., Tavazzi, I., & Offord, E. (2001). Comparison of the Antioxidant Activity of Commonly Consumed Polyphenolic Beverages (Coffee, Cocoa, and Tea) Prepared per Cup Serving. *Journal of Agricultural and Food Chemistry, 49*, 3438–3442.
278. Riederer, P., Sofic, E., Rausch, W. D., Schmidt, B., Reynolds, G. P., Jellinger, K., & Youdim, M. B. H. (1989). Transition metals, ferritin, glutathione, and ascorbic acid in Parkinsonian brains. *Journal of Neurochemistry, 52*, 515–520.
279. Rietveld, A., & Wiseman, S. (2003). Antioxidant Effects of Tea: Evidence from Human Clinical Trials. *Journal of Nutrition, 133*, 3285S–3292S.
280. Rivera, L., Moron, R., Sanchez, M., Zarzuelo, A., & Galisteo, M. (2008). Quercetin amelioratesmetabolic syndrome and improves the inflammatory status in obese Zucker rats. *Obesity, 16*, 2081–2087.
281. Roberts, E. A. H. (1949). The fermentation process in tea manufacture. Condensation of catechin and its relation to the chemical change in fermentation. *Biochemical Journal, 45*(5), 538–542.
282. Roberts, E. A. H. (1957). Oxidative condensation of flavanols in tea fermentation. *Chemistry & Industry, 41*, 1355–1356.
283. Roberts, E. A. H., Cartwright, R. A., & Oldschool, M. (1957). The phenolic substances of manufactured tea. I. Fractionation and paper chromatography of water-soluble substances. *Journal of the Science of Food and Agriculture, 8*, 72–80.
284. Robertson, A. (1992). The chemistry and biochemistry of black tea production, the non-volatiles. In: K. C., Wilson & M. N., Clifford (Eds.), *Tea: Cultivation to Consumption* (pp. 555–601). London, UK: Chapman and Hall.
285. Rosen, C. J., & Bouxsein, M. L. (2006). Mechanisms of disease: is osteoporosis the obesity of bone? *Natural Clinical Practice Rheumatology, 2*(1), 35–43.
286. Rossi, L., Mazzitelli, S., Arciello, M., Capo, C. R., & Rotilio, G. (2008). Benefits from dietary polyphenols for brain aging and Alzheimer's disease. *Neurochemical Research, 33*, 2390–2400.
287. Roth, A., Schaffner, W., & Hertel, C. (1999). Phytestrogen kaempferol (3,4',5,7- tetrahydroxyflavone) protects PC12 and T47D cells from β-amyloid-induced toxicity. *Journal of Neuroscience Research, 57*, 399–404.
288. Ruan, J. Y., Ma, L. F., & Shi, Y. Z. (2013). Potassium management in tea plantations: its uptake by field plants, status in soils, and efficacy on yields and quality of teas in China. *Journal of Plant Nutrition and Soil Science, 176*, 450–459.

289. Ruan, J. Y., Gerendas, J., Hardter, R., & Sattelmacher, B. (2007). Effect of root zone pH and form and concentration of nitrogen on accumulation of quality-related components in green teat. *Journal of the Science of Food and Agriculture, 87*, 1505–1516.

290. Sadava, D., Whitlock, E., & Kane, S. E. (2007). The green tea polyphenol, epigallocatechin-3-gallate inhibits telomerase and induces apoptosis in drug resistant lung cancer cells. *Biochemistry and Biophysics Research Communications, 360*(1), 233–237.

291. Sahu, S. C., & Gray, G. C. (1993). Interactions of flavonoids, trace metals, and oxygen: nuclear DNA damage and lipid peroxidation induced by myricetin. *Cancer Letters, 70*, 73–79.

292. Salahuddin, M., Jalalpure, S. S., & Gadge, N. B. (2010). Antidiabetic activity of aqueous bark extract of Cassia glauca in streptozotocin induced diabetic rats. *Canadian Journal of Physiology and Pharmacology, 88*, 153–160.

293. Samir, N., Bachani, M., Harshavardhanna, D., & Steiner, J. P. (2012). Catechins protect neurons against mitochondrial toxins and HIV proteins via activation of the BDNF pathway. *Journal of Neurovirology, 18*(6), 445–455.

294. Sasazuki, S., Kodama, H., Yoshimasu, K., Liu, Y., Washio, M., Tanaka, K., Tokunaga, S., Kono, S., Arai, H., Doy, Y., Kawano, T., Nakagaki, O., Takada, K., Koyanagi, S., Hiyamuta, K., Nii, T., Shirai, K., Ideishi, M., Arakawa, K., Mohri, M., & Takeshita, A. (2000). Relation between green tea consumption and the severity of coronary atherosclerosis among Japanese men and women. *Annals of Epidemiology, 10*, 401–408.

295. Satav, J. G., & Katyare, S. S. (2004). Effect of streptozotocin-induced diabetes on oxidative energy metabolism in rat liver mitochondria – a comparative study of early and late effects. *Indian Journal of Clinical Biochemistry, 19*, 23–31.

296. Sato, F., Matsukawa, Y., Matsumoto, K., Nishino, H., & Sakai, T. (1994). Apigenin induces morphological differentiation and G2-M arrest in rat neuronal cells. *Biochemical and Biophysical Research Communications, 204*, 578–584.

297. Scholz, E., & Bertram, B. (1995). *Camellia sinensis* (L.): O. Kuntze Der Teestrauch. *Zeitschrift fur Phytotherapie, 17*, 235–250.

298. Schroeter, H., Boyd, C., Spencer, J. P., Williams, R. J., Cadenas, E., & Rice-Evans, C. (2002). MAPK signaling in neurodegeneration: influences of flavonoids and of nitric oxide. *Neurobiology of Aging, 23*, 861–880.

299. Seeram, N. P., Henning, S. M., Niu, Y., Lee, R., Scheuller, H. S., & Heber, D. (2006). Catechin and caffeine content of green tea dietary supplements and correlation with antioxidant capacity. *Journal of Agricultural and Food Chemistry, 54*, 1599–1603.

300. Sellamuthu, P. S., Muniappan, B. P., Perumal, S. M., & Kandasamy, M. (2009). Antihyperglycemic effect of mangiferin in streptozotocin induced diabetic rats. *Journal of Health Sciences, 55*, 206–214.

301. Serafini, M., Ghiselli, A., & Ferro-Luzzi, A. (1996). In vivo antioxidant effect of green and black tea in man. *European Journal of Clinical Nutrition, 50*, 28–32.

302. Severson, R. K., Nomura, A. M., Grove, J. S., & Stemmermann, G. N. (1989). A prospective study of demographics, diet, and prostate cancer among men of Japanese ancestry in Hawaii. *Cancer Research, 49*, 1857–1860.

303. Shahidi, F., & Wanasundara, P. K. J. (1992). Phenolic antioxidants. *Critical Reviews in Food Science and Nutrition, 32*, 67–103.

304. Shimmyo, Y., Kihara, T., Akaike, A., Niidome, T., & Sugimoto, H. (2008). Flavonols and flavones as BACE-1 inhibitors: Structure–activity relationship in cell-free, cell based and in silico studies reveal novel pharmacophore features. *Biochimica et Biophysica Acta (BBA): General Subjects, 1780*(5), 819–825.
305. Shimoi, K., Saka, N., Kaji, K., Nozawa, R., & Kinae, N. (2000). Metabolic fate of luteolin and its functional activity at focal site. *BioFactors, 12*, 181–186.
306. Shin, J. S., Kim, K. S., Kim, M. B., Jeong, J. H., & Kim, B. K. (1999). Synthesis and hypoglycemic effect of chrysin derivatives. *Bioorg. Medicinal Chemistry Letters, 9*, 869–874.
307. Sies, H., Schewe, T., Heiss, C., & Kelm, M. (2005). Cocoa polyphenols and inflammatorymediators. *American Journal of Clinical Nutrition, 81*, 304S–312S.
308. Singh, A., Naidu, P. S., & Kulkarni, S. K. (2003). Reversal of aging and chronic ethanolinduced cognitive dysfunction by quercetin a bioflavonoid. *Free Radical Research, 37*, 1245–1252.
309. Siriwardhana, N., Kalupahana, N. S., Cekanova, M., LeMieux, M., Greer, B., & Moustaid-Moussa, N. (2013). Modulation of adipose tissue inflammation by bioactive food compounds. *The Journal of Nutritional Biochemistry, 24*, 613–23.
310. Sonoda, T., Nagata, Y., Mori, M., Miyanaga, N., Takashima, N., Okumura, K., Goto, K., Naito, S., Fujimoto, K., Hirao, Y., Takahashi, A., Tsukamoto, T., Fujioka, T., & Akaza, H. (2004). A case-control study of diet and prostate cancer in Japan: possible protective effect of traditional Japanese diet. *Cancer Science, 95*, 238–242.
311. Sridevi, M., Chandramohan, G., & Pugalendi, K. V. (2007). Protective effect of Solanum surattense leaf-extract on blood glucose, oxidative stress and hepatic marker enzymes in STZ-diabetic rats. *Asian Journal of Biochemistry, 1*, 117–121.
312. Stadler, K., Jenei, V., Bolcshazy, G. V., Somogyi, A., & Jakus, J. (2003). Increased nitric oxide levels as an early sign of premature aging in diabetes. *Free Radical Biology and Medicine, 35*, 1240–1251.
313. Stangl, V., Dreger, H., Stangl, K., & Lorenz, M. (2007). Molecular targets of teapolyphenols in the cardiovascular system. *Cardiovascular Research, 73*(2), 348–58.
314. Steffen, Y., Gruber, C., Schewe, T., & Sies, H. (2008). Mono-O-methylated flavanols and other flavonoids as inhibitors of endothelial NADPH oxidase. *Archives of Biochemistry and Biophysics, 469*, 209–219.
315. Stewart, N., Hicks, G. G., Paraskevas, F., & Mowat, M. (1995). Evidence for a second cell cycle block at G2/M by p53. *Oncogene, 10*, 109–115.
316. Stodt, U. W., Blauth, N., Niemann, S., Stark, J., Pawar, V., Jayaraman, S., Koek, J., & Engelhardt, U. H. (2014). Investigation of processes in black tea manufacture through model fermentation (oxidation) experiments. *Journal of Agricultural and Food Chemistry, 62*, 7854–7861.
317. Subash-Babu, P., Ignacimuthu, S., & Prince, P. S. M. (2008). Restoration of altered carbohydrate and lipid metabolism by hyponidd, a herbal formulation in streptozotocin-induced diabetic rats. *Asian Journal of Biochemistry, 2*, 90–98.
318. Suganuma, M., Okabe, S., Sueoka, N., Sueoka, E., Matsuyama, S., Imai, K., Nakachi, K., & Fujiki, H. (1999). Green tea and cancer chemoprevention. *Mutation Research, 428*, 339–344.
319. Sugimura, T. (2000). Nutrition and dietary carcinogens. *Carcinogenesis, 21*, 387–395.

320. Sumpio, B. E., Cordova, A. C., Berke-Schlessel, D. W., Qin, F., & Chen, Q. H. (2006). Green Tea, the "Asian Paradox," and Cardiovascular Disease. *Journal of the American College of Surgeons, 202*(5), 814–825.

321. Suri, S., Taylor, M. A., Verity, A., Tribolo, S., Needs, P. W., & Kroon, P. A. (2008). A comparative study of the effects of quercetin and its glucuronide and sulfate metabolites on human neutrophil function in vitro. *Biochemical Pharmacology, 76*, 645–653.

322. Takahashi, H., Nomata, K., Mori, K., Matsuo, M., Miyaguchi, T., Noguchi, M., & Kanetake, H. (2004). The preventive effect of green tea on the gap junction intercellular communication in renal epithelial cells treated with a renal carcinogen. *Anticancer Research, 24*, 3757–3762.

323. Tanaka, T., Mine, C., Inoue, K., Matsuda, M., & Kouno, I. (2002). Synthesis of theaflavin from epicatechin and epigallocatechin by plant homogenates and role of epicatechin quinone in the synthesis and degradation of theaflavin. *Journal of Agricultural and Food Chemistry, 50*(7), 2142–2148.

324. Tang, N., Wu, Y., Zhou, B., Wang, B., & Yu, R. (2009). Green tea, black tea consumption and risk of lung cancer: A meta-analysis. *Lung Cancer, 65*, 274–283.

325. Tarahovsky, Y. S., Kim, Y. A., Yagolnik, E. A., & Muzafarov, E. N. (2014). Flavonoid–membrane interactions: involvement of flavonoid–metal complexes in raft signaling. *Biochim Biophys Acta, 1838*, 1235–1246.

326. Taubenberger, J. K., Reid, A. H., Janczewski, T. A., & Fanning, T. G. (2001). Integrating historical, clinical and molecular genetic data in order to explain the origin and virulence of the 1918 Spanish influenza virus. *Philosophical Transactions of the Royal Society of London B: Biological Sciences, 356*, 1829–1839.

327. Taubert, D., Berkels, R., Roesen, R., & Klaus, W. (2003). Chocolate and blood pressure inelderly individuals with isolated systolic hypertension. *JAMA, 290*, 1029–1030.

328. Taubert, D., Roesen, R., Lehmann, C., Jung, N., & Schomig, E. (2007). Effects of low habitual cocoa intake on blood pressure and bioactive nitric oxide: a randomized controlled trial. *JAMA, 298*, 49–60.

329. Tewes, F. S., Koo, L. C., Meisgen, T. S., & Rylander, R. (1990). Lung Cancer Risk and Mutagenicity of Tea. *Environmental Research, 52*, 23–33.

330. Thomas, M. J. (1995). The role of free radicals and antioxidants: how do we know that they are working. *Critical Reviews in Food Science and Nutrition, 35*, 21–39.

331. Tijburg, L. B. M., Matter, T., Folts, J. D., Weisgerber, U. M., & Katan, M. B. (1997). Tea flavonoids and cardiovascular diseases: A review. *Critical Reviews in Food Science and Nutrition, 37*, 771–785.

332. Tiwari, R. P., Bharti, S. K., Kaur, H. D., Dikshit, R. P., & Hoondal, G. S. (2005). Synergistic antimicrobial activity of tea and antibiotics. *Indian Journal of Medical Research, 122*, 80–84.

333. Tokimitsu, I. (2004). Effects of tea catechins on lipid metabolism and body fat accumulation. *Bio Factors, 22*, 141–143.

334. Tota, S., Awasthi, H., Kamat, P. K., Nath, C., & Hanif, K. (2010). Protective effect of quercetin against intracerebral streptozotocin induced reduction in cerebral blood flow and impairment of memory in mice. *Behavioral Brain Research, 209*, 73–79.

335. Tribolo, S., Lodi, F., Connor, C., Suri, S., Wilson, V. G., & Taylor, M. A. (2008). Comparative effects of quercetin and its predominant human metabolites on adhesion molecule expression in activated human vascular endothelial cells. *Atherosclerosis, 197*, 50–56.

336. Tu, Y. C., Lian, T. W., Yen, J. H., Chen, Z. T., & Wu, M. J. (2007). Antiatherogenic effects of kaempferol and rhamnocitrin. *Journal of Agricultural and Food Chemistry, 55*, 9969–9976.

337. Tuorkey, M. J. (2015). Molecular targets of luteolin in cancer. *European Journal of Cancer Prevention,* doi: 10.1097/CEJ.0000000000000128

338. USFDA. (2006). U.S. Food and Drug Administration. Accessible on: http://www.accessdata.fda.gov/drugsatfda_docs/nda/2006/021902s000TOC.cfm.

339. Uchiyama, S., Taniguchi, Y., Saka, A., Yoshida, A., & Yajima, H. (2011). Prevention of diet-induced obesity by dietary black tea polyphenols extract in vitro and in vivo. *Nutrition (Burbank, LosAngeles County, Calif.), 27*(3), 287–292.

340. Ukil, A., Maity, S., & Das, P. K. (2006). Protection from experimental colitis by theaflavin-3,3'-digallate correlates with inhibition of IKK and NF-kappa-B activation. *British Journal of Pharmacology, 149*(1), 121–131.

341. Valko, M., Leibfritz, D., Moncol, J., Cronin, M. T., Mazur, M., & Telser, J. (2007). Free radicals and antioxidants in normal physiological functions and human disease. *The International Journal of Biochemistry & Cell Biology, 39*, 44–84.

342. Villar, I. C., Galisteo, M., Vera, R., O'Valle, F., Garcia-Saura, M. F., & Zarzuelo, A. (2004). Effects ofthe dietary flavonoid chrysin in isolated rat mesenteric vascular bed. *Journal of Vascular Research, 41*, 509–516.

343. Villar, I. C., Vera, R., Galisteo, M., O'Valle, F., Romero, M., & Zarzuelo, A. (2005). Endothelialnitric oxide production stimulated by the bioflavonoid chrysin in rat isolatedaorta. *Planta Medica, 71*, 829–834.

344. Vinson, J. A., & Zhang, J. (2005). Black and green tea equally inhibits diabetic cataracts in a streptozotocin-induced rat model of diabetes. *Journal of Agricultural and Food Chemistry, 53*, 3710–3713.

345. Vlassara, H., Bucala, R., & Striker, L. (1994). Pathogenic effects of advanced glycozylation: Biochemical, biologic and clinical implications for diabetes and aging. *Laboratory Investigation, 70*, 138–151.

346. Wanasundara, U. N., & Shahidi, F. (1998). Antioxidant and pro-oxidant activity of green tea extracts in marine oils. *Food Chemistry, 63*, 335–342.

347. Wang, C., & Kurzer, M. S. (1997). Phytestrogen concentration determines effects on DNA synthesis in human breast cancer cells. *Nutrition and Cancer, 28*, 236–247.

348. Wang, C. F., & Kurzer, M. S. (1998). Bioactive natural products. *Nutrition and Cancer, 31*, 90–100.

349. Wang, W., Heideman, L., Chung, C. S., Pelling, J. C., Koehler, K. J., & Birt, D. F. (2000). Cell-cycle arrest at G2/M and growth inhibition by apigenin in human colon carcinoma cell lines. *Molecular Carcinogenesis, 28*(2), 102–10.

350. Wang, Z. Y., Agarwa, R., Bickers, D. R., & Mukhtar, H. (1991). Protection against ultraviolet I3 radiation-induced photocarcinogenesis in hairless mice by green tea poly-phenols. *Carcinogenesis, 12*, 1527–1530.

351. Weisburger, E. J. H. (1994). Effects of green and black tea on hepatic xenobiotic metabolizing systems in the male rat. *Xenobiotica, 24*, 119–127.

352. Winterbone, M. S., Tribolo, S., Needs, P. W., Kroon, P. A., & Hughes, D. A. (2009). Physiologically relevant metabolites of quercetin have no effect on adhesion molecule orchemokine expression in human vascular smooth muscle cells. *Atherosclerosis, 202*, 431–438.

353. WHO (World Health Organization) (2003). Global Cancer Rates Could Increase by 50% to 15 million by 2020. Available at URL: http://www.who.int/mediacenter/news/releases/2003/pr27/en/print.html/Accessed on 13–11–2008.

354. Wu, A. H., Arakawa, K., Stanczyk, F. Z., Berg, D. V. D., Koh, W. P., & Yu, M. C. (2005). Tea and circulating estrogen levels in postmenopausal Chinese women in Singapore. *Carcinogenesis, 26*, 976–980.

355. Wu, A. H., Yu, M. C., Tseng, C. C., Hankin, J., & Pike, M. C. (2003a). Green tea and risk of breast cancer in Asian Americans. *International Journal of Cancer, 106*, 574–579.

356. Wu, C. D., & Wei, G. X. (2002). Tea as a functional food for oral health. *Nutrition, 18*, 443–444.

357. Wu, C. H., Lu, F. H., Chang, C. S., Chang, T. C., Wang, R. H., & Chang, C. J. (2003b). Relationship among habitual tea consumption, percent body fat, and body fat distribution. *Obesity Research, 11*, 1088–1095.

358. Wu, D. G., Yu, P., Li, J. W., Jiang, P., Sun, J., Wang, H. Z., & Bie, P. (2014). Apigenin potentiates the growth inhibitory effects by IKK-beta-mediated NF-kappaB activation in pancreatic cancer cells. *Toxicology Letters, 224*(1), 157–164.

359. Wu, M. J., Weng, C. Y., Ding, H. Y., & Wu, P. J. (2005). Anti-inflammatory and antiviral effects of Glossogyne tenuifolia. *Life Sciences, 76*, 1135–1146.

360. Xiao, H. B., Lu, X. Y., Sun, Z. L., & Zhang, H. B. (2011). Kaempferol regulates OPN–CD44 pathway to inhibit the atherogenesis of apolipoprotein E-deficient mice. *Toxicology and Applied Pharmacology, 257*, 405–411.

361. Xiao, N., Mei, F., Sun, Y., Pan, G., Liu, B., & Liu, K. (2014). Quercetin, luteolin, and epigallocatechin gallate promote glucose disposal in adipocytes with regulation of AMP-activated kinase and/or sirtuin 1 activity. *Planta Medica, 80*(12), 993–1000.

362. Xu, Y., Ho, C. T., Amin, S. G., Han, C., & Chung, F. L. (1992). Inhibition of tobacco-specific nitrosamine-induced lung tumorigenesis in A/J mice by green tea and its major polyphenol as antioxidants. *Cancer Research, 52*(14), 3875–3879.

363. Yamaguchi, R., Tatsumi, M. A., Kato, K., & Yoshimitsu, U. (1988). Effect of metal salts and fructose on the autoxidation of methyl linoleate in emulsions. *Agricultural and Biological Chemistry, 52*, 849–850.

364. Yang, C. S., & Wang, Z. Y. (1993). Tea and cancer. *Journal of the National Cancer Institute, 85*, 1038–1049.

365. Yang, C. S., Hong, J., Hou, Z., & Sang, S. (2004). Green tea polyphenols: antioxidative and prooxidative effects. *Journal of Nutrition, 134P*, 3181S.

366. Yang, C. S., Liao, J., Yang, G. Y., & Lu, G. (2005). Inhibition of lung tumorigenesis by tea. *Experimental Lung Research, 31*, 135–144.

367. Yang, C. S., Wang, X., Lu, G., & Picinich, S. C. (2009). Cancer prevention by tea: animal studies, molecular mechanisms and human relevance. *Nature Reviews Cancer, 9*(6), 429–439.

368. Yang, Z. F., Bai, L. P., Huang, W. B., Li, X. Z., Sui-Shan Zhao, S. S., Zhong, N. S., & Jiang, Z. H. (2014). Comparison of in vitro antiviral activity of tea polyphenols against influenza A and B viruses and structure–activity relationship analysis. *Fitoterapia, 93*, 47–53.
369. Yang, Z. G., Jia, L. N., Shen, Y., Ohmura, A., & Kitanaka, S. (2011). Inhibitory effects of constituents from Euphorbia lunulata on differentiation of 3 T3-L1 cells and nitric oxide production in RAW264.7 cells. *Molecules, 16*(10), 8305–18.
370. Yang, Z. Y., Tu, Y. Y., Xia, H. L., Jie, G. L., Chen, X. M., & He, P. M. (2007). Suppression of free-radicals and protection against H_2O_2-induced oxidative damage in HPF-1 cell by oxidizedphenolic compounds present in black tea. *Food Chemistry, 105*, 1349–56.
371. Yanishlieva, N. V., & Marinova, E. M. (2001). Stabilization of edible oils with natural antioxidants. *European Journal of Lipid Science and Technology, 103*, 752–767.
372. Yao, L. H., Jiang, Y. M., & Caffin, N. (2006). Phenolic compounds in tea from Australian supermarkets. *Food Chemistry, 96*, 614–620.
373. Yao, S., Pan, M. H., Lo, C. Y., Lai, C. S., & Ho, C. T. (2013). Black tea: chemical analysis and stability. *Food & Function, 4*, 10–18.
374. Yao, Y., Han, D. D., Zhang, T., & Yang, Z. (2010). Quercetin improves cognitive deficits in rats with chronic cerebral ischemia and inhibits voltage-dependent sodium channels in hippocampal CA1 pyramidal neurons. *Phytotherapy Research, 24*, 136–140.
375. Ye, J., Fan, F., Xu, X., & Liang, Y. (2013). Interactions of black and green tea polyphenols with whole milk. *Food Research International, 53*(1), 449–455.
376. Yen, G. C., & Chen, H. Y. (1995). Antioxidant activity of various tea extract in relation to their antimutagenicity. *Journal of Agricultural and Food Chemistry, 43*, 27–32.
377. Yilmazer-Musa, M., Griffith, A. M., Michels, A. J., Schneider, E., & Frei, B. (2012). Grape seed and tea extracts and catechin 3-gallates are potent inhibitors of aamylase and a-glucosidase activity. *Journal of Agricultural and Food Chemistry, 60*(36), 8924–8929.
378. Ying, H. Z., Liu, Y. H., Yu, B., Wang, Z. Y., Zang, J. N., & Yu, C. H. (2013). Dietary quercetin ameliorates nonalcoholic steatohepatitis induced by a high-fat diet in gerbils. *Food and Chemical Toxicology, 52*, 53–60.
379. Yokozawa, T., Nakagawa, T., & Kitani, K. (2002). Antioxidative activity of green tea polyphenol in cholesterol-fed rats. *Journal of Agricultural and Food Chemistry, 50*, 3549–3552.
380. Youdim, M. B. H., & Riederer, P. (2004). Iron in the brain, normal and pathological. In: Adelman, G., & Smith, B. (eds.). *Encyclopedia of Neuroscience. Nature Reviews Neuroscience, 5*(11), 863–73.
381. Yuksel, Z., Avci, E., & Erdem, Y. K. (2010). Characterization of binding interactions between green tea flavanoids and milk proteins. *Food Chemistry, 121*(2), 450–456.
382. Yung, L. M., Leung, F. P., Wong, W. T., Tian, X. Y., Yung, L. H., Chen, Z. Y., Yao, X. Q., & Huang, Y. (2008). Tea polyphenols benefit vascular function. *Inflammopharmacology, 16*, 230–234.
383. Zanoli, P., Avallone, R., & Baraldi, M. (2000). Behavioral characterization of the flavonoids apigenin and chrysin. *Fitoterapia, 71*, 117–123.

384. Zarzuelo, A., Jimenez, I., & Gamez, M. J. (1996). Effects of luteolin 5-O-betarutinoside in streptozotocin-induced diabetic rats. *Life Sciences, 58*, 2311–2316.
385. Zhang, M., Holman, C. D. A. J., Huang, J. P., & Xie, X. (2007). Green tea and the prevention of breast cancer: A case control study in southeast China. *Carcinogenesis, 28*, 1074–1078.
386. Zhang, X. H., Zou, Z. Q., Xu, C. W., Shen, Y. Z., & Li, D. (2011). Myricetin induces G2/M phase arrestin HepG2 cells by inhibiting the activity of the cyclin B/Cdc2 complex. *Molecular Medicine Reports, 4*, 273–277.
387. Zhao, M., Ma, J., Zhu, H. Y., Zhang, X. H., Du, Z. Y., Xu, Y. J., & Yu, X. D. (2011). Apigenin inhibits proliferation and induces apoptosis in human multiple myeloma cells through targeting the trinity of CK2, Cdc37 and Hsp90. *Molecular Cancer, 10*, 104.
388. Zheng, C. H., Zhang, M., Chen, H., Wanga, C. Q., Zhang, M. M., Jiang, J. H., Tian, W., Lv, J. G., Li, T. J., Zhu, J., & Zhou, Y. J. (2014). Luteolin from Flos Chrysanthemi and its derivatives: New small molecule Bcl-2 protein inhibitors. *Bioorganic & Medicinal Chemistry Letters, 24*, 4672–4677.
389. Zheng, X. Q., Jin, J., Chen, H., Du, Y. Y., Ye, J. H., Lu, J. L., Lin, C., Dong, J. J., Sun, Q. L., Wu, L. Y., & Liang, Y. R. (2008). Effect of ultraviolet B irradiation on accumulation of catechins in tea (*Camellia sinensis* (L) O. Kuntze). *African Journal of Biotechnology, 7*, 3283–3287.
390. Zhong, L., Golberg, M. S., Gao, Y. T., Hanley, J. A., Parent, M. E., & Jin, F. (2001). A population-based case-control study of lung cancer and green tea consumption among women living in Shanghai, China. *Epidemiology, 12*, 695–700.

CHAPTER 11

POTENTIAL HEALTH BENEFITS OF LEMON GRASS—A REVIEW

NIGHAT ZIAUDIN and HASNAIN NANGYAL

CONTENTS

11.1 Introduction 283
11.2 Lemon Grass: Botanical Description and Diversity 284
11.3 Lemon Grass: Importance and Economic Status
in Agriculture 285
11.4 Lemon Grass: Biochemistry and Nutrition Properties 285
11.5 Lemon Grass: Potential Health Benefits 286
11.6 Conclusions 290
11.7 Summary 291
Keywords 291
References 291

11.1 INTRODUCTION

General proposal to the general population is to expand the admission of sustenance rich in cancer prevention agent mixes because of their understood sound impacts. Reactive oxygen species (ROS) (e.g., superoxide radicals, hydroxyl radical and per hydroxyl radicals) have been connected with carcinogenesis, coronary heart illnesses. The cell reinforcement end specifically radical-intervened oxidative responses may be utilized in a system for counteractive action of maturing related maladies and well-being issues. This has prompted to hunt down cell reinforcement standards, the recognizable proof of common assets, and the confinement of dynamic cancer prevention agent atoms. Cell reinforcements have been identified in various farming and sustenance items including oats, organic products, vegetables and oil seeds [24].

Plants are still a potential of therapeutic mixes. Generally, plants are utilized as part of oral wellbeing and to treat numerous sicknesses. According to World Wellbeing Association (WHO) definition, *"restorative plant is a plant that can be utilized for remedial purposes as well as its mixes can be utilized as a pioneer as part of the blend of semi engineered drugs* [5]".

Plants with therapeutic quality can be utilized both as part of sorted out and disorderly shape. Plants have strong remedial specialists, because of the vicinity of dietary (minerals and vitamins) and non-nutritious (sulfides, dynamic phytochemicals including lignans, filaments, plant sterols, polyphenolics, carotenoids, flavonoids, coumarins, terpenoids, curcumins, saponins and phthalates) segment, thus can be advanced as "useful nourishment" [34].

In this chapter, authors have discussed biochemical properties and health benefits of lemon grass (*Cymbopogon citrates*).

11.2 LEMON GRASS: BOTANICAL DESCRIPTION AND DIVERSITY

Lemon grass (*Cymbopogon citrates*) belongs to family poaceae and is a local herb from India and has been developed in tropical and sub-tropical countries. It is broadly utilized as a medication. Lemon grass oil is utilized as a pesticide and additive, and has antifungal properties [10].

The concentrate of its leaves as a beverage is useful in remembering the stomach torment. Regardless of its uses, lemon grass helps in treatment of hot conditions, unwinding of mental anxiety and in numerous physiological angles because of vicinity of dynamic fixings, for example, myrcene (an antibacterial and torment reliever), citronellol, geraniol and citronellal [4]. *C. citratus* plays its part as a bug repellent and carminative operators.

The most vital of bioactive constituents are alkaloids, saponins, tannins, flavonoids, phlobotannins and heart glycoside [1, 2].

11.3 LEMON GRASS: IMPORTANCE AND ECONOMIC STATUS IN AGRICULTURE

Lemon grass (*Cymbopogon citratus*) has been expended in different structures. Because of the creation of lemon grass oil as a noteworthy segment, two of the species *C. citrate* and *C. flexosusare* for the most part are called lemon grass [23]. It is utilized as a treatment of different infirmities like hacks, infection, elephantiasis influenza, opthalmia, gingivitis, clogging, jungle fever, migraine, pneumonia, vascular disarranges, stomach throb and looseness of the bowels. It has been contemplated as vasorelaxing, mitigating, diuretic and as a cure in treating the ringworm infestation, gastrointestinal unsettling influences, fevers and hypertension because of high cancer prevention agent levels [32]. Lemon grass typically is utilized as a domestic treatment of antibacterial gastrointestinal issue, neurological infections as an antispasmodic, pain relieving, narcotic, hostile to pyretic and diuretic [5].

11.4 LEMON GRASS: BIOCHEMISTRY AND NUTRITION PROPERTIES

Concentrate of leaves of lemon grass has diuretic, vasorelaxing and calming cure in treatment of numerous sicknesses. In developing countries with high prescription cost, individuals utilize the separates of this plant for treatment of distinctive infections [27]. Lemon grass is a society solution for hack, influenza, cerebral pain and numerous vascular issues. Numerous natural dynamic substances have been disengaged from the *C. citratus*. Most critical is citral, which helps in absorption and in addition to remember fits, muscle issues and migraine. The oil from this plant has antibacterial and antifungal benefits [30]. Lemon grass has several therapeutic qualities because of vicinity of some compound substances, which can create a distinct physiological activity in human body.

The most critical of these bioactive constituents are: alkaloids, saponins, tannins, flavonoids, phlobotannins and heart glycoside [2].

11.5 LEMON GRASS: POTENTIAL HEALTH BENEFITS

Numerous restorative plants contain expansive measure of organic dynamic mixes. For example, polyphenols can assume an imperative part in adsorbing and killing the free radicals, extinguishing singlet and triplet oxygen, or deteriorating peroxides. A large portion of these phytochemicals have critical cancer prevention agent limits that are connected with lower events and lower death rates of few human diseases [13].

11.5.1 ANTI-INFLAMMATORY PROPERTIES

Dissolvable concentrates, polyphenol rich extractants and citral secludes are boss segments of lemon grass showing calming exercises. Lemon grass tea possesses analgesic activity due to presence of terpenes (especially myrcene). Hyperalgesia is induced by both carrageenan and PGE2. Oral application of decoction obtained from lemon grass also possesses anti-inflammatory properties [7]. Likewise, watery extracts devoid of lipid and fundamental oil and polyphenol portions (phenolic acids, flavonoids and tannins) of lemon grass leaves were studied for their mitigating properties. Flavonoids and tannins divisions displayed better calming strength because of vicinity of luteolinglycosides. Phenolic acids then again showed tasteful hindrance of PGE2 generation. The evaluation of the fundamental oil containing citral has been explored for such inertia. Therefore, ethanolic lemon grass remove (half) is strong enough against aggravation brought about by LPS instigated by murine alveolar macrophages. The system activity is by repressing the emission of NO and expert incendiary cytokine tumor putrefaction component TNF-α [27]. Citral and other monoterpenes from lemon grass exhibited *in-vivo* anti-inflammatory activity using carrageenan induced paw edema and peritonitis in a model rat study. Paw edema was reportedly reduced by application of citral (100 and 200 mg/kg of the body weight) and peritonitis was also reduced as leukocyte conversion to peritoneal cavity [29]. In addition, citral is dose dependent in reducing COX-2 mRNA, protein expression and activated peroxisome proliferator-activated receptor (PPARα and γ) in LPS induced U937 human macrophage like cells [3]. The PPARα and γ belong to a group of nuclear receptor proteins that play essential role in regulation of cell development, differentiation and metabolism by functioning as transcription factor [19].

11.5.2 ANTI-OXIDANT PROPERTIES

Oxidation is an essential procedure in human cells, tissue and frameworks prompting development of ROSs, which incorporate hydrogen peroxide (H_2O_2), superoxide anion and free radicals. Because of its reactivity, ROSs can harm biochemical segments like cell film, cell lipids, proteins and DNA [11]. Also, ROSs act as real inducer of a few wellbeing issues like atherosclerosis, rheumatoid joint pain and muscle demolition. Others are waterfalls, sure neurological issue, tumor and maturing. Cancer prevention agents must be available in the body to offer defensive system against harming impacts of oxidation procedure brought on by these radicals [18, 32].

Scientists have distinguished cell reinforcement possibilities of lemon grass concentrates and have recorded their capacities to decrease ROSs. Such systems incorporate hindrance of lipoperoxidation and decolorization of 2,2-diphenyl-1-picrylhydrazyl (DPPH) [22]. Infusions and decoctions arranged from lemon grass have indicated cell reinforcement properties by rummaging superoxide anion, restraining lipoperoxidation and decolorizing DPPH. These impacts are higher in mixture than decoction [9]. Thus, lemon grass implantation showed more grounded cancer prevention agent exercises in connection to different concentrates of methanolic, 80% fluid ethanol and decoction. Further studies found that tannin and flavonoid portions of sans oil imbuement concentrate were most dynamic hostile to oxidative operators compared with phenolic acid divisions [17]. Fundamental oil from lemon grass showed cancer prevention agent property by DPPH rummaging test. The outcomes demonstrated that both leaves and stalk extricates have radical searching capacity in a dosage subordinate way [22].

11.5.3 ANTI-NOCICEPTIVE PROPERTIES

The likelihood of lemon tea having anti-nociceptive impacts has been studied in the past. Reports have demonstrated almost no positive activities of lemon grass separates, thus invalidating the cases in society medications [8, 20, 31]. However, later studies showed positive results. According to Viana et al. [31], lemon grass tea had anti-nociceptive property as shown by positive results [15]. Vital oil of lemon grass, which contained citral and no myrcene, was explored for their anti-nociceptive exercises utilizing three exploratory models in mice. The hot plate test demonstrated that reaction to boost by the mice was expanded by crucial oil regulated intra-peritoneal (I.P.), while writhing by acidic corrosive impelling demonstrated that intra-peritoneal and oral organization of essential oil created hindrance of stomach contraction in a dosage subordinate way. In the formalin test, licking time was definitely restrained by vital oil directed I.P. at both first and second period of the trials [30]. They reported that opioid receptors are included in the anti-nociceptive activity, since opponent naloxone hindered the impacts of the vital oils under trials. Same gathering of agents indicated that the disparity could be due to contrasts in plant chemo types. According to Quintans-Junior [29], citral from lemon grass has been observed to be hostile to nociception properties. It was assumed that citral is equipped to inhibit for displaying fringe anti-nociceptive property of writhing and nociception [18].

11.5.4 ANTI-FUNGI PROPERTIES

The activity of fundamental oils from lemon grass decoction against both pathogenic and eatable organisms is of massive commitment. Lemon grass oil demonstrated an inhibition thus promising prospect among few vital oils for the development of parasites cells, which are ensnared in discharging mycotoxins amid capacity of grains and other nourishment items [16, 25]. Here, the synergistic impacts of oil parts indicated both synergistic and adversarial impacts among diverse oils under study [25, 31]. Fundamental oil portion of lemon tea has been accounted to be hostile for contagious impact against filamentous organisms of distinctive classes, thus demonstrating its effectiveness against

Potential Health Benefits of Lemon Grass—A Review

both sickness and non-pathogenic growth. Also, the oil is equipped to inactivate the ailment due to yeast cells (*Candida* spp.) by hindering their development [12].

11.5.5 CYTOTOXICITY AND ANTI-MUTAGENICITY

A few studies (both *in-vivo* and *in-vitro*) have explored cytotoxicity and mutagenicity impacts of lemon grass concentrate with a specific goal to affirm the health benefits of lemon grass tea. All phenolic mixes separated from methanolic concentrate of lemon grass were nontoxic to human lung fibroblasts even at high focus (1 mM) [9]. In another study, grown-up rats subjected to oral utilization of lemon grass tea for 2 months did not show any adverse impact on both the rodent and their subsequent posterity [21]. Rodent on lemon grass myrene diet did not create resilience dissimilar to analgestic drug morphine. Insurance of mitochondria layer focusing on murine alveolar macrophages was supposedly restored by 5% ethanol concentrate of lemon grass thus exhibiting cytoprotective property [30]. Lemon grass separate got sing 80% ethanol did not demonstrate any matagenic properties in Salmonella change test. Indeed, the concentrate could counter synthetic transformation in Salmonella typhimurium strains TA98 and TA100 [32]. Additionally, any adverse impact to chromosome instigated by mitomycin C in human lymphocytes was hindered by lemon grass.

11.5.6 ANTI-MICROBIAL PROPERTIES

Hostile to bacterial movement in concentrates of plant materials has been explained from different sources with promising results. This study has additionally been researched in the unpredictable oil segment of the watery concentrate of lemon grass. Among the major bioactive mixes recognized in the oil were α-citral (geranial) and β-citral (neral) segments. These parts show their antibacterial action by repressing the development of both Gram positive and Gram negative microscopic organisms. However, the third segment myrcene has no antibacterial action separately yet do improve movement when consolidated with others.

11.5.7 ANTI-OBESITY AND ANTIHYPERTENSIVE ACTIVITY

A few studies have been completed on the possibility of lemon grass separates as a well-spring of hypolipidemic and hypoglycemic substances, which might bring down the dangers of hypertension and heftiness. Research reports have demonstrated that citratus watery concentrates when bolstered to rats at 500 mg/kg/day were able to prompt critical lessening in hypoglycemic record not withstanding counter regulatory factors such as catecholamine, cortisol and glucagon. Hypolipidemic impact was observed with discernible diminishment in low thickness lipid levels in the circulatory system. The instrument, by which the tea adequately performs these impacts, stayed subtle however a few specialists have related it with expanded insulin blend and discharge (hyperinsulinemia) or expanded fringe glucose use [2, 6]. The vicinity of against hypertensive mixes (for example, flavonoids and alkaloids) has been accounted to help with the hypoglycemic properties displayed by lemon grass watery concentrate due to presence of key oil and different extractants [26, 28]. Thus, lemon grass concentrates were useful in decreasing cholesterol levels in the circulatory system, due to the vicinity of an endogenous ligand of focal sort benzodiazepine receptors known as endozepineocta-deca-neuropeptide (ODN), which are inhibitors of nourishment admission in small creatures [13].

11.6 CONCLUSIONS

Most studies on lemon grass tea and extracts were conducted through infusion, decoction or organic solvent extraction without considering effects of other factors, such as method of cultivation, harvesting time, controlled oxidation/fermentation, roasting/frying and withering conditions. Previous works have reported that these factors can affect composition, physicochemical and biological properties of tea from other sources. There are possibilities that properties of lemon grass tea might also be affected if studies consider these factors.

11.7 SUMMARY

The use of herbal preparations has remained the main approach of folk medicine to the treatment of ailments and debilitating diseases. Initial intensive research on lemon grass extracts (tea) may have shown conflicting evidences. However, the resurgence of claims by folk medicine practitioners has necessitated further inquiry into the efficacy of the tea. Lemon grass tea contains a few biological compounds in its decoction, mixture and crucial oil extractes. Hostile to oxidant, mitigating, against bacterial, hostile to corpulence, anti-nociceptive, anxiolytic and antihypertensive proofs of lemon grass tea were unmistakably illustrated to back introductory pharmacological cases. Lemon grass tea is non-lethal, non-mutagenic and gets wide acceptance among prescription experts in developing nations.

KEYWORDS

- **decoction**
- **infusion**
- **lemon grass**

REFERENCES

1. Abdul Fattah, S. M., Abosree, Y. H., Bayoum, H. M., & Eissa, H. A. (2010). The use of lemongrass extracts as antimicrobial and food additive potential in yogurt. *Journal of American Science, 6*, 582–594.
2. Adegbegi, A J., Usunomena, U., Lanre, A. B., Amenze, O., & Gabriel, O. A. (2012). Comparative Studies on the Chemical Composition and Antimicrobial Activities of the ethanolic extract of lemon grass leaves and stem. *Asian Journal of Medical Sciences, 4*, 145–114.
3. Adejuwon, A. A., & Esther, O. A. (2007). Hypoglycemic and hypolipidemic effects of fresh leaf aqueous extract of *Cymbopogon citratus* Stapf in rats. *Journal of Ethnopharmacology, 112*, 440–444.
4. Anibijuwon, I. I., & Ojo, O. O. (2010). Studies on Extracts of Three Medicinal Plants of South-Western Nigeria: *Hoslundia opposita, Lantana camara* and *Cymbopogon citratus*; *Advances in Natural and Applied Sciences, 4*(1), 93–98

5. Behboud, J. A., Ebaci, Babak, M. A., & Hassanzad, Z. (2012). Antibacterial Activities of Lemon Grass Methanol Extract and Essence on Pathogenic Bacteria. *American-Eurasian Journal of Agricultural & Environmental Science*, *12*(8), 1042–1046.
6. Carbajal, D., Casaco, A., Arruzazabala, L., Gonzalez, R., & Tolon, Z. (1989). Pharmacological study of *Cymbopogon citratus* leaves. *Journal of Ethnopharmacology*, *25*, 103–107.
7. Carlini, E. A., Conta¬, J. D. P., Silva, F. A. R., Da Silvera, F. N. G., Frochtengarte, M. L., & Bueno, O. F. A. (1986). Pharmacology of lemongrass (*Cymbopogon citrates* Stapf.). I. Effects of teas prepared from leaves on laboratory animals. *Journal of Ethnopharmacology*. *17*, 37–64.
8. Cheel, J., Theoduloz, C., Rodrguez, J., & Schmeda, H. G. (2005). Free radical scavengers and antioxidants from lemongrass (*Cymbopogon citratus* (DC) Stapf.). *Journal of Agricultural and Food Chemistry*, *53*, 2511–2517.
9. Deepa, G., Aditya, M., Nishtha, K., & Thankamani, M. (2012). Comparative analysis of phytochemical profile and antioxidant activity of some Indian culinary herbs. *Research Journal of Pharmaceutical, Biological and Chemical Sciences*, *3*, 485–490.
10. Devasagayam, T. P. A., Tilak, J. C., Boloor. K. K., Sane, Ketaki, S., Ghaskadbi, Saroj, S., & Lele, R. D. (2004). Free radicals and antioxidants in human health: Current status and future prospects. *Journal of Association of Physicians of India, 52*, 794–804.
11. Dharmendra, S., Suman, P. S. K., Atul, P. K., Subhash, C. G., & Sushil, K. (2001). Comparative antifungal activity of essential oils and constituents from three distinct genotypes of *Cymbopogon* spp. *Current Science*, *80*, 1264–1266.
12. Djerijdane, A., Yousafi, N., Nadjemi, B., Boutassouna, D., & Stocker, P. (2006). Antioxidant activity of some Algerian medicinal plants extracts containing phenolic compounds. *Food Chemistry*, *97*, 650–660.
13. Do Rego, J. C., Orta, M. H., Leprince, J., Tonon, M. C., Vaudry, H., & Costentin, J. (2007). Pharmacological characterization of the receptor mediating the anorexigenic action of the octadecaneuropeptide: evidence for an endozepinergic tone regulating food intake. *Neuro psycho pharmacology*, *32*, 1641–1648.
14. Fandohan, P., Gnonlonfin, B., Laleye, A., Gbenou, J. D., Darboux, R., & Moudachirou, M. (2008). Toxicity and gastric tolerance of essential oils from Cymbopogoncitratus, *Ocimum gratissimum* and *Ocimum basilicumin* Wistar rats. *Food and Chemical Toxicology, 46*, 2493–2497.
15. Figueirinha, A., Pararhos, A., Perez-Alonso, J. J., Santos Buelga, C., & Batista, M. T. (2008). *Cymbopogon citratus* leaves: Characterization of flavonoids by HPLC–PDA–ESI/MS/MS and an approach to their potential as a source of bioactive polyphenols. *Food Chemistry*, *110*, 718–728.
16. Finkel, T. (1998). Oxygen radicals and signaling. *Current Opinion in Cell Biology*, *10*, 248–253.
17. Leite, J. R., Seabra, M. L., Maluf, E., Assolant, K., Suchecki, D., Tufik, S., Klepacz, S., Calil, H. M., & Carlini, E. A. (1986). Pharmacology of lemon grass (*Cymbopogon citratus* Stapf), III: Assessment of eventual toxic, hypnotic and anxiolytic effects on humans. *Journal of Ethnopharmacology*, *17*, 75–83.
18. Lucia, M. O., Formigoni, S., Lodder, H. M., Filho, O. G., Ferreira, T. M. S., & Carlini, E. A. (1986). Pharmacology of lemon grass (*Cymbopogon citratus* Stapf), II:

Potential Health Benefits of Lemon Grass—A Review 293

Effects of daily two-month administration in male and female rats and in offspring exposed in utero. *Journal of Ethnopharmacology, 17,* 65–74.

19. Mirghani, M. E. S., Liyana, Y., & Parveen, J. (2012). Bioactivity analysis of lemongrass (*Cymbopogan citratus*) essential oil. *International Food Research Journal, 19,* 569–575.

20. Mohd, F. A., Kofi, R. A., & A. G. G. Candlish. (2011). Mutagenic and cytotoxic properties of three herbal plants from Southeast Asia. *Tropical Biomedicine, 24*(2), 49–59.

21. Netzel, M. G., Netzel, Q. Tian, S. Schwartz, B, Wigdhal K., & Parveen, Z. (2007). Native austraian food a novel source of antioxidant for food. *Innovative Food Sciences and Emerging Technologies, 8,* 339–346.

22. Nguefack, J., Tamgue, O., Dongmo, J. B. L., Dakolea, C. D., Leth, V., Vismer, H. F., Zollo, P. H. A., & Nkengfack, A. E. (2012). Synergistic action between fractions of essential oils from *Cymbopogon citratus, Ocimum gratissimum* and *Thymus vulgaris* against Penicilliumexpansum. *Food Control, 23,* 377–383.

23. Omotade. I. O. (2009). Chemical profile and antimicrobial activity of *Cymbopogon citratus* leaves. *Journal of Natural Products, 2,* 98–103.

24. Onabanjo, A. O., Agbaje, E. O., & Odusote, O. O. (1993). Effects of Aqueous Extracts of *Cymbopogon citratus* in Malaria. *Journal of Proto zoological Research, 3,* 40–45.

25. Quintans-Júnior, L. J., Guimarães, A. G., Santana, M. T., Araújo, B. E. S., Moreira, F. V., Bonjardim, L. R., Araújo, A. A. S., Siqueira, J. S., Antoniolli, A. R., Botelho, M. A., Almeida, J. R. G. S., & Santos, M. R. V. (2011). Citral reduces nociceptive and inflammatory response in rodents. *Brazilian Journal of Pharmacognosy, 21,* 497–502.

26. Rathabai, V., & Kanimozhi. D. (2013). Phytochemical Screening In-Vitro Antioxidant and Antimicrobial activity of ethanolic extract of *Cymbopogon citrates* L. *International Journal of Research in Pharmaceutical and Biomedical Sciences, 4*(3), 234–236.

27. Souza, F. M. L., Lodder, H. M., Gianotti, F., O., Ferreira, T. M., & Carlini, T. A. (1986). Pharmacology of lemon grass (*Cymbopogon citratus* Stapf.), II: Effects of daily two month administration in male and female rats and in offspring exposed in utero. *Journal of Ethnopharmacology, 17,* 65–74.

28. Thannickal, V. J., & Fanburg, B. L. (2000). Reactive oxygen species in cell signaling. *American Journal of Physiology Lung Cellular and Molecular Physiology, 279,* 1005–1028.

29. Tiwari, M., Dwivedi, U. N., & Kakkar, P. (2010). Suppression of oxidative stress and pro-inflammatory mediators by *Cymbopogon citratus* Stapf extract in lipopolysaccharide stimulated murine alveolar macrophages. *Food and Chemical Toxicology, 48,* 2913–2919.

30. Vanisha, S. N., & Hema, M. (2012). Potential Functions of Lemon Grass (*Cymbopogon citratus*) in Health and Disease. *International Journal of Pharmaceutical & Biological Archives, 3,* 1035–1043.

31. Viana, G. S. B., Vale, T. G., Pinho, R. S. N., & Matos, F. J. A. (2000). Anti-nociceptive effect of the essential oil from *Cymbopogon citratus* in mice. *Journal of Ethnopharmacology, 70,* 323–327.

32. Vinitketkumnuen, U., Puatanachokchai, R., Kongtawelert, P., Lertprasertsuke, N., & Matsushima, T. (1994). Antimutagenicity of lemon grass (*Cymbopogon citratus* Stapf) to various known mutagens in Salmonella mutation assay. *Mutation Research, 341,* 71–77.

GLOSSARY

Glossary of Terms in Machine Vision System
(https://en.wikipedia.org/wiki/Glossary_of_machine_vision)

Angular resolution describes the resolving power of any image forming device such as an optical or radio telescope, a microscope, a camera, or an eye.

Aperture refers to the diameter of the aperture stop of a photographic lens. The aperture stop can be adjusted to control the amount of light reaching the film or image sensor.

Aspect ratio of an image is its displayed width divided by its height (usually expressed as "$x:y$").

Barcode (also **bar code**) is a machine-readable representation of information in a visual format on a surface.

Bitmap is a data file or structure representing a generally rectangular grid of pixels, or points of color, on a computer monitor, paper, or other display device.

Blob discovery is an inspecting an image for discrete blobs of connected pixels (e.g., a black hole in a gray object) as image landmarks. These blobs frequently represent optical targets for machining, robotic capture, or manufacturing failure.

Camera is a device used to take pictures, either singly or in sequence. A camera that takes pictures singly is sometimes called a photo camera to distinguish it from a video camera.

Camera Link is a serial communication protocol designed for computer vision applications based on the National Semiconductor interface Channel-link. It was designed for the purpose of standardizing scientific and industrial video products including cameras, cables and frame grabbers. The standard is maintained and administered by the Automated

Imaging Association or (AIA): the global machine vision industry's trade group.

Charge-coupled device (CCD) is a sensor for recording images, consisting of an integrated circuit containing an array of linked or coupled, capacitors. CCD sensors and cameras tend to be more sensitive, less noisy, and more expensive than CMOS sensors and cameras.

CIE 1931 Color Space: In the study of the perception of color, one of the first mathematically defined color spaces was the **CIE XYZ color space** (also known as **CIE 1931 color space**), created by the International Commission on Illumination (CIE) in 1931.

CMOS (complementary metal-oxide semiconductor) is a major class of integrated circuits. CMOS imaging sensors for machine vision are cheaper than CCD sensors but noisier.

C-Mount is a standardized adapter for optical lenses on CCD - cameras.

CoaXPress (CXP) is an asymmetric high speed serial communication standard over coaxial cable. CoaXPress combines high speed image data, low speed camera control and power over a single coaxial cable. The standard is maintained by JIIA (Japan Industrial Imaging Association).

Color blindness is also known as color vision deficiency in humans, It is the inability to perceive differences between some or all colors that other people can distinguish.

Color is the perception of the frequency (or wavelength) of light. It can be compared to how pitch (or a musical note) is the perception of the frequency or wavelength of sound.

Color temperature (White light) is commonly described by its color temperature. A traditional incandescent light source's color temperature is determined by comparing its hue with a theoretical, heated black-body radiator. The lamp's color temperature is the temperature in kelvins at which the heated black-body radiator matches the hue of the lamp.

Color vision (CV) is the capacity of an organism or machine to distinguish objects based on the wavelengths (or frequencies) of the light they reflect or emit.

Glossary

Computer vision is an interdisciplinary field related to artificial intelligence, machine learning, robotics, signal processing, geometry, etc. The purpose of computer vision is to program a computer to "understand" a scene or features in an image. It is the study and application of methods which allow computers to "understand" image content.

Contrast is the difference in visual properties that makes an object (or its representation in an image) distinguishable from other objects and the background.

CS-Mount is same as C-Mount but the focal point is 5 mm shorter. A CS-Mount lens will not work on a C-Mount camera. CS-mount is a 1-inch diameter, 32 threads/inch mounting thread.

Data matrix is a two dimensional Barcode.

Defocus refers to a translation along the optical axis away from the plane or surface of best focus. In general, defocus reduces the sharpness and contrast of the image.

Depth of field (DOF) is the distance in front of and behind the subject (in optics), which appears to be in focus.

Depth perception (DP) is the visual ability to perceive the world in three dimensions. It is a trait common to many higher animals. Depth perception allows the beholder to accurately gauge the distance to an object.

Diaphragm is a thin opaque structure with an opening (aperture) at its center (in optics). The role of the diaphragm is to stop the passage of light, except for the light passing through the aperture.

Edge detection (ED) marks the points in a digital image at which the luminous intensity changes sharply. It also marks the points of luminous intensity changes of an object or spatial-taxon silhouette.

Electromagnetic interference: Radio Frequency Interference (RFI) is electromagnetic radiation which is emitted by electrical circuits carrying rapidly changing signals, as a by-product of their normal operation, and which causes unwanted signals (interference or noise) to be induced in other circuits.

Field of view (FOV) is the part which can be seen by the machine vision system at one moment. The field of view depends from the lens of the system and from the working distance between object and camera.

FireWire is a personal computer (and digital audio/video) serial bus interface standard, offering high-speed communications. It is often used as an interface for industrial cameras.

Focus: An image, or image point or region, is said to be in focus if light from object points is converged about as well as possible in the image; conversely, it is out of focus is light is not well converged. The border between these conditions is sometimes defined via a circle of confusion criterion.

Frame grabber is an electronic device that captures individual, digital still frames from an analog video signal or a digital video stream.

Fringe Projection Technique is a 3D data acquisition technique employing projector displaying fringe pattern on a surface of measured piece, and one or more cameras recording image(s).

Gamut or **color gamut** is a certain *complete subset* of colors, in color reproduction, including computer graphics and photography, the

Graphical User Interface (or GUI, sometimes pronounced "gooey") is a method of interacting with a computer through a metaphor of direct manipulation of graphical images and widgets in addition to text.

Grayscale digital image is an image in which the value of each pixel is a single sample. Displayed images of this sort are typically composed of shades of gray, varying from black at the weakest intensity to white at the strongest, though in principle the samples could be displayed as shades of any color, or even coded with various colors for different intensities.

Histogram (Color) is a representation of the distribution of colors in an image, derived by counting the number of pixels of each of given set of color ranges in a typically two-dimensional (2D) or three-dimensional (3D) color space.

Histogram is a graphical display of tabulated frequencies in statistics. A histogram is the graphical version of a table which shows what proportion of cases fall into each of several or many specified categories. The

Glossary 299

histogram differs from a bar chart in that it is the *area* of the bar that denotes the value, not the height, a crucial distinction when the categories are not of uniform width. The categories are usually specified as non-overlapping intervals of some variable. The categories (bars) must be adjacent.

HSV color space (HSV: Hue, Saturation, Value model; also called HSB: Hue, Saturation, Brightness) defines a color space in terms of three constituent components: Hue, the color type (such as red, blue, or yellow); Saturation, the "vibrancy" of the color and colorimetric purity; and Value, the brightness of the color.

Image file formats provide a standardized method of organizing and storing image data. This article deals with digital image formats used to store photographic and other image information. Image files are made up of either pixel or vector (geometric) data, which is rasterized to pixels in the display process, with a few exceptions in vector graphic display. The pixels that make up an image are in the form of a grid of columns and rows. Each of the pixels in an image stores digital numbers representing brightness and color.

Incandescent light bulb generates light using a glowing filament heated to white-hot by an electric current.

ISO 9000 is a family of standards for quality management systems. ISO 9000 is maintained by ISO, the International Organization for Standardization and is administered by accreditation and certification bodies.

JPEG is a most commonly used standard method of lossy compression for photographic images.

Kell factor is a parameter used to determine the effective resolution of a discrete display device.

Laser is a device that emits light through a specific mechanism for which the term laser is an acronym: light amplification by stimulated emission of radiation.

Lens Controller is a device to control a motorized (ZFI) lens. Lens controllers may be internal to a camera, a set of switches used manually, or a sophisticated device that allows control of a lens with a computer.

Lens is a device that causes the light either to converge and concentrate or to diverge, usually formed from a piece of shaped glass. Lenses may be combined to form more complex optical systems as a Normal lens or a Telephoto lens.

Lighting refers to either artificial light sources such as lamps or to natural illumination.

Machine Vision (MV) is the application of computer vision to industry and manufacturing.

Metrology is the science of measurement. There are lots of applications for machine vision in metrology.

Motion perception (MP) is the process of inferring the speed and direction of objects and surfaces that move in a visual scene given some visual input.

Neural network (ANN) is an interconnected group of artificial neurons that uses a mathematical or computational model for information processing based on a connectionist approach to computation. In most cases an ANN is an adaptive system that changes its structure based on external or internal information that flows through the network.

Normal lens (or entrocentric lens) is a lens that generates images that are generally held to have a "natural" perspective compared with lenses with longer or shorter focal lengths. Lenses of shorter focal length are called wide-angle lenses, while longer focal length lenses are called telephoto lenses.

One-dimensional, ID

Optical character recognition (OCR) involves computer software designed to translate images of typewritten text (usually captured by a scanner) into machine-editable text, or to translate pictures of characters into a standard encoding scheme representing them in (ASCII or Unicode).

Optical resolution describes the ability of a system to distinguish, detect, and/or record physical details by electromagnetic means. The system may be imaging (e.g., a camera) or non-imaging (e.g., a quad-cell laser detector).

Glossary 301

Q-factor of a resonant cavity is equal to the ratio of the resonant frequency to the bandwidth of the cavity resonance. The average lifetime of a resonant photon in the cavity is proportional to the cavity's Q. If the Q factor of a laser's cavity is abruptly changed from a low value to a high one, the laser will emit a pulse of light that is much more intense than the laser's normal continuous output. This technique is known as Q-switching.

Region of interest (ROI) is a selected subset of samples within a dataset identified for a particular purpose.

RGB color model utilizes the additive model in which red, green, and blue light are combined in various ways to create other colors.

Separate video (S-Video) is also known as **Y/C** (or *erroneously*, S-VHS and super video). It is an analog video signal that carries the video data as two separate signals (brightness and color), unlike **composite video** that carries the entire set of signals in one signal line. S-Video, as most commonly implemented, carries a high-band width 480i or 576i resolution video (standard definition video). It does not carry audio on the same cable.

Shutter is a device that allows light to pass for a determined period of time, for the purpose of exposing the image sensor to the right amount of light to create a permanent image of a view.

Shutter speed is the time for which the shutter is held open during the taking an image to allow light to reach the imaging sensor. In combination with variation of the lens aperture, this regulates how much light the imaging sensor in a digital camera will receive.

Smart camera is an integrated machine vision system which, in addition to image capture circuitry, includes a processor, which can extract information from images without need for an external processing unit, and interface devices used to make results available to other devices.

Spatial-Taxons are information granules, composed of non-mutually exclusive pixel regions, within scene architecture. They are similar to the Gestalt psychological designation of figure-ground, but are extended to include foreground, object groups, objects and salient object parts.

Structured light 3D scanner refers to the process of projecting a known pattern of illumination (often grids or horizontal bars) on to a scene. The way that these patterns appear to deform when striking surfaces allows vision systems to calculate the depth and surface information of the objects in the scene.

SVGA (Super Video Graphics Array) is a broad term that covers a wide range of computer display standards.

Telecentric lens is a compound lens with an unusual property concerning its geometry of image-forming rays. In machine vision systems, telecentric lenses are usually employed in order to achieve dimensional and geometric invariance of images within a range of different distances from the lens and across the whole field of view.

Telephoto lens is the one whose focal length is significantly longer than the focal length of a normal lens.

Thermography is a type of Infrared imaging.

Three-dimensional (3D) computer graphics is a three-dimensional representation of geometric data that is stored in the computer for the purposes of performing calculations and rendering 2D images. Such images may be for later display or for real-time viewing.

Three-dimensional (3D) scanner is a device that analyzes a real-world object or environment to collect data on its shape and possibly color. The collected data can then be used to construct digital, three dimensional models useful for a wide variety of applications.

TIFF (Tagged Image File Format) is a file format for mainly storing images, including photographs and line art.

Two-dimensional (2D) computer graphics is a computer-based generation of digital images—mostly from two-dimensional models (such as 2D geometric models, text, and digital images) and by techniques specific to them.

USB (Universal Serial Bus) provides a serial bus standard for connecting devices, usually to computers such as PCs, but is also becoming commonplace on cameras.

VESA (Video Electronics Standards Association) is an international body, founded in the late 1980s by *NEC Home Electronics* and eight other video display adapter manufacturers. The initial goal was to produce a standard

Glossary   303

for 800×600 SVGA resolution video displays. Since then VESA has issued a number of standards, mostly relating to the function of video peripherals in IBM PC compatible computers.

VGA (Video Graphics Array) is a computer display standard first marketed in 1987 by IBM.

Wide-angle lens is a lens whose focal length is shorter than the focal length of a normal lens, in photography and cinematography.

X-rays is a form of electromagnetic radiation with a wavelength in the range of 10 to 0.01 nm, corresponding to frequencies in the range 30 to 3000 PHz (10^{15} hertz). X-rays are primarily used for diagnostic medical and industrial imaging as well as crystallography. X-rays are a form of ionizing radiation and as such can be dangerous.

Y-cable is a self-describing name of a type of cable containing three ends of which one is a common end that in turn leads to a split into the remaining two ends. When looked upon, a Y-cable can resemble the Latin letter "Y". Y-cables are typically, but not necessarily, short (less than 12 inches), and often the ends connect to other cables. Uses may be as simple as splitting one audio or video channel into two, to more complex uses such as splicing signals from a high density computer connector to its appropriate peripheral.

Zoom lens is a mechanical assembly of lenses whose focal length can be changed, as opposed to a prime lens, which has a fixed focal length.

INDEX

A

Abiotic factors, 44, 123, 168, 181
Absorption, 14, 80, 132, 212, 285
Acid
  consumption, 168
  mine drainage (AMD), 154, 155, 168, 169, 171
  mine water, 150
  thinning, 217
Acidianus, 152
Acidiphilum, 152, 166
Acidithiobacillus
  ferrooxidans, 150, 152, 154, 163, 166, 167, 172
  thiooxidans, 150, 152, 154, 163, 166, 172
Active
  contour algorithm, 9
  oxygen method (AOM), 233
  site, 78, 79, 86, 87, 92, 97, 204
Additive
  enzyme, 92
  model, 301
Adenine nucleotide translocase (ANT), 253
Adenosine triphosphate (ATP), 241
Adequate exploratory efforts, 170, 172
Adipocytes, 235
Adipocyte-triggered inflammatory response, 239
Adsorption, 73, 211
Aerobic, 165
Aerva lanata, 133
Agar, 91, 136, 138
Agricultural, 5, 25, 43, 73, 119, 123, 124, 148
  crops, 124
  industrial waste, 73
  production, 123, 124

Agriculture, 5, 107, 108, 118, 120, 123–125, 134, 184
Agrobacterium, 122
Agrochemicals, 109, 110, 119, 123, 124
Agronomy, 108
Agyrosis, 132
Air-lift percolator, 153, 156
Air-liquid contact, 156
Albumic, 116
Alcaligenes eutrophus, 153
Aldehyde group, 210, 217
Aldehydes, 200
Alginate, 91
Alkaline conditions, 200, 206
Alkylation process, 91
Alternative current (AC), 60, 41, 192, 201, 203, 205, 206, 208, 210, 214
Alzheimer disease, 231, 242
Amalgamation, 185, 186
Ameliorate, 242
Amide, 91
Amino acids, 44, 118, 185, 237
Ammonium sulfate, 74, 102, 165, 202
Amorphous, 205, 206, 208, 212, 214
Amyloidprecursor protein (APP), 253
Amylopectin, 201, 203, 205–210, 214, 217
  content, 207, 208
  oxidation, 206
  region, 201
Amylose, 192, 197, 201, 204–210, 213–215, 217
  amylopectin, 204
  amylose interactions, 204
  content, 192, 205
  digestion, 201
  leaching, 208–210, 214
  lipid, 215
Amyotrophic lateral sclerosis, 253

Analgestic drug morphine, 289
Androgens, 248
Angiogenesis, 242, 244
Angiosperms, 134
Angular resolution, 295
Anti-allergic, 235, 241
Anti-alzheimer property, 242
Anti-anxiolytic effect, 256
Anti-arteriosclerotic, 256
Anti-atherogenic, 241, 256
Anti-atherosclerotic, 241, 256
Antibacterial, 132, 138, 141, 142, 284, 285, 289
  action, 142, 289
  activity, 132, 138, 141
  bioassay, 138
  gastrointestinal issue, 285
Antibiosis production, 184
Antibiotics, 132, 134, 254
Anticancer
  activity, 240
  reagents, 134
Anti-carcinogen, 237, 241, 243–245
Anti-carcinogenic, 237, 241, 243–245
Anti-diabetic, 235, 241, 255
Anti-diabetogenic activity, 236
Anti-estrogenic, 241, 256
Antifungal benefits, 285
Anti-hepatotoxic, 235
Antihyperglycemic, 255
Antihyperlipidemic, 241
Anti-hyperproteinemic, 255
Anti-inflammatory, 235–241, 243, 252, 255, 256, 286
  activities, 240, 286
  effects, 238, 240
  properties, 240, 286
Antimicrobial, 44, 49, 132, 138, 142, 241, 254, 256
Anti-mutagenic, 236, 239, 241
Anti-mutagenicity, 239
Anti-neoplastic, 235
Anti-nociceptive, 288, 291
Anti-nociceptive activity, 288
Anti-obesity, 238, 239, 241
Anti-osteoporotic, 235, 241

Antioxidant, 113, 116, 117, 188, 230, 231, 233–235, 237, 239–244, 253, 255, 256
  activity, 117, 188, 233, 234, 244, 256
  properties, 116, 230, 233, 234, 243
  property, 233, 253
Anti-oxidative, 238
Anti-platelet, 235
Anti-proliferative, 236
Antispasmodic, 285
Anti-thrombotic, 241
Anti-tumor, 231, 236, 238, 256
Anti-tumorigenic, 231, 236, 256
Anti-viral, 240, 241
Anxiolytic, 241, 291
Aortarelaxation, 236
Aperture, 295, 297, 301
Apigenin, 232, 235, 236
Apoptosis, 235, 236, 242, 244, 250
  human carcinoma cells, 236
Machine vision system applications, 16
  horticultural products, 18, 19
  dairy, 19
  beverages, 19
  meat/poultry/fish, 20
  egg sorting, 20
  bakery products/snacks/confectionaries, 20, 21
  grains/cereals, 21, 22
Aqueous phases, 111
Arabidopsis, 186
Argyria, 132
Arsenopyrite, 149, 154
Arteriosclerosis, 242
Artificial
  intelligence, 297
  light sources, 300
  neurons, 300
Asparagus, 121
Aspergillus
  flavus, 46, 133
  japonicus, 78, 80
  niger, 77, 79, 83, 84, 132, 136, 154
  oryzae, 73, 75, 77–80, 82–89, 94, 101–103
Asymmetric, 296

Index 307

Atherosclerosis, 252, 254, 287
Atlantic salmon filets, 23
Atmospheric
    gases, 156
    pressure, 35, 38–42, 45–47, 49, 50
    pressure plasma jet, 35, 49
Atomization spraying, 117
Automated Train Examiner (ATEx), 16
Automatic
    observation system, 8
    portioners, 20
Auxin, 121
Avalanche, 35, 38
Azadirachta indica, 133

B

Bacillus
    licheniformis, 133, 154
    stearothermophilus, 74, 86
    subtilis, 48, 90
Bacteria, 47, 64, 74, 80–82, 86, 90, 94,
    132–134, 138, 149–153, 156, 161,
    162, 165–167, 172, 184, 249
Bacterial
    culture, 138
    movement, 289
    strains, 137
    suspension, 137
Bactericidal nature, 118
Balanced protein, 194
Ball grid arrays (BGAs), 16
Barcode, 295, 297
Barium chloride dehydrate, 137
Barren solution, 158, 167
Barytes, 170
Bauxite, 170
Beef
    carcasses, 9
    cuts, 9
Benzotropolone ring structure, 238
Benztropolone rings, 240
Beta-galactosidase, 102
Bioactive, 116, 117, 134, 135, 285, 286,
    289

Biochemical changes (seed priming),
    194
    DNA, 194–195
    drying-back/seed longevity, 196
    effects on
    enzymes, 196
    protein synthesis, 195
    RNA, 195
Biochemical reactions, 71, 157
Bio-degradable, 108, 110, 125
Bio-extraction, 149, 165, 170, 171
Bio-geotechnology, 150
Bio-hydrometallurgical
    processes, 170, 172
    techniques, 170
Biohydrometallurgy, 150, 172
Bio-leaching, 148–153, 157, 161–164,
    166, 169–172
    environments, 153
Biological, 14, 71, 108, 122, 132–134,
    141, 148, 155, 161, 164, 166, 171,
    188, 235, 236, 240, 290, 291
    activities, 235, 236
    control agents, 134
    engineering, 108
    functions, 236
    priming, 188
    processes, 164
    properties, 290
Bio
    mining, 148–150, 161, 171
    oxidation, 149–151, 162, 167, 171
    preparation treatment, 184
    reactor leaching, 168, 171
    safety chamber, 138
    stimulators, 183
    synthetic, 131, 136
    technology, 108, 120, 122–124, 131,
    149, 171, 173
Black
    body radiator, 296
    tea, 230
Blob discovery, 295
Blood pressure, 237, 252
Brassica vegetables, 240
Breakdown viscosity, 215

Breast cancer, 241, 242, 247–249, 255
Bresanla ham, 45
Brix value, 13
Broccoli, 235
Bromine, 202
Bronchogenic cancer, 244
B-spline algorithm, 10
Buffer
　lactose acetate, 103
　solution, 74, 79, 89, 94
Bulgaricus bacteria, 80
Bulk fertilizers, 119
Burn treatment, 132

C

Cadmium, 148, 167
Calcium, 85, 91, 94, 96, 97, 101, 118,
　165, 194, 253
　alginate, 91, 94, 96, 97, 101
　nitrate, 165
Calibration, 8, 18, 19, 22, 23
Camellia sinesis, 229
Camera, 7, 9, 22, 295
Camptothecin, 236
Cancer, 230, 231, 235–237, 239–252,
　255, 256, 283–287
Capping, 123
Carbon
　dioxide, 156, 157, 165
　nano-fibers surface, 122
　nano-tube, 120, 124
Carbonyl groups, 202, 206, 207, 232
Carboxyl groups, 91, 202, 205–207, 211,
　216, 217
Carboxylation, 186, 235
Carcinogenesis, 241, 283
Cardio-protective, 241
Cardiovascular disease, 230
Catalyze, 71, 118
Catechin gallate, 238
Catechins, 231–234, 237, 238, 243, 244,
　256
Catecholamine, 290
Cattail millet, 192
Cavity resonance, 301

Cell
　cycle arrest, 235, 236, 242
　cytoskeleton, 185
　development, 287
　division, 185, 188, 243
Cell
　elimination, 244
　film, 186, 287
　growth, 236, 242
　lipids, 287
　membrane, 66, 122, 188
　wall, 66, 122, 123
Cereal, 21, 45, 46, 119, 192, 194, 201,
　203, 213
　crops, 194
　starch, 194, 201, 203, 213
Cerebral pain, 285
Certification bodies, 299
Chalcone
　flavanone isomerase, 235
　synthase, 234
Chalcopyrite, 163, 171
Chamomile, 235
Charged coupled device (CCD), 5, 7, 9,
　18, 22, 296
Chelating agents/property, 87, 232
Chelation, 73
Chemolithoautotrophs, 165, 173
Chemolithotroph, 167
Chitosan, 90–96, 98, 102, 103, 108, 109,
　116, 118, 124
Chlorophyll, 230
Chocolate biscuits, 20
Cholesterol, 240, 255, 290
Chromite, 170
Chromium, 148
Chronic
　diseases, 231, 235, 243
　myelogenous leukemia, 242
Chrysin, 232, 236, 237
Cigarette smoking, 250, 252
Cinematography, 303
Circulatory system, 290
Citratus watery concentrates, 290
Citronellol, 284
Citrus, 18

Index 309

Classification of,
  machine vision system, 11, 15
    hyperspectral imaging, 15
    infrared, 13, 14
    multispectral imaging, 15
    ultraspectral, 16
    ultraviolet, 13
    visible spectral, 11–13
    X-ray, 14, 15
Clean technology, 148, 173
Clogging, 285
Clouding agent, 216
Clustering, 10
Coaxial cable, 296
CoaXPress (CXP), 296
Cold
  jet plasma, 49
  plasma, 34, 36, 37, 40, 41, 45–49
  smoked salmon, 45
Color
  analysis, 24
  based recognition system, 21
  blindness, 296
  calibration, 22, 23
  coordinate systems, 11
  development, 21
  distinctions, 13
  distribution, 9
  gamut, 298
  grading, 8
  measurement, 19, 22, 24, 25
  models, 21, 22
  ranges, 298
  reproduction, 298
  scanners, 8
  space, 296
  temperature, 296
  vision (CV), 296
    deficiency, 296
Colorimeters, 13
Colorimetric
  measurements, 8
  purity, 299
Commercial
  scale plants, 156

units, 151
Commercial-scale plants, 157
Compatible computers, 303
Competent authority, 10
Complementary metal oxide semicon-
  ductor (CMOS), 7, 9, 296
Complex
  algorithms, 20
  microbial interaction, 153
Composite
  materials, 121
  video, 301
Compressed sterile air, 156
Computational model, 300
Computer
  connector, 303
  graphics, 298, 302
  hardware/software, 23
  vision, 4, 10, 17–19, 22, 23, 26, 295,
    297, 300
    algorithms, 19
    system, 4, 10, 19, 23, 26
Concentric lamellae, 113
Confocal microscopy images, 10
Confusion criterion, 298
Conical flask, 136
Conventional
  colorimeter, 24
  hybrid breeding, 197
  methods, 122, 148, 151, 172
  mining processes, 151
  starch modifications, 217
Conveyor belt, 9, 20, 21
Cooling phase, 64
Copper, 40, 41, 85, 102, 132, 148–151,
  157–159, 161, 167, 170, 171
Copper-nickel sulfide, 171
Coriandrum sativum, 133
Corn
  oil, 243
  starch, 204, 214
Coronary heart disease, 237, 243, 252,
  283
Cortisol, 290
Coupling energy, 34
Covalent

binding, 73
crosslinking, 200, 217
Cross-linked starch, 200, 201, 205, 207, 209, 211, 212, 215–217
Cross-linker dosages, 200
Cross-linking, 90, 197, 198, 200, 201, 203, 205–212, 214–216
Cross-segment container, 180
Crucial
distinction, 299
oil extractes, 291
Cryochemical synthesis, 133
Crystalline, 208, 212–214
array, 214
domains, 208
Crystallinity, 208, 214
Crystallite, 210, 214
Crystallization, 72, 202, 204
Cultivation, 135, 290
Curve conformity, 76
Cyanidation, 159
Cytokine tumor putrefaction, 286
Cytoprotective property, 289
Cytotoxicity, 289

D

Dairy
industry, 72, 103
products, 19, 101, 103
Date codes, 13
Debilitating diseases, 291
Decoction, 286–288, 290, 291
Decolorization, 287
Decomposition, 72, 73, 89, 101, 150
Defocus, 297
Degree of,
crystallinity, 214
substitution, 202
Dehydration, 188
Dementia, 231, 253
Denaturation
energy, 81, 82
factors, 83, 87
process, 96
Depth of field (DOF), 297

Depth perception (DP), 297
Desmodium trifolium, 133
Detectors, 14
Deterioration, 123
Detritus, 20
Diabetes, 240, 243, 252, 254–256
mellitus, 254
Diabetic, 230, 240, 255
Dialysis, 74–76
Diaphragm, 297
Dielectric barrier discharge, 35, 45, 46, 49
Diffusion, 10, 15, 37, 66, 138, 188
Digestibility, 204
Digestion, 181, 201, 204, 230, 238
Digestive tract, 244
Digital
camera, 7, 8, 19, 23, 301
image, 4, 5, 297–299, 302
processing technology, 4
imaging method, 24
numbers, 299
still frames, 298
video stream, 298
Dilution-to-extinction approach, 135
Direct
air nozzles, 21
electric resistance heating, 59
manipulation, 298
plasma diagnostics, 42
Dissolution, 117, 149, 156, 162, 164, 166, 168
Dithiothreitol (DTT), 123
Double distilled water, 136
Drought-tolerant crops, 192
Dry-fermented sausage, 243
Drying, 23, 76, 180, 181, 187, 188, 216, 230
Dump leaching, 157, 159, 168
Dyslipidemia, 240

E

Economic, 25, 118, 148, 152, 170
Ecosystem homeostasis, 124
Eco-toxicity, 119

Index 311

Edge detection (ED), 297
Edible films, 216
Electric
conductivity, 59, 61, 63, 64, 67
current, 35, 59, 60, 66, 299
field intensity, 61
field, 35, 38, 42, 59–61, 66
heterogeneity, 60
resistance, 59, 60, 65
Electro conductive heating, 59, 68
Electrodes, 35, 38–43, 58, 61, 63, 65, 67
Electromagnetic radiation, 34, 297, 303
Electron
density, 35, 37, 42
temperature, 37, 38, 42
Electrons, 34–38, 166
Electro-osmosis, 66
Electrophoresis, 77, 78, 83, 84
Electroporation, 66
Electrostatic
deposition, 113
repulsion, 211, 217
Elephantiasis influenza, 285
Elevation, 209, 210, 215
Elution volume, 84
Emission, 22, 118, 286, 299
Emulsifier, 112, 116, 216
Emulsion, 63, 109–112, 115–117
Emulsion evaporation methods, 117
Encapsulated fertilizers, 118
Encapsulating materials, 108
Encapsulation, 73, 119, 125, 216, 238
Endophyte, 134–136
Endophytic
fungal, 139, 140, 141
fungi, 132, 134–136, 138, 141
relationship, 134
Endosperm, 192
Endothelium, 236
Energy metamorphosis enzyme value, 102
Enterococcus sp, 137, 141
Enzymatic activity, 74, 76, 80, 82, 86–88, 91–98, 100
Enzyme, 71–73, 86, 87, 89, 91, 99, 148, 188, 204, 237–239, 253

complex, 78
conversion, 217
denaturation energy value, 81
immobilizing granules, 97
molecules, 87, 94
stability, 78, 79, 83
Epicatechin, 231, 233, 238
Epichlorohydrin, 197, 217
Epidemophyton floccosum, 254
Epigallocatechin, 231, 233, 238, 257
Erlenmeyer flasks, 156
Escherichia coli, 47, 137, 141, 254
Esophageal cancer, 247, 249, 250
Estrogen, 248, 255
Estrogenic, 241
Ethanol, 115, 135, 234, 287, 289
Eukaryotic
micro-organisms, 132
organisms, 134
Exopolymeric substance (EPS), 162, 163
Extracellular signal-regulated kinase (ERK), 236, 242
Extraction, 9, 72, 73, 77, 150–152, 158, 159, 164, 166–169, 171

F

Fat globules, 63, 238
Fermentation, 72, 73, 230, 238, 254, 256, 290
Fermented tea, 230
Ferric, 161–163, 165, 166, 253
ions, 161, 163, 165
precipitation, 166
Ferroplasma, 152
Ferrous, 161, 163, 165–167, 232
Fertilizers, 118
Fibroblasts, 236, 289
Ficus panda, 132, 135, 136, 141
Field of view (FOV), 298
Final
inspection cells, 16
viscosity, 215
FireWire, 298
Fixed focal length, 303
Flatbed scanner, 9, 19, 24

Flavanols, 232, 237
Flavones, 234, 235
Flavonoids, 231–233, 235, 245, 257, 284–287, 290
Flora, 45, 123
Focal
adhesion kinase (FAK), 236
length, 300, 302, 303
point, 297
Food and Agriculture Organization (FAO), 229
Food
industry process, 197
material, 58, 61
pack checks, 16
particles, 60
preservation technology, 49
processing, 4, 5, 22, 25, 43, 49, 67
products, 16, 22, 25, 44, 194
quality, 10
supply, 123
Fossilized tissues, 134
Fouling, 60
Fourth state of matter, 34, 49
Frame grabber, 298
Free
enzyme, 73, 93–96, 98, 99, 101–103
radical, 34, 231, 240, 255, 286, 287
Freeze damage detection, 13
Friedreich ataxia, 253
Fringe projection technique, 298
Functional groups, 201, 203, 206, 207, 211, 216, 232, 234
Functional properties of,
pearl millet starch, 217
least gelation concentration, 221
pasting properties, 224–226
swelling/solubility aspects, 217–220
swelling volume, 221–222
X-ray diffraction (XRD) studies, 223–224
Fungal, 73, 109, 134–136, 139, 140, 254
Fungi, 72, 102, 133–135, 149, 153, 154, 172, 184

Fusarium oxysporum, 133

G

Galacto-oligosaccharides, 90
Galactose, 73, 85, 87–90, 96, 97, 101, 102, 242
Galactose sugar, 87, 88
Galena, 163
Gallocatechin, 233, 238
Galvanic
conversion, 161, 172
interaction, 164
Gamma rays, 14
Gap-junctional intercellular communication (GJIC), 237
Garlic oil, 124
Gas diffusion, 168
Gel
filtration, 76–78, 84, 85, 102
punching machine, 138
Gelatin, 91, 98, 116
Gelatinization, 199, 200, 202, 204, 207, 209, 210, 213, 215
Gene
delivery, 122
expression, 122, 123
transformation, 122
Genetic
engineering, 124, 172, 197, 217
material, 124
modification, 108, 197
mutation, 172, 244
Gentle hydration method, 113
Geometry, 34, 40, 169, 297, 302
Geomicrobiology, 173
Geraniol, 284
Germination, 110, 119, 179–187
Gigabit Ethernet, 9
Gingivitis, 285
Glandular stomach, 244
Gliding arc cold plasma system, 49
Glucose, 89, 90, 102, 136, 202, 204, 235, 236, 254, 255, 290
Glucosuria, 254
Glutathione reductase (GR), 186

Index   313

Gluteraldehyde, 90, 91, 95, 98, 101
Glycerophospholipids, 112
Glycolysis, 120
Glycosidic
  bond, 204
  linkage, 215, 217
  linkages, 215
Gold, 108, 123, 124, 132, 149, 150, 157–159, 161, 171
Grade ores, 152, 157, 159, 171
Grain yield, 118
Grains, 5, 13, 21, 44, 192, 193, 288
Gram
  negative, 44, 152, 241, 289
  positive, 44, 241, 289
Granular
  integrity, 216
  resistance, 215
  state, 200
Granules, 91–93, 95, 97, 194, 200, 201, 204, 210, 212, 301
Graphical
  images, 298
  user interface, 298
Greedy algorithms, 9
Green
  pepper, 235
  synthesis, 142
  tea, 230
Green tea polyphenols (GTPs), 243, 248, 250, 251, 253–255
Ground electrode, 40, 41
Gutbuster, 121

H

Hairline cracks, 20
Hallucinogens, 133
Halopriming, 180, 182, 188
Health benefits of tea, 275
  anti-diabetic effects, 268–269
  effects of tea on different cancer types, 257
    breast cancer, 262–263
    esophageal cancer, 263–264
    lung cancer, 258–262

    oral cancer, 264
    prostate cancer, 265
    skin cancer, 264–265
    stomach cancer, 263
  effects on
    anti-microbial activity, 268
    cardiovascular diseases, 266
    neurodegenerative diseases, 266–267
Heart disease, 242, 244, 252
Heat
  channeling, 60
  dissipation, 60
  transfer, 59, 64
Heavy
  light particles, 34
  metal, 148, 153, 161, 167, 169, 244
  particles, 34, 36, 37
Heftiness, 290
Hegni (Niger), 192
Helicobactor pylori, 249
Hepatic lipid accumulation, 239
Herbicidal agents, 119, 120
Herbicide, 109, 110, 120, 123, 124
Heterocyclic ring, 232
Heterogeneous, 63, 164, 239
  system, 164
  texture, 63
Heterotrophs, 152
High
  adhesiveness, 183
  blood pressure, 252
  cholesterol level, 252
  level processing, 10
  mammographic density, 248
  specific surface area, 119
  speed, 25, 296
  temperature plasma, 35
  voltage electrode, 41
Higher membrane hydrophilicity, 115
Histogram, 7, 298, 299
Homogeneity, 23, 181
Homogeneous, 38, 63
Horticultural crops, 180
Host plant, 134

314 Engineering Interventions in Agricultural Processing

Hosting molecules, 122
Hot percolation treatment, 136, 138
Human
    health, 229, 238
    influenza-A, 254
    inspection, 25
    papilloma viruses, 254
    prostate carcinoma cells, 236
Hunter colorimeter, 24
Hunterlab colorimeter, 24
Huntington disease, 253
Hydrogen
    bond, 87, 194, 200, 201, 211, 217
    peroxide, 202, 243, 287
Hydrolysis, 89, 90, 92, 98–101, 103,
    201, 204, 238
Hydrolyzing enzyme, 204
Hydrometallurgy, 149, 173
Hydrophilic, 112, 205, 212
    effect, 112
    nano-silica, 120
    residues, 112
    tail, 114
Hydropriming, 180, 181, 188
Hydroxyl, 91, 186, 200, 202, 206, 215,
    217, 232, 233, 238, 243, 283
Hyperalgesia, 286
Hypercholestoremia, 254
Hyperinsulenemia, 255
Hyperspectral imaging, 13–15, 26
Hypertension, 240, 252, 254, 285, 290
Hypertriglyceridemia, 254
Hypochlorite, 201–203, 206, 207, 214
    oxidation, 201–203
Hypoglycemic, 255, 290
    properties, 290
    record, 290
Hypolipidemic, 290
    impact, 290

I

Illuminants, 22
Illumination, 6, 22, 23, 300, 301
Image
    acquisition, 5–7, 9, 15, 18, 22–24, 26
    analysis, 4, 6, 9, 24, 25
    capture circuitry, 301
    color, 9
    data, 296, 299
    file formats, 299
    files, 299
    illumination, 6
    information, 10, 299
    landmarks, 295
    processing, 7, 10, 24
    rays, 302
    sensor, 295, 301
    texture information, 23
    video processing toolbox, 25
Imaging sensor, 16, 296, 301
Imbibition, 181, 183, 185, 187, 211, 212
Immaculate replication, 187
Immobilization, 73, 90–92, 94–97, 99
    activity, 91
    process, 73, 90–92, 94, 95, 97
Immobilized enzyme, 73, 91–103
Immobilizing substance, 96, 97
Immunological techniques, 153
Incandescent light bulb, 299
Incubation, 82, 92, 95, 96, 102, 103,
    136, 137, 139
Industrial, 4, 16, 18, 35, 58, 63, 123,
    134, 148, 149, 151, 197, 216, 295, 298
    cameras, 298
    ohmic cookers, 63
    waste products, 151
Inexhaustible, 72
Influenza, 254, 285
Infrared imaging, 302
Infusion, 239, 287, 290, 291
Inhibiting effect, 87, 88
Inhibition, 68, 73, 86, 87, 96, 97, 101,
    138, 141, 167, 168, 235, 236, 240–
    242, 244, 288
Inhibitors, 96, 167, 169, 238, 241, 290
Inoculum preparation, 137
Inoculums, 156
In-situ leaching (ISL), 158
Inspection, 14, 16, 18, 20, 21, 25
Insulin
    deficiency, 254

Index

sensitivity, 237
Integration, 49, 122, 131
Intensive mineral investigation, 170
Inter-electrode separation, 39
Interferometer type, 16
Intergranular binding forces, 211
Intermediate level processing, 10
Intermittent fermentation processes, 73
Ionization, 37, 38
Ionized gaseous matter, 34
Ionizing radiation, 183, 303
Iron, 85, 102, 113, 148, 152, 162, 163, 165–167, 170, 194, 240, 243, 252, 253
Irradiation, 23, 200, 237
Ischemic attack, 254
Isolation, 73

J

Joule heating, 59, 68
Juice box containers, 13
Jungle fever, 285

K

Kaem, 232, 241, 242
Kaempferol, 232, 241, 242, 257
Kaolin, 119
Kell factor, 299
Kelvins, 296
Kidney, 244
Kinetic parameters, 156
Kinetin, 181
Klebsiella pneumonia, 141
Kluyveromyces
lactis, 81, 94, 96, 101
marxianus, 98, 101
Knudsen number, 38, 49

L

Label inspection, 16
Lactase deficiency, 101
Lactobacillus delbrueckii ssp, 80
Lactose, 72, 73, 89, 90, 98–103
Lamellae, 113
Laminar air flow chamber, 135
Landfill space, 148

Laser, 8, 299
ablation, 133
Lateral roots, 184
Leach liquor, 156, 168
Leachate, 158, 167
Leaching techniques, 161
commercial scale, 165
dump leaching, 167
heap leaching, 167
in-situ leaching, 166
reactor leaching, 168–169
vat leaching, 167–168
laboratory scale, 161
manometric technique, 163
percolator leaching, 164
shake flask technique, 164
stationary flask technique, 163, 164
pilot plant, 164
agitation tank leaching process, 165
column leaching, 165
Lead, 35, 87, 122, 148, 157, 170, 171
Leaf area index, 184
Lean grade ores, 148
Least gelation concentration (LGC), 211, 217
Lemon grass health benefits, 302
anti-fungi properties, 304, 305
anti-inflammatory properties, 302, 303
anti-microbial properties, 305
anti-nociceptive properties, 303
anti-obesity/antihypertensive activity, 306
anti-oxidant properties, 303
cytotoxicity/anti-mutagenicity, 305
Lens, 299, 300
Lentils, 8, 9
Leptospirillium spp, 152
Lettuce, 14, 18, 43, 44, 47, 48, 194
Leukemia cells, 236
Leukocyte conversion, 286
Levansucrase enzyme, 90
Light

intensity, 237
system, 6
Lighting, 300
Linear structure, 206
Line-scan type, 7
Lipid peroxidation, 186, 238
Lipids, 44, 115, 120, 186, 194, 207, 208, 211, 231, 240
Lipoperoxidation, 287
Lipophilic
molecules, 112
nano-silica, 120
substitution, 217
Lipophilicity, 240
Liposome layer, 112
Liposomes, 110, 112–115, 125
Lithography, 133
Liver, 57, 231, 244
Liver cancer, 231
Lixivient, 162
Local thermodynamic equilibrium, 35, 36
Logarithm, 81, 83, 84
Long wavelength infrared (LWIR), 14, 15
Low
enzymatic efficiency, 102
level processing, 10
shear stress resistance, 194, 202
temperature plasma, 35, 38
Luminous intensity, 297
Lung, 231, 241, 242, 244–246, 289
cancer, 231, 241, 242, 244–246
Luteolin, 232, 235
Luteolinglycosides, 286
Lymphocytes, 289

M

Machine vision system components, 6
camera hardware interface, 9
image analysis, 9
high level processing, 11
intermediate level processing, 11
low level processing, 11
image illumination, 6, 7

digital camera, 7 8
scanner, 8, 9
vision camera, 7
Macro-porous material, 117
Magnesium, 85, 102, 132, 165
Magnetic resonance imaging (MRI), 14, 15
Mahangu, 192
Mammary
cancer, 231
gland, 244
Manganese, 85, 86, 102, 148, 170, 171
Manometric technique, 153, 155
Marbling, 20
Marine oil, 243
Marker gene, 123
Maturity index (IM), 18
Measurement, 5, 13, 19, 20, 42, 156, 157, 207, 300
Meat, 13, 20, 43–46, 57, 58, 62–68, 199, 248
Meat Development Association (MDA), 57, 242
Mechanical, 4, 35, 66, 112, 117, 132, 201, 303
agitation, 157
processes, 112
Mechanical-optical-electronic-software system, 4
Medicinal plant, 141
Mercaptoethanol, 86–88, 102
Mercury, 148
Mesenchymal stem cells (MSCs), 235
Mesoporous silica nano-particles (MSNs), 117, 122, 123
Metabolic, 148, 166, 180, 183, 187
Metabolism, 71, 166, 235, 240, 253, 287
Metal, 42, 73, 85–87, 132, 148–153, 157, 159, 161–171, 203, 232, 253, 296
bioextraction, 168
leaching, 165
pollution, 148
remediation, 170
solubilization, 150
sulfide oxidation, 161
sulfide, 161–163, 166, 167, 169

values, 149
Metallospaera, 152
Metallurgy, 170
Metaphor, 298
Methanolic, 287, 289
Methylxanthines, 234
Metrology, 300
Michaelis constant, 88
Micro hollow cathode discharge, 35
Microbes, 45–48, 66, 132, 133, 138, 153, 156, 184
Microbial
  activity, 150, 152, 167
  diversity, 153
  inactivation, 44
  leaching, 149, 151, 161, 169, 170, 173
  metal mobilization processes, 149
  oxidation, 173
  safety, 45
Microbiological
  investigations, 156
  security, 64
  solubilization processes, 149, 173
Micrococcus, 137
Micro-emulsions (MEs), 110–112
Microinjection method, 122
Micro-macronutrients, 108
Micrometer range, 39
Micronutrients, 118
Microorganisms, 44, 45, 48, 65–67, 72, 118, 134, 149, 150, 152, 153, 156, 157, 159, 161, 163, 167, 169, 172, 184, 256
Micro-porous material, 117
Micro-projectile, 122
Microscope, 122, 141, 213, 295
Microscopic organisms, 184, 289
Mid wavelength infrared (MWIR), 14, 15
Mid-level processing techniques, 11
Mild vortexing, 122
Millet, 192, 194, 196, 204, 205, 208–210, 212, 214–216
  starch, 194, 204, 205, 208–210, 212, 214–216

Mineralogy, 149
Minerals, 118, 149, 154, 157–159, 161, 162, 164–171, 194, 252, 284
Minimum breakdown voltage, 38, 39
Mining, 148, 149, 151, 152, 157, 158, 164, 165, 170–172
Minolta
  colorimeter, 23
  readings, 23
Mitigating, 124, 285, 286, 291
Mitochondria, 289
Mitochondrial membrane permeabilization (MMP), 241, 253
Mitogen-activated protein kinase (MAPK), 242, 253
Moisture content (MC) , 14, 66, 204, 205, 214
Molding flash detection, 16
Molecular weight, 83, 84, 90, 102, 202, 215
Monochrome and polychrome cameras, 7
Monodispersed, 141
Monosaccharides, 72
Moraceae, 135
Morphotaxonomic, 136
Motion perception (MP), 300
Mucuna bean, 202, 215
Muller Hinton Agar (MHA), 138
Multifunctional reagent, 200
Multiple
  correlation coefficients, 18, 19
  myeloma cells, 236
Multiple-linear regression (MLR), 18, 19
Multispectral sensor, 15
Multivariate statistics, 23
Murine alveolar macrophages, 286, 289
Mushroom, 18
Mutagenesis, 241
Mutagenic, 241, 245, 291
Mutagenicity impacts, 289
Mutation, 230, 244
Mycotoxins, 288
Myocardial infarction, 254
Myrcene, 284, 286, 288, 289

Myricetin, 241

N

Nano-based
  delivery systems, 124
  materials, 108, 125
  products, 124
Nano-capsule, 116, 122, 125
Nano-crystals, 117
Nano-delivery systems classification, 118
  emulsions, 119
    micro-emulsions, 119–120
    nano-emulsions, 120
    lipid-based nano-particles, 123
    polymeric nano-particles, 124–125
    nano-porous materials, 125
    nano-crystals, 125
  vesicular delivery system, 120
    liposomes, 120–121
    niosomes, 121–123
Nano-devices, 108
Nano-emulsions (NEs), 110, 112, 119, 120, 124
Nano-fertilizer, 118, 119, 125
Nano-fiber, 122, 125
Nano-formulation, 109, 121, 122
Nano-herbicide molecules, 120
Nano-materials, 108–110, 119, 120, 124, 125, 131, 132
Nano-particle, 108–112, 115–120, 122–125, 131–133, 137–142
Nano-particle applications, 126
  delivery of
    agrochemicals, 127
    genetic materials into plant tissues, 130–131
    plant growth regulators, 129–130
  efficient delivery of fertilizers, 126–127
  nano-herbicides, 127–128
  nano-pesticides, 128–129
Nano-pesticide, 120, 121, 125

Nano-scale, 120, 122, 125
Nano-silica component, 124
Nanosilver production, 133
Nano-size, 120–122, 125
Nano-structure lipid carriers (NLCs), 115, 116
Nano-technology, 108, 111, 119, 124, 125, 131
Nano-zeolite, 118
Naringenin chalcone, 234
National Semiconductor interface, 295
Native starch, 194, 197–199, 201–203, 205, 207–216
Natural
  biochemical cycles, 148
  microbial process, 148
  product, 133, 134
  restoration, 124
Near infrared, 5, 13, 14
  spectroscopy, 5
  system (NIRS), 13
Negligible water dissolvability, 183
Neural
  biomolecules, 252
  network (ANN), 10, 23, 300
Neurocognitive disorder, 253
Neuro-degeneration, 253
Neurodegenerative
  disease, 252, 253
  disorders, 241
Neurological
  disorders, 239
  infections, 285
  issue, 287
Neuropathy, 254
Neuro-protective, 241, 242, 253
Neutral species, 34, 35
Nickel, 148, 151, 167, 171
Niosomes, 110, 113, 115
Nitrogen, 13, 38, 39, 44, 118, 184, 243, 252
Nitrosamines, 243
Nociception, 288
Non-biodegradable products, 123
Non-carbohydrate substances, 208
Non-covalent interactions, 200, 217

Index

Non-enzymatic oxidation, 161
Non-equilibrium plasma, 34, 37
Non-fermented tea, 230
Non-homogeneity, 23
Non-local thermodynamic equilibrium
  plasma, 50
Non-thermal plasma (NTP), 34–37,
  43–45, 49
Non-waxy starch, 200, 217
Non-woven fabrics, 16
Nuclear, 14, 26, 35, 122, 236, 242, 287
  magnetic resonance (NMR), 14, 15,
  26, 239, 240
Nucleic acids, 44, 187
Nucleotides, 123

O

Obesity, 231, 252, 255
Ohmic heating principles, 67
  advantages/disadvantages, 67, 68
  electric conductivity, 69–72
  electric field intensity, 69
  system performance coefficient, 72
  thermal power, 68, 69
Oil seeds, 179, 284
Oligomeric derivatives, 237
Oligosaccharides, 90, 188
Online poultry inspection, 20
Oolong tea, 230
Opioid receptors, 288
Opthalmia, 285
Optical
  character recognition (OCR), 300
  resolution, 300
  targets, 295
Optimal temperature, 80, 82, 83, 94, 95
Orange juice, 22, 45, 113
Organic
  products, 284
  solvent extraction, 290
Oscillatory movement, 60
Osmo-molding, 182
Osmopriming, 180, 182, 185, 186, 188
Osmotic
  arrangement, 183

potential, 180, 181, 187
  weight, 180
Osteoporosis, 235
Ovarian cancer, 241, 242
Overall acceptability (OA), 19
Overlapping rings, 138
Oxidation, 44, 148, 149, 151, 161–169,
  186, 197–207, 211, 215–217, 230–
  232, 238–240, 243, 287, 290
Oxidative
  damage, 231, 237, 239, 252, 253,
  255–257
  deformation, 206
  reaction, 215
  stress, 240, 242, 243, 252, 253, 255,
  256
  treatment, 214
Oxidized
  agent, 162, 163, 202, 203, 206, 216
  starch, 201–203, 209–212, 214, 217
Oxygen, 40, 44, 156, 157, 164, 165, 167,
  169, 184, 186, 202, 204, 231, 232,
  234, 243, 252, 286

P

Packaging, 21, 43
Pancreas, 244
Pancreatic
  beta cell integrity, 240
  cancer, 236
Parasites cells, 288
Parasitism, 184
Partial differential equation, 10
Partial least square (PLS), 8, 18, 19
Particulate foods, 43
Pasteurization value, 64, 65, 68
Pasting
  property, 201
  temperature, 215
Pathogen, 109, 134, 135
Pathogenic microbial strains, 132
Paw edema, 286
Peak viscosity, 202
Pearl millet starch modification, 204

chemical modification of pearl millet starch, 207
  cross linking of starch, 207
  dual modification of starch, 213–214
  oxidation of starch, 211–213
Penicillium chrysogenum, 83, 84
Pennisetum glaucum, 192, 193
Peptide bonds, 87, 91
Peritoneal cavity, 286
Peritonitis, 286
Permanent image, 301
Permanganate, 202
Peroxisome proliferator-activated receptor, 287
Pesticide, 109, 110, 118–121, 123, 124, 284
Pharmacological activities, 142, 235
Phenol, 91, 232
Phenolic
  acids, 286
  compound, 230, 233, 255
Phosphate, 101, 165, 200, 207, 215, 216
Phospholipid, 112, 113, 115
Phosphorus, 118, 197, 207
  oxychloride, 197
Photography, 298, 303
Photoprotector, 237
Physical
  priming, 180, 188
  remediation technologies, 170
Physicochemical, 19, 148, 194, 197, 201, 290
Plant
  cells, 108, 109, 122, 123
  growth regulators, 108, 121, 124
  root, 118, 121, 184
  staining agents, 108
  tissues, 122, 124
  transformation, 108, 124
  vascular system, 124
Plasma, 33–37, 40–42, 46, 48
  cholesterol level, 252
  corona discharge, 50
  discharges, 38

electrodes, 34
intensity, 34
jet, 40–42, 46
needle, 35
treatment, 44, 45, 49
Plasmon peak, 139
Plastic container, 58
Plastid, 185
Plethora, 134
Plumule length, 182, 183
Pneumonia, 285
Polar head, 112, 114
Polyacrylamide gel, 77, 83, 84, 91
Polycationic carriers, 94
Polydipsia, 254
Poly-ethoxy glycol (PEG), 124
Polyethylene glycol (PG), 182
Polygonal starch granules, 194, 217
Polymer, 110, 116, 200, 205
Polymerase chain reaction (PCR), 153
Polymeric nano-particles (PNs), 116, 117
Polymers, 108, 116, 200, 233, 239
Polynomial algorithms, 23
Polyphenoels, 233
Polyphenol
  antioxidants, 117
  oxidases, 230
Polyphenols, 229, 230, 232–234, 239, 244, 254, 255, 286
Polysaccharides, 108
Polysiloxane-polyvinyl alcohol, 94, 96, 97
Polyuria, 254
Polyvinyl alcohol, 101
Pore enlargement, 123
Pork surface, 45
Positive
  charge, 74, 94
  ions, 34
Potassium, 85, 102, 118, 165
Potato, 13, 18, 21, 22, 24, 197, 203, 213, 215
Potential drug delivery agent, 116
Poultry, 20, 43, 48, 199
Pre-sowing seed treatments, 180

Prime lens, 303
Prokaryotes, 134
Proliferation, 236, 239, 242–244, 248, 255
Pro-myelocitic leukemia, 242
Prostacyclin production, 238
Prostate cancer, 231, 236, 241, 246, 247, 251
Protection, 94, 96, 119, 134, 231, 240, 241, 251, 253, 255
Protective
    function, 135
    wax layer, 120
Protein, 13, 45, 71, 74–77, 83–85, 87, 91, 96, 108, 116, 117, 123, 180, 185, 186, 194, 211, 212, 231, 233, 238, 241, 242, 244, 248, 253, 255, 287
    Bax, 242
    bundles, 77
Protozoa, 153, 172, 184
Pseudomonas
    aeruginosa, 66
    spp, 153
    stutzeri, 133
Pyrite, 149, 162, 166

Q

Quecetinmediated cardioprotection, 240
Quercetin, 116, 232, 240, 243, 255

R

Radiation, 35, 183, 299
Radiative de-excitation rate, 37
Radicle distension, 180, 183
Radio frequency interference (RFI), 297
Radio telescope, 295
Raffinose, 188
Raw
    agricultural products, 44
    extract, 77, 136–138
    image data, 11
Reaction speed, 81, 88
Reactive oxygen species (ROS), 235, 242, 243, 283
Refractory

gold ores, 151, 171
    ores, 152
Region of interest (ROI), 9, 10, 301
Resolution video, 301, 303
Resonant
    cavity, 301
    photon, 301
Respirometer, 155
Retinopathy, 254
Retrogradation, 194, 197, 202, 203, 206, 218
Reverse micelles, 133
Rheumatoid joint pain, 287
Rhizobacteria, 184
Rhizomucor sp, 74, 78, 79, 85, 88
Rhizosphere, 121, 184
Rhodobacter sphaeroides, 153
Ring test, 19
Ringworm infestation, 285
Robotics, 297

S

Salmonella typhimurium strains, 289
Salting out, 117
Sanyo, 192
Saponins, 121
Saturation, 13, 48, 73, 74, 92, 102
Saw blades, 16
Scanner, 8, 9, 300–302
Scilab software, 25
Screening, 73, 102, 103, 172
Seafood, 20
Sediments, 154, 169, 170
Seed
    germination, 181, 182, 184, 185
    preparation, 180
    priming types, 190
        biochemical priming, 193, 194
        halopriming, 192
        hydropriming, 190–191
        osmopriming, 191–192
        physical priming, 193
        solid matrix priming, 192, 193
    priming, 180, 187, 188
    surface, 183

Seedling, 179–183
Segmentation, 5, 9–11, 24
  algorithms, 9
  process, 5
Selective nano-materials, 124
Semi-fermented tea, 230
Semi-liquid state, 200
Semipermeable film, 180
Sensors, 13
Serial communication, 295, 296
Serum cholesterol, 243
Shake flask techniques, 153
Shelf life, 116
Short Wavelength Infrared (SWIR), 14, 15, 26
Short-wave near infrared (SW-NIR), 14
Shutter, 301
Silica
  Gel, 90
  material, 92
Silver, 132, 133, 136–142, 150
  nanoparticles, 132, 133, 136–142
  nitrate, 138, 142
Simple stepwise expandability, 152
Skin
  brain, 244
  cancer, 250, 251, 255
  photo aging, 251
  thyroid, 236
Slow wound healing, 254
Sludges, 169, 170
Small cell lung carcinoma (SCLC), 244
Smart
  camera, 301
  delivery fertilizers, 118
Snacks, 20
Sodium
  alginate, 116
  chloride, 76, 182
  hypochlorite, 135, 201
  trimetaphosphate, 197
  tripolyphosphate, 200
Soil
  accumulation, 148
  bacteria, 184
  biocontrol agents, 184

humidity, 237
quality, 118, 123
rhizobacteria, 184
Solid lipid nano-particles (SLNs), 115, 116
Solid matrix priming (SMP), 180, 183, 188
Solubility, 72, 111, 117, 119, 198, 207, 209, 210, 216
Solubilization, 162, 163, 166
Solubilized amylose, 214
Sonication, 116, 121
Sonicators, 112
Sophisticated device, 299
Sorghum, 180, 192, 194
Soybean oil, 243
Spatial-taxon silhouette, 297
Spatial-taxons, 301
Spectrum, 12, 13, 15
Sphalerite, 163, 171
Spherical starch granules, 194, 218
Sphingolipids, 112
Splicing signals, 303
Spline
  algorithm, 10
  interpolation, 10
Squamous cell carcinoma, 250
Stabilizer, 198, 216
Standard encoding scheme, 300
Staphylococcus
  albus, 141
  saprophyticus, 153
Starch, 97, 101, 109, 117, 122, 181, 192, 194, 197–218, 238
  gelation, 211
  granule integrity, 215
  granules, 97, 200, 201, 204, 207, 209, 210, 212, 214, 215
  matrices, 208
  modification, 194, 197
  molecule, 198, 200–203, 205, 206, 209–212, 215, 217
  phosphate, 200
Starch-lipid interaction, 211, 218
Starch-starch interaction, 218
Stationary flask technique, 153, 156

Index

Sterile scalpel, 135
Sterilization, 41, 43, 156
Sterilizing, 43, 49
Stirred tank process, 157
Stockpiling, 186
Streptococcus thermophilus, 80, 82, 86
Streptomyces spp, 153
Streptomycin sulfate, 138, 136
Striking surfaces, 302
Substancepoly acryl amide, 98
Sulfates, 161, 164
Sulfide, 149–151, 154, 159, 162, 164, 166–169
Sulfobacillus, 152, 154, 163
Sulfolobus, 166
Sulfur compounds, 162, 165
Sulphur-oxidizing microbes, 163
Superoxide dismutase (SOD), 186
Surface
    inspection, 16
    plasmon resonance (SPR), 139, 140
    sterilization, 37, 43
    tension, 119, 167, 169
Surfactant, 111, 113, 115–117, 167
Super video graphics array (SVGA), 302, 303
Swelling, 198–200, 204, 207–212, 215, 216, 218
    activity, 207
    behavior, 207–210
    power, 208, 209, 216
    volume, 211, 218
System performance coefficient (SPC), 64

T

Thiobacillus. ferrooxidans, 150, 154
Tea polyphenols, 246
    flavones, 248
        apigenin, 249
        chrysin, 250
        luteolin, 249
    flavanols, 251
        catechins, 251
        myricetin, 255

quercetin, 254
theaflavins, 252
thearubigins, 253
Telecentric lens, 302
Telephoto lens, 300, 302
Temperature holding phase, 64
Textile fabric, 132
Textural features, 9
Theaflavins (TFs), 233, 238, 239, 257
Thearubigins, 233, 238, 239, 257
Theobromine, 234
Theophylline, 234
Thermal, 34–38, 48, 59, 60, 65, 73, 80, 83, 94–96, 102, 117, 122, 132, 133, 194, 202
    decomposition, 133, 194, 202
    lag, 59
    optimum degree, 80
    plasma, 34–36, 38
    temperatures, 95
    treatment, 117
Thermionic tube devices, 7
Thermodynamic equilibrium, 34
Thermography, 302
Thermophiles, 166
Thermophilic bacteria, 173
Thermothrix, 152
Thiobacillus, 150
Thiobacillus ferrooxidans, 150
Thiobacilluscaldus, 166
Thiosulfate, 162, 173
Threonine, 194, 242
Thresholding process, 9
Tagged Image File Format (TIFF), 9, 302
Titanium, 120, 124, 132
Tobacco, 184, 250, 252
Tolerance, 96, 167, 169, 198, 199
Tomatoes, 45, 47, 49
Topoisomerases, 235
Toxic, 118, 123, 132, 133, 167–169, 202, 242, 244
    chemicals, 123
    substances, 132, 167
Toxicity, 132, 148
Trace nutrients, 118

Traditional
  delivery system, 124
  heating, 59, 60, 65, 66
  incandescent light source, 296
  mining techniques, 171, 172
Transferosomes, 110, 115
Transformation process, 73
Transgenic vehicle, 122
Translate images, 300
Transmission electron microscopy
  (TEM), 132, 137, 141
Triacylglycerides, 243, 255
Tricarboxylic acid (TCA), 235
Trichoderma, 184
Trichophyton
  mentagrophytes, 254
  rubrum, 254
True optimal temperature, 83
Tuber, 201, 203, 213
  crops, 201
  starch, 201, 203, 213
Turbidity, 137
Tyrosinase inhibitory activity, 237

U

Ultrasonication, 122
Ultraviolet
  irradiation, 133
  light (UV), 13, 44, 46, 132, 137–139,
  237, 251, 255
  rays, 14, 250
Unsaturated fats, 186
Uranium, 157, 158
Urokinase enzyme, 255
Universal Serial Bus (USB), 7, 9, 302
Uterus, 244

V

Vascular endothelial growth factor
  (VEGF), 236
Vascular smooth muscle cells (VSMCs),
  237
Vasodilatory activity, 235
Vasoprotector, 237
Vector graphic display, 299

Video Electronics Standards Association
  (VESA), 302, 303
Vesicular
  delivery system, 110
  nanodelivery systems, 110
Video graphics array (VGA), 303
Vibrio cholera, 137
Video
  camera, 295
  peripherals, 303
Virtual grader, 18
Viscosities, 198–200, 202, 203, 214–217
Viscous mass, 214
Visible light (VIS), 7, 9, 11, 15, 112, 138
Vision systems, 20, 21, 302
Vitamins, 108, 284
Volatilization, 118
Voltage, 38–42, 44, 58, 59, 61–63, 66,
  67, 253
Voltage-dependant anion channel
  (VDAC), 253

W

Wafer dicing, 16
Warburg instrument, 155, 173
Water
  binding capacity, 211, 218
  treatment, 132
Wavelengths, 11, 13, 18, 19, 22, 296
Waxy millet, 215
Weeds, 21, 109, 119, 120, 124
Wetting, 181
Wheat sprouts, 235
White tea, 230
Wide-angle lenses, 300, 303
Wine production, 13
Wood quality inspection, 16
World Health Organization (WHO), 244,
  284

X

Xenobiotic, 123, 124
X-rays, 14, 218, 303
  diffraction (XRD), 213, 214, 218
  imaging technologies, 14

Index   325

radiography, 14

Y

Y-cable, 303

Z

Zeolites, 118
Zinc, 132, 148, 151, 157, 167, 170, 171, 194
Zooglearamigera, 153
Zoom lens, 303

PGSTL 11/07/2017